Power Systems

Electrical power has been the technological foundation of industrial societies for many years. Although the systems designed to provide and apply electrical energy have reached a high degree of maturity, unforeseen problems are constantly encountered, necessitating the design of more efficient and reliable systems based on novel technologies. The book series Power Systems is aimed at providing detailed, accurate and sound technical information about these new developments in electrical power engineering. It includes topics on power generation, storage and transmission as well as electrical machines. The monographs and advanced textbooks in this series address researchers, lecturers, industrial engineers and senior students in electrical engineering. ** Power Systems is indexed in Scopus**

More information about this series at http://www.springer.com/series/4622

Radu-Emil Precup · Tariq Kamal ·
Syed Zulqadar Hassan
Editors

Solar Photovoltaic Power Plants

Advanced Control and Optimization Techniques

Springer

Editors
Radu-Emil Precup
Faculty of Automation and Computers
Politehnica University of Timişoara
Timişoara, Romania

Syed Zulqadar Hassan
School of Electrical Engineering
Chongqing University
Chongqing, China

Tariq Kamal
Department of Electrical
and Electronics Engineering
Sakarya University
Serdivan/Sakarya, Turkey

ISSN 1612-1287 ISSN 1860-4676 (electronic)
Power Systems
ISBN 978-981-13-6150-0 ISBN 978-981-13-6151-7 (eBook)
https://doi.org/10.1007/978-981-13-6151-7

Library of Congress Control Number: 2018968094

This Springer imprint is published by the registered company Springer Nature Singapore Pte Ltd.
The registered company address is: 152 Beach Road, #21-01/04 Gateway East, Singapore 189721, Singapore

Preface

Growing adversarial environmental impact, escalating energy demand, ever-expanding utilization of fossil fuels coupled with rising production costs have brought substantial attention to sustainable development worldwide. In this context, global efforts have been made to promote utilizing more renewable energy resources, among which solar photovoltaic contributes is one of the most promising clean energy sources to the world energy consumption. Approximately, 90% of the global installed photovoltaic systems are integrated with grid. Megawatt photovoltaic power plants are generally preferred to install in remote areas due to the requirement of wide land area. Usually, a medium voltage network is adopted to transfer power to load centers. Therefore, compact, reliable, dynamic control and system stability, photovoltaic power plant planning and optimal sitting and utilization have become increasingly important. This book discusses control and optimization techniques (i.e., improved fuzzy control, artificial intelligence, back-propagation neural network, adaptive neuro-fuzzy control, sliding-mode control, predictive control, backstepping, secant incremental gradient based on Newton–Raphson, cuckoo search algorithm, particle swarm optimization, and gray wolf optimizer) in the broadest sense, covering new theoretical results and the applications of newly developed methods in photovoltaic systems. Going beyond classical control techniques, it promotes the use of control and optimization strategies with improved efficiency, based on linearized models and purely continuous (or discrete) models and proved by appropriate performance indices. These new strategies not only enhance the performance of the photovoltaic systems, but also decrease the cost per kilowatt-hour generated.

The material of the book is organized into the following eleven chapters. All the chapters have been included in this book after a rigorous review process. Special importance was given to chapters offering novel control and optimization techniques in solar photovoltaic systems. The contributed chapters provide new ideas and approaches, clearly indicating the advances made in control system analysis and simulation with respect to the existing state-of-the-art.

Inverter (DC–AC converter) is the key interface between the solar photovoltaic array and mains in the grid-integrated photovoltaic system. The inverter must follow the frequency and voltage of the grid and extract maximum power from the solar photovoltaic. Therefore, the quality of the output current in an inverter integrated photovoltaic systems is an important standard. Chapter "Adaptive Control Techniques for Three-Phase Grid-Connected Photovoltaic Inverters" provides the development of model reference adaptive control techniques for grid-connected photovoltaic inverter systems under uncertain parameters and disturbances. The control objectives are analyzed based on the photovoltaic inverter output requirement. The ability to compensate grid-side harmonic disturbances and asymptotic adaptive disturbance rejection is enhanced.

The output power of a photovoltaic system depends on solar radiation that falls on its PN junction as well as the percentage of solar radiation it converts into electricity (conversion efficiency). Since there is always a unique maximum power point on each power–voltage curve, maximum power point tracking units should be utilized in photovoltaic sources to increase their efficiency. Chapter "Application of Sliding-Mode Control for Maximum Power Point Tracking of PV Systems" presents one-loop and two-loop sliding-mode control schemes to increase the efficiency in photovoltaic systems. A maximum power point searching unit is utilized in the searching loop, and a tracking controller is utilized in the other loop to extract the maximum photovoltaic power photovoltaic power source.

In photovoltaic systems, DC bus voltage balancing is critical. Fluctuations in bus voltage cause power imbalance that originates from different sources of disturbances such as sudden change in load and/or weather parameters. Such a power imbalance results in extra energy. Chapter "Predictive Control of Four-Leg Converters for Photovoltaic Energy Systems" is devoted to predictive control of four-leg converters for photovoltaic systems. The predictive current control enables grid-connected operation, whereas predictive voltage control is used for stand-alone operation of photovoltaic energy systems. The predictive control strategies fulfill the control requirements concerning output current control, load voltage control, balancing of DC link capacitor voltage, and neutral-leg switching frequency minimization. Chapter "A Novel Maximum Power Point Tracking Method for Photovoltaic Application Using Secant Incremental Gradient Based on Newton Raphson" discusses some common algorithms dedicated to maximum power point tracking of the photovoltaic system such as perturb and observe, particle swarm optimization and gray wolf optimizer. The chapter also develops a new maximum power point tracking method for photovoltaic application using secant incremental gradient based on Newton–Raphson method. The proposed method has better performance in achieving global maximum power point with more tracking efficiency and convergence speed versus classical methods.

Solar photovoltaic experiences some deficiencies and some fundamental problems when utilized as a stand-alone energy source. In this context, solar photovoltaic is integrated with certain power sources and/or storage systems in a hybrid power system to increase the reliability. Chapter "Study on Control of Hybrid Photovoltaic-Wind Power System Using Xilinx System Generator" describes a photovoltaic–wind hybrid power system using a Xilinx system generator. Maximum power point tracking techniques are adopted in order to extract the maximum energy from the renewable energy sources. The virtual flux-oriented control scheme is adopted to control the grid-connected three-phase inverter based on the backstepping approach.

Over the last few years, fuzzy, neural networks, and other artificial intelligence techniques have contributed substantially in the modeling, control, and optimization of solar photovoltaic systems. Chapter "Artificial Intelligence for Photovoltaic Systems" presents an overview on the applications of artificial intelligence techniques in photovoltaic systems. Particular attention is devoted to forecasting and modeling of meteorological data, basic modeling of solar cells, and sizing of photovoltaic systems. A comparison between conventional techniques and the added benefits of using machine learning methods is given. Similarly, Chapter "Applications of Improved Versions of Fuzzy Logic Based Maximum Power Point Tracking for Controlling Photovoltaic Systems" reviews the applications of different conventional and improved fuzzy logic-based maximum power point tracking techniques in photovoltaic systems. Based on simulation and experimental results, the chapter provides a comparative study considering the main assessment criteria such as fast convergence, conversion efficiency, algorithm's complexity, and practical implementation to figure out the relative merits and limitations of the available maximum power point tracking techniques.

Chapter "A New Method for Generating Short-Term Power Forecasting Based on Artificial Neural Networks and Optimization Methods for Solar Photovoltaic Power Plants" introduces the application of artificial neural networks and particle swarm optimization to generate short-term power forecasting for solar photovoltaic plants. Power prediction is estimated using real-time data of 1 MW photovoltaic power plant in use. Estimation power data are compared with real-time data, and the precision of the proposed method is demonstrated. Chapter "Evaluation on Training Algorithms of Back Propagation Neural Network for a Solar Photovoltaic Based DSTATCOM System" suggests a back-propagation neural network control algorithm based on fast Fourier transform control algorithm for distribution static compensator integrated solar photovoltaic systems. Harmonic elimination in terms of accuracy, number of iterations (epochs), and training time has been improved in the proposed algorithm.

Chapter "Power Extraction from PV Module Using Hybrid ANFIS Controller" presents the implementation of a hybrid adaptive neuro-fuzzy inference system controller for maximum power extraction from PV module. This chapter also provides the effect of load impedance and converter topologies on adaptive neuro-fuzzy inference system controller design. The hardware results are very promising and show that the adaptive neuro-fuzzy inference system control system performance is better than other conventional control systems in terms of efficiency, stability, and precision. Chapter "An Online Self Recurrent Direct Adaptive Neuro-Fuzzy Wavelet Based Control of Photovoltaic Systems" focuses on a new wavelet-based online direct adaptive neuro-fuzzy control of photovoltaic systems. The conversion efficiency and output power are better than the well-known used traditional and intelligent maximum power point tracking controllers.

Timişoara, Romania
Serdivan/Sakarya, Turkey
Chongqing, China
December 2018

Radu-Emil Precup
Tariq Kamal
Syed Zulqadar Hassan

Acknowledgements

We would like to express our sincere thanks to all authors and reviewers who have dedicated their valuable time and hard work to this book. Special thanks to Prof. Luis Fernández Ramírez for his advice and help during the writing of this book. Finally, thanks to Executive Editor Dr. Christoph Baumann and Project Coordinator Mr. Ramamoorthy Rajangam for their great efforts and support during the implementation of this project.

Contents

Editorial Advisory Board

List of Reviewers

Abdelhalim Zaoui Electromagnetic Systems Laboratory, Polytechnic Military School, Algiers, Algeria
Abdelouahed Mesbahi Laboratory of Energy and Electrical Systems, Hassan II University of Casablanca, ENSEM, Casablanca, Morocco
Adrian Nedelcu National Institute for Research and Development in Electrical Engineering, ICPE-CA, Bucharest, Romania
Ahmet Dogan Department of Electrical Engineering, Faculty of Engineering, Erciyes University
Aishwarya Kokare Electrical Engineering Department, G. H. Raisoni Institute of Engineering and Technology, Pune, India
Ali Muhtar Department of Electrical Engineering and Information Technology, Faculty of Engineering, Universitas Gadjah Mada, Yogyakarta, Indonesia
Alper Kerem Osmaniye Korkut Ata University, KVHS, Department of Electrical and Energy, Osmaniye, Turkey
André P. Meurer Federal University of Santa Maria, Santa Maria—RS, Brazil
Andrej Trpovski TUM CREATE, Singapore

Siriroj Sirisukprasert Department of Electrical Engineering, [illegible], Thailand
Srivani S. G. Iyengar Department of Electrical and Electronics Engineering, R. V. College of Engineering, Bengaluru, India
Sukumar Mishra Department of Electrical Engineering, Indian Institute of Technology Delhi, New Delhi, India
Tamer Khatib Department of Energy Engineering and Environment, An-Najah National University, Nablus [illegible]
Tiong Teck Teo School of Electrical and Electronic Engineering, [illegible] University, Singapore
Ugur Demirdelen Department of [illegible] and Electronics, [illegible] Science and Technology University, Turkey
Tunku Muhammad Nizar Tunku Mansur School of [illegible] Engineering, Universiti Malaysia Perlis, [illegible], Malaysia
Uğur Arifoğlu Department of Electrical and Electronics Engineering, [illegible] Engineering, Sakarya University, Turkey
Ujjwal B. Vyas Department of Electrical and Electronics Engineering, R. V. College of Engineering, Bengaluru, India
Wagner Brignol Center of Excellence in Energy and [illegible], [illegible] University of South Minas, Brazil
Wanshi Hong University of Virginia, USA
Wilson D. Pavón Vallejos Department of Electrical Engineering, Universidad Politécnica Salesiana, Quito, Ecuador
Yang Du Department of Electrical and Electronic Engineering, Xi'an Jiaotong-Liverpool University, China
Yassine Sayouti LIE, ATT Hassan II University, [illegible], Morocco
Yi-nan Chen Department of Electrical Engineering, [illegible], Section 3, [illegible], Taiwan
Yi-Wei Kuo Department of Electrical Engineering, National [illegible] University of Science and Technology, [illegible], Taiwan
Youness Chaddadi Sidi Mohamed [illegible] Faculty of Science and Technology, Renewable Energy and Intelligent Systems Laboratory, Morocco

Adaptive Control Techniques for Three-Phase Grid-Connected Photovoltaic Inverters

Wanshi Hong, Gang Tao and Hong Wang

Abstract This chapter presents a framework of model reference adaptive control (MRAC) techniques for three-phase grid-connected photovoltaic (PV) inverter systems with uncertain parameters and disturbances. Such adaptive controllers are employed to achieve two main goals: (i) the asymptotic tracking for the output of a time-varying reference signal by the PV system with high-order dynamics and parameter uncertainties, which cannot be achieved by some conventional control techniques, and (ii) the asymptotic rejection of a practical class of unknown high-order harmonic signal disturbances, which is crucial for desired PV system operations. In this chapter, a full PV inverter system dynamic model is derived, and adaptive control design conditions are verified for such system models. An MRAC based disturbance rejection scheme is also developed for the PV inverter system with parameter and disturbance uncertainties. Desired system performances are ensured analytically and simulation results are listed to verify the result. This study shows the potential advantages of using adaptive control techniques for PV inverter systems, for ensuring desired PV system stability, output tracking, and disturbance rejection properties.

Nomenclature

$\delta(t)$	System disturbance term
ω	Grid fundamental frequency
$\omega(t)$	Compact form of signals
$\Psi(t)$	Estimate of high frequency gain matrix
ρ_i	Control relative degrees of ith control input
$\Theta(t)$	Parameter estimates

W. Hong · G. Tao (✉)
University of Virginia, Charlottesville, VA 22904, USA
e-mail: gt9s@virginia.edu

W. Hong
e-mail: wh3cv@virginia.edu

H. Wang
Pacific Northwest National Laboratory, Richland, WA 99352, USA
e-mail: hong.wang@pnnl.gov

R.-E. Precup et al. (eds.), *Solar Photovoltaic Power Plants*, Power Systems,
https://doi.org/10.1007/978-981-13-6151-7_1

Θ^*	Compact form of nominal parameters
$\tilde{\Theta}(t)$	Difference between parameter and its estimates
$\varepsilon(t)$	Estimation error
$\xi(t)$, $\zeta(t)$	Estimation signals
$\xi_m(s)$	Left interactor matrix
A, B, B_d, C	Inverter system parameters
a_{ij}^*	jth coefficient of $d_i(s)$
C_f	Filter capacitance
$d_i(s)$	Polynomial at ith diagonal term at the left interactor matrix
$d_{a,b,c}$	Three phase duty cycle
$e(t)$	Tracking error
$G(s)$	Transfer matrix of inverter system
$i_{g,a,b,c}$	Grid side three-phase current
I_{gd}, I_{gq}	Inverter d-q axis output current
I_{gd}^*, I_{gq}^*	Reference d-q axis current signals
$i_{i,a,b,c}$	Inverter side three-phase current
I_{PV}, U_{PV}	Input current and voltage of the inverter system
$K_1(t)$, $K_2(t)$, $K_{3f}(t)$	Adaptive control parameter estimate
K_1^*, K_2^*, K_{3f}^*	Nominal control parameter
K_p	High frequency gain matrix
L_6	Observability matrix of inverter system
L_f	Inverter side filter inductance
L_g	Grid side filter inductance
$m(t)$	Normalize matrix
P	Active power
P^*	Maximum power point
Q	Reactive power
$r(t)$	Reference input
R_f, R_g, R_c	Filter ESR
S_6	Controllability matrix of inverter system
S_p, Γ_p, K_s	Gain matrices
T	DQZ transformation
t	Time variable
$U_{a,b,cN}$	Grid voltage in a-b-c axis
u_{abc}	Control signal in a-b-c axis
$U_{c,a,b,c}$	Three-phase capacitance voltage
$U_{d,qN}$	Grid voltage in d-q axis
$U_{d1N}, U_{d,q6s,cN}$	Grid side voltage magnitudes
$W_m(s)$	Reference model
x	State representation
$x_{a,b,c}$	State representation in a-b-c reference frame
$x_{d,q,o}$	State representation in d-q-o reference frame
$y(t)$	Inverter output signal
$y_m(t)$	Reference output signal.

1 Introduction

This section introduces the photovoltaic (PV) inverter and provides background about renewable energy and some conventional control methods for PV inverter systems. It then describes the state-of-the-art research into the PV inverter control problem.

1.1 Research Motivation

Solar energy is inexhaustible and eternal. Most energy on Earth comes directly or indirectly from the sun. Every year, about 1.8×10^{18} kWh of energy is radiated from the sun to Earth, that is about 10 thousand times more than Earth's power consumption. The use of photovoltaic (PV) energy has become a trend around the world. By the end of 2015, the USA ranked the fourth for solar energy usage in the world. By the end of 2016, USA had 40 GW of installed PV capacity, which is almost twice as much as the PV capacity for the previous year [1]. From February 2016 to January 2017, utility-scale solar power generated 35.5 TWh, or 0.92% of total U.S. electricity demand. All these figures show the important strategic position of solar energy in the field of power generation.

PV Inverter

A PV inverter is a crucial part of the power system because it converts the direct current (DC) of the PV power generation devices (such as solar panels) into an acceptable utility frequency alternating current (AC) for grid-connected or off-grid users [2]. Hence, PV inverters are the core of any PV power generation system (grid-connected or off-grid). The quality of the output current of a PV inverter is an important inverter standard, so the control strategy for inverter systems has been studied to guarantee the desired output quality [3, 4]. Moreover, to have a PV inverter work most efficiently, the output current of the inverter should follow the reference currents that are obtained from the Maximum Power Point Tracking (MPPT) module (to be introduced in Sect. 2).

Major Technical Problems for PV Inverter Systems

With the continuously increasing demand for solar energy over the past decades, the grid-connected PV inverter system control problem has been of a great research interest. However, the large variations in weather mean sunlight intensity is often uncertain and the distributed installation of solar panels makes it difficult to detect system damages. These factors will lead to difficulties in controlling the PV inverters to assure that they work efficiently. As a result, improving the reliability of PV inverter systems has been a major research task. The main technical problems are as follows:

- *Randomness of the energy source.* For a renewable power source, random effects such as temperature, environment, or light intensity can influence the inverter system output. Therefore, the controller for the inverter should have the ability to make the inverter output track a given time-varying reference signal, which is obtained from the MPPT module.

- *Harmonic pollution.* For inverter systems, the power of electronic devices might cause the output current harmonics to increase. Moreover, when a polluted grid is connected to a PV inverter system, i.e., the grid-side current contains harmonic components, this will influence the performance of the PV inverter system. Thus, the PV inverter system should have the ability to reduce high-order harmonic influences.
- *Islanding effect.* An islanding effect occurs when the grid is powered off but the PV system does not detect this and continues working alone. This condition endangers power company customers and workers. Other protections, such as the isolation of AC and DC units, should also be applied to prevent the inverter from causing damage to the grid. Thus, PV systems should have anti-islanding protection modules [5].
- *Parameter uncertainty.* System uncertainties have major influences on PV inverter systems due to the unreliable characteristics of the solar energy source. A major challenge in grid integration for distributed PV systems is the unknown system uncertainties. Adaptive control technology has the desired potential to solve these problems.

Conventional Control Methods and Their Drawbacks

A Proportional and Integral (PI) controller [6, 7] and a PI controller combined with a Proportional and Resonant (PI + PR) controller are most commonly used by the photovoltaic industry. However, some weak points still remain in these conventional controllers: (i) The classical PI controller does not have the ability to deal with harmonic effects from the grid. (ii) Although a PI + PR controller can reduce the effect of the harmonics, for three-phase inverter systems the controller structure is rather complicated [3]. They may not work for distributed PV systems whose orders are high. (iii) Tracking of a time-varying reference signal cannot be achieved only by using classical controllers, which is a major issue for PV systems because the reference signal is generated from the online updating MPPT module. This drawback will cause some major problems in achieving integrated control of distributed PV inverter systems, which require control cooperation. (iv) The PV system uncertainties cannot be effectively handled by conventional methods.

System Uncertainty Issue

System uncertainties have major influences on a PV inverter system because of the unreliable characteristics of the renewable energy source. That is, (i) the uncertainty of system parameters, such as inverter system resistance or inductance; (ii) the uncertainty of the output voltage of the photovoltaic generation devices, such as solar panels; and (iii) the uncertainty of system loads and faults (for distributed PV inverter systems, the distance among the solar panels can lead to different sunlight intensity for each inverter system, and there are uncertain factors such as load variation and failure or damage of system components). A major challenge in grid integration for distributed PV systems is the system uncertainty problem, but the adaptive control technique has the desired potential to solve the main uncertainty issues.

Hence, advanced adaptive control techniques that can effectively deal with system uncertainties are needed to improve PV inverter system reliability.

Adaptive Control

In this chapter, the ability of adaptive control techniques to meet the desired control objectives of PV inverter systems is studied. Adaptive control is particularly aimed at the design of controllers for the system that has uncertainties [8] (including environmental, structural, and parameter uncertainties). These sources of uncertainty are common to power systems. Payload variation or component aging cause parametric uncertainties. In power systems, PV and other types of renewable energy-based power systems, in particular the randomness of input power sources such as solar energy and wind power also leads to parametric uncertainties. Component failure in a power system leads to structural uncertainties, and external weather influences and harmonic effects are typical environmental uncertainties. Adaptive control has been successful in addressing new challenging problems and offering encouraging solutions, when dealing with uncertainties that often appear in automobile engines, electronic devices, and other industrial processes.

Unlike other control methods like proportional-integral-derivative (PID) control [9], robust, optimal, or nonlinear control [10] methods, which require a certain knowledge of system parameters, adaptive controllers do not require such knowledge and they use online performance error information to adapt parameter uncertainties [11, 12].

1.2 Literature Review

Many research approaches have been taken to improve the reliability of PV inverter systems. In this chapter, we introduce some exciting methods that address the improvement of the control strategy of PV inverter systems.

In [3], the authors developed PI + PR control techniques for PV inverter systems to eliminate the high-order harmonic effects. The function of a PR controller is to eliminate the harmonics in a particular frequency, in which frequency the harmonic component cannot be reduced by the PI controller. In this research, a specific method of modeling the inverter has been applied. In [4], the authors constructed classical controllers for the inverter using LCL filters. This method eliminated the current harmonics and the basic control objectives of an inverter system. However, for the three-phase inverter, the topology of the control structure is rather complicated because of the complexity of the three-phase system, and the uncertainty problem for the PV inverter systems has not been considered.

In [13], the authors used an LCL filter in the inverter system to physically eliminate high-order harmonics, derived an inverter system model with an LCL filter, and developed a controller for this type of inverter system. In this chapter, the significance of the LCL filter is mentioned, because an LCL filter can improve the efficiency of filtering and also reduce the volume of the filter. The authors used a classical control method for the inverter control problem, and they did not consider the harmonic problem for this system. Moreover, with this improvement for the PV inverter systems, the controller will be harder to design because the order of the filter system has increased.

In [14], three-phase inverter control using a nonlinear Kalman filter is introduced. This paper develops a nonlinear approach to the inverter system using a Kalman filter. To simplify the control problem, in this research the authors proposed to directly control the power to track the maximum power point P^*.

In [15–17], some adaptive approaches to the PV inverter control problem are given. In [17], an adaptive control scheme was used to help predict the system parameters when the system is working in a polluted environment. However, the authors still used the classical PI + PR controller for control of the main circuit. In [15], an adaptive droop control was also used to help predict system parameters. A simple adaptive control design was developed in [16] by only considering the off-grid case without an LCL filter in the inverter system.

1.3 Technical Contributions

From the discussion above, it can be seen that it is crucial to develop a reliable controller to deal with high-order harmonics disturbances and system uncertainties, so as to achieve the output tracking of a time-varying signal for PV inverter systems. This need motivated our research on multivariable model reference adaptive control (MRAC) techniques for PV inverter systems. The main contributions of this work are the following:

- The control objectives are analyzed based on the PV inverter output requirement.
- The ability to compensate grid-side harmonic disturbances and asymptotic adaptive disturbance rejection is enhanced.
- A state feedback output tracking MRAC scheme is developed with some key techniques established for PV inverter systems.

The remainder of this chapter is organized as follows. Section 2 first introduces the details of three-phase grid-connected PV inverter systems, then derives a model for three-phase PV inverter systems, and analyzes control objectives for three-phase PV inverter systems. Section 3 develops an adaptive control design for the PV inverter system and analyzes the stability of the inverter system relative to the proposed adaptive control design. Section 4 presents a simulation study using the developed adaptive controller to verify the desired system performance. Section 5 summarizes this chapter and discusses the future topics related to this research.

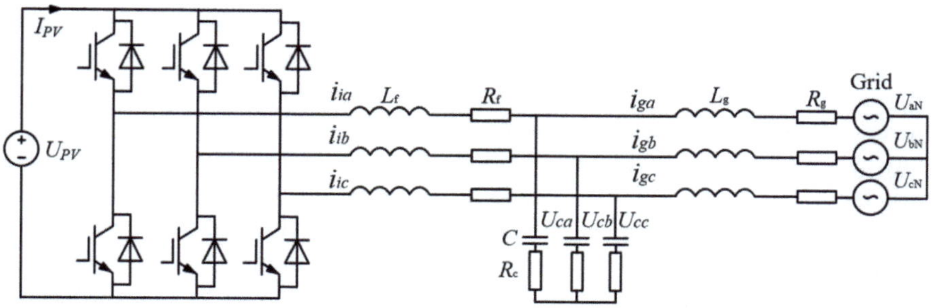

Fig. 1 Inverter circuit structure [18]

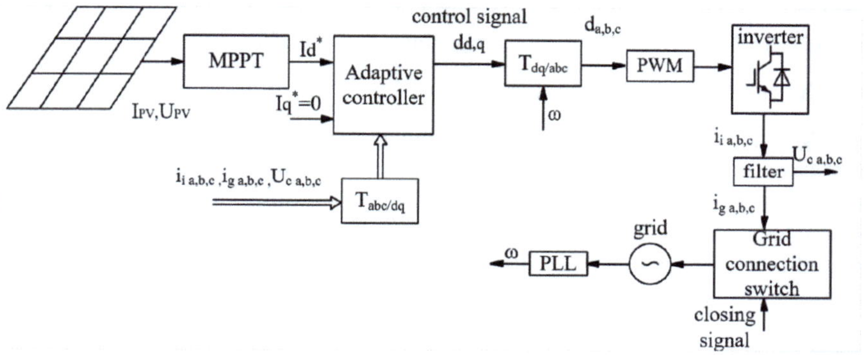

Fig. 2 Adaptive PV inverter control system structure [18]

2 PV Inverter System Modeling

The function of a three-phase inverter is to manipulate the input DC voltage and current with switching signals to change it into the desired three-phase AC current. Figure 1 shows the circuit structure of the three-phase grid-connected PV inverter system. The solar panel generates current and voltage (I_{PV} and U_{PV}), which are the input of this inverter system. The current and voltage go through a series of insulated gate bipolar transistors (IGBTs) and are converted to inverter side current $i_{i,a,b,c}$. Passing the LCL filter (represented by L_f, L_g and C_f, R_f, R_g, and R_c, which are the Equivalent Series Resistance) the inverter-side current is filtered and the inverter system produces the output $i_{g,a,b,c}$.

The adaptive control system structure is shown in Fig. 2, where the subscripts a, b, c in the signals $d_{a,b,c}$, $i_{i,a,b,c}$, $U_{c,a,b,c}$ etc., denote three components of each signal. Our adaptive controller controls the switching state of the IGBTs to have the PV inverter system constantly operating at the maximum power point P^*. To achieve this goal, our adaptive controller should correspond with each module in the inverter system. The reference current signals I^*_{gd}, I^*_{gq} are for the inverter output current I_{gd}, I_{gq} to track. In this design, we assume that $I^*_{gq} = 0$ for all time for simplicity, which means that the reference output power contains no reactive power. For applications, the reactive power can be set to nonzero elements. In such cases, I_{gq}^* can be time-varying as well. The control strategy to track $I_{gq}^*(t)$ is similar to the tracking of I_{gd}^* introduced in this chapter. The control signal u_{abc} represents the duty cycles of the pulse-width modulation (PWM) waveform [19], which generates switching signals to the IGBTs inside the inverter and produces the inverter-side currents $i_{i,a,b,c}$. In this process, the DC-side current is changed to the desired AC output current. Passing through the LCL filter, the final outputs $i_{g,a,b,c}$ are obtained. Before the grid connection, an outer control loop is required to confirm whether the phase and magnitude of the inverter output voltage are identical to the grid side voltage. Thus, a closing signal generation module is required. Our work was to design the adaptive controller (shown in Fig. 2) to track the maximum power point and cooperate with other system components shown in Fig. 2.

Maximum Power Point Tracking

The MPPT module is a crucial part of the inverter system. For different solar panels that have different insolation, there is a maximum power point within the allowed range of the voltage (i.e., dependent on the variation of the sunlight intensity). The goal of MPPT is to find the maximum power point P^* which is used in the latter part of the system to calculate the reference current for the adaptive controller. This process constantly updates the maximum power point P^*.

LCL Filter

The conventional three-phase grid-connected inverter uses an LC filter or just an L filter. However, with the power level of the inverter advanced to a new level, the power electronic devices require a lower switching frequency to eliminate the power loss, which will lead to an increase in the high-order harmonics of the grid side. As a result, to meet the total harmonic distortion (THD) standard, the inductance will become very high if one only uses the L filter. This will lead to a series of problems that will cause not only higher cost and and require a larger size for the system, but also increase the inductance affecting the system dynamically. Recently, the replacement of the L filter by an LCL filter has been one of the most modern solutions for solving the above problems. LCL filters show better performance in reducing the high-order harmonics with a lower total inductance.

Grid-Connecting Process

As one of the crucial technical aspects of the inverter control problem, the inverter system first should generate voltage that has some related magnitude, the same phase, and the same frequency as the grid side. The grid-connecting process follows two steps:

Step 1 The inverter should first start working in isolation from the grid using a transformer to generate three-phase voltage that has the same magnitude as the grid side voltage. The maximum tolerance error for the output voltage is 10% of the rated voltage [20].

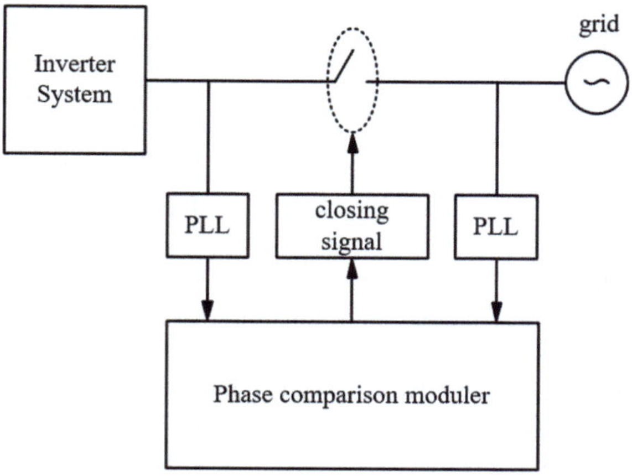

Fig. 3 Scheme diagram for the inverter-grid connecting process

Step 2 With the inverter generating the same magnitude of the three-phase voltage, the phase tracking method is used to generate the closing signal [21, 22]. The basic principle is to connect the inverter to the grid at the moment when the inverter system output has the same frequency and the same phase as the grid.

With the above steps accomplished, the inverter system can be successfully connected to the grid. A block diagram showing the control of the grid-connection process is provided in Fig. 3. In this chapter, we are mainly considering the current control problem for the grid-connected system, which occurs after this grid connection process is accomplished.

Inverter System Modeling

We can write the three-phase circuit dynamic equations from Fig. 1 as follows:

$$\frac{\mathrm{d}}{\mathrm{d}t}\begin{bmatrix} I_{ia} \\ I_{ib} \\ I_{ic} \end{bmatrix} = \frac{u_{PV}}{\sqrt{3}L_f}\begin{bmatrix} d_a \\ d_b \\ d_c \end{bmatrix} - \frac{R_f}{L_f}\begin{bmatrix} I_{ia} \\ I_{ib} \\ I_{ic} \end{bmatrix} - \frac{R_c}{L_f}\begin{bmatrix} I_{ia}-I_{ga} \\ I_{ib}-I_{gb} \\ I_{ic}-I_{gc} \end{bmatrix} - \frac{1}{L_f}\begin{bmatrix} U_{ac} \\ U_{bc} \\ U_{cc} \end{bmatrix} \tag{1}$$

$$\frac{\mathrm{d}}{\mathrm{d}t}\begin{bmatrix} I_{ga} \\ I_{gb} \\ I_{gc} \end{bmatrix} = \frac{1}{L_g}\begin{bmatrix} U_{ac} \\ U_{bc} \\ U_{cc} \end{bmatrix} + \frac{R_g}{L_g}\begin{bmatrix} I_{ia} \\ I_{ib} \\ I_{ic} \end{bmatrix} + \frac{R_c}{L_g}\begin{bmatrix} I_{ia}-I_{ga} \\ I_{ib}-I_{gb} \\ I_{ic}-I_{gc} \end{bmatrix} - \frac{1}{L_g}\begin{bmatrix} U_{aN} \\ U_{bN} \\ U_{cN} \end{bmatrix} \tag{2}$$

$$\frac{\mathrm{d}}{\mathrm{d}t}\begin{bmatrix} U_{ac} \\ U_{bc} \\ U_{cc} \end{bmatrix} = -\frac{1}{C_f}\begin{bmatrix} I_{ia}-I_{ga} \\ I_{ib}-I_{gb} \\ I_{ic}-I_{gc} \end{bmatrix}, \tag{3}$$

where $[d_a, d_b, d_c]^T = u(t)$ is the control vector, which are the duty cycles of the PWM module.

To achieve maximum power point tracking, we need to change (1)–(3) into d-q-0 axis by applying the following DQZ transformation T [23–25]

$$T = \sqrt{\frac{2}{3}} \begin{bmatrix} \cos(\omega t) & \cos\left(\omega t - \frac{2\pi}{3}\right) & \cos\left(\omega t + \frac{2\pi}{3}\right) \\ -\sin(\omega t) & -\sin\left(\omega t - \frac{2\pi}{3}\right) & -\sin\left(\omega t + \frac{2\pi}{3}\right), \\ \frac{1}{\sqrt{2}} & \frac{1}{\sqrt{2}} & \frac{1}{\sqrt{2}} \end{bmatrix}, \tag{4}$$

$$\begin{bmatrix} x_d \\ x_q \\ x_o \end{bmatrix} = T \begin{bmatrix} x_a \\ x_b \\ x_c \end{bmatrix} \Leftrightarrow \begin{bmatrix} x_a \\ x_b \\ x_c \end{bmatrix} = T^{-1} \begin{bmatrix} x_d \\ x_q \\ x_o \end{bmatrix} = T^T \begin{bmatrix} x_d \\ x_q \\ x_o \end{bmatrix}, \tag{5}$$

where $TT^T = I$. Because no neutral line connection existed, based on the properties of this three-phase circuit topology, all terms related to 0 will be zero [23].

Thus, the state-space model for d-q-0 axis will contain six states. This process also reduces the number of unknown parameters to be estimated. Note that an accurate phase tracking is for this coordinate transformation process. More studies about phase tracking can be found in [21, 22].

State-Space Equation

The state-space model in d-q-0 axis is

$$\dot{x}(t) = Ax(t) + Bu(t) + B_d\delta(t),\ y(t) = Cx(t), \tag{6}$$

where

$$x = \left[I_{id}\ I_{iq}\ I_{gd}\ I_{gq}\ U_{dc}\ U_{qc}\right]^T,\ u = \left[d_d\ d_q\right]^T, \tag{7}$$

$$A = \begin{bmatrix} -\frac{R_f+R_c}{L_f} & \omega & \frac{R_c}{L_f} & 0 & -\frac{1}{L_f} & 0 \\ -\omega & -\frac{R_f+R_c}{L_f} & 0 & \frac{R_c}{L_f} & 0 & -\frac{1}{L_f} \\ \frac{R_c+R_g}{L_g} & 0 & -\frac{R_c}{L_g} & \omega & \frac{1}{L_g} & 0 \\ 0 & \frac{R_c+R_g}{L_g} & -\omega & -\frac{R_c}{L_g} & 0 & \frac{1}{L_g} \\ \frac{1}{C_f} & 0 & -\frac{1}{C_f} & 0 & 0 & \omega \\ 0 & \frac{1}{C_f} & 0 & -\frac{1}{C_f} & -\omega & 0 \end{bmatrix}, \tag{8}$$

$$B = \begin{bmatrix} \frac{U_{PV}}{\sqrt{3}L_f} & 0 & 0\,0\,0\,0 \\ 0 & \frac{U_{PV}}{\sqrt{3}L_f} & 0\,0\,0\,0 \end{bmatrix}^T,\ B_d = \begin{bmatrix} 0\,0 & -\frac{1}{L_g} & 0 & 0\,0 \\ 0\,0 & 0 & -\frac{1}{L_g} & 0\,0 \end{bmatrix}^T, \tag{9}$$

$$C = \begin{bmatrix} C_1 \\ C_2 \end{bmatrix} = \begin{bmatrix} 0\,0\,1\,0\,0\,0 \\ 0\,0\,0\,1\,0\,0 \end{bmatrix},\ \delta = \left[U_{dN}\ U_{qN}\right]^T, \tag{10}$$

where $u(t) = [d_d, d_q]^T$ is the control vector, and δ is the grid side voltage. A, B, B_d and C are the parameter matrices that contain all the system parameters. This state-space model is the plant to construct adaptive control design.

Inverter Requirements

To obtain the control objectives, we need to analyze the requirements for the inverter output:

(i) The inverter voltage should have the same phase and frequency as that of the grid-side voltage.

(ii) The inverter system should be robust for uncertainties. This requirement is crucial especially for the PV inverters, because the randomness of solar energy may cause more uncertainty situations.

(iii) The harmonic components of the output inverter current should stay below a certain rate. Harmonics such as the 11, 13 and 17 and 19th..., can directly be eliminated from the LCL filter, but lower-order harmonics, like the 5 and 7th, cannot be fully eliminated by filters. In this research, we mainly consider the harmonics from the grid side. The disturbance $\delta(t)$ in (6) can be expressed as

$$\delta(t) = \begin{bmatrix} U_{dN}(t) \\ U_{qN}(t) \end{bmatrix} = \begin{bmatrix} U_{d1N}+U_{d6cN}\cos(6\omega t)+U_{d6sN}\sin(6\omega t) \\ U_{q6cN}\cos(6\omega t)+U_{q6sN}\sin(6\omega t) \end{bmatrix}, \tag{11}$$

where U_{d1n} represents the grid fundamental voltage [13, 23]. By performing the *DQZ* transformation, the 5th- and 7th-order harmonics on the *a-b-c* axis can be expressed as a summation of sinusoid signals in the 6th order. From the control perspective, the elimination of the high-order harmonics is equivalent to rejecting the grid-side disturbance $\delta(t)$.

(iv) The inverter output power should meet some requirements. The inverter power expressions are

$$P = \frac{2}{3}(I_{gd}U_{dN} + I_{gq}U_{qN}) = \frac{2}{3}I_{gd}U_{d1N} \tag{12}$$

$$Q = \frac{2}{3}(I_{gq}U_{dN} + I_{gd}U_{qN}) = \frac{2}{3}I_{gq}U_{d1N}, \tag{13}$$

where P is the active power and Q is the reactive power. For this research, we assume that the inverter system does not generate reactive power, i.e., $I^*_{gq} = 0$. In practice, we can choose other $I^*_{gq}(t)$ values to stabilize the output voltage, if needed. The reference signal is chosen to be

$$I^*_{gd}(t) = \frac{2P^*(t)}{3U_{d1N}}, \; I^*_{gq} = 0, \tag{14}$$

where P^* is from MPPT module.

The first requirement (i) can be achieved by obtaining the frequency and phase of the system from the phase lock loop (PLL) module. The results of studies of the adaptive PLL module have been recently reported [17–21]. As for the controller developed in this paper, requirements (ii)–(iv) are major considerations. To meet the above inverter requirements, we propose the control objectives described below.

Control Objectives

From the above discussion, we can see that building a model reference adaptive controller is important for achieving efficient control of the PV inverter system given the presence of system uncertainties (characterized by the unknown parameters in

(A, B, B_d, C)) and unknown disturbances (characterized by $\delta(t)$ in (11) with unknown magnitudes). Thus, we propose an adaptive controller design for the inverter system (6) to meet the following control objectives:

(a) All closed-loop signals are bounded.
(b) The inverter output $y(t) = [I_{gd}\ I_{gq}]^T$ achieves asymptotic tracking of the reference signal $y_m(t)$, where

$$y_m(t) = [I^*_{gd}(t) I^*_{gq}(t)] \tag{15}$$

with $I^*_{gd}(t) = \frac{2P^*(t)}{3U_{d1N}}$ and $I^*_{gq} = 0$ from the inverter requirement.
(c) The inverter system operates normally under the effects of grid-side voltage distortions.

The physical meaning of the stated control objectives is that the adaptive controller should work at P^* under the uncertainties caused by component aging and the randomness of the solar energy, and it should eliminate the effect of high order harmonic disturbances. Moreover, the proposed controller will make the inverter system stable. With the control objectives stated, a model reference adaptive controller with disturbance rejection ability is to be developed to achieve these control objectives.

3 Adaptive Control of PV Inverter Systems

This section first provides some necessary design conditions and their verification, then derives the nominal control design with all parameters known. Finally, the adaptive control design is presented.

3.1 Verification of Control Design Conditions

To begin the adaptive control design, we need to verify that the system (6) meets some basic design conditions.

Basic Design Conditions for MRAC

The multi-input multi-output (MIMO) system for the MRAC design is given by

$$\dot{x} = Ax + Bu, \ y = Cx, \tag{16}$$

where $A \in R^{n\times n}$, $B \in R^{n\times M}$ and $C \in R^{M\times n}$. The objective of MRAC is to have the output signal $y(t)$ track the given reference signal $y_m(t)$, from the model reference system:

$$y_m(t) = W_m(s)[r](t), \tag{17}$$

where $r(t)$ is a chosen reference input signal, and $W_m(s)$ is an $M \times M$ stable transfer matrix. As a notation, $W_m(s)[r](t) \triangleq \mathscr{L}^{-1}[W_m(s)r(s)]$. $W_m(s)$ is chosen to be $W_m(s) = \xi_m^{-1}(s)$, where $\xi_m(s)$ is the left interactor matrix of $G(s) = C(sI - A)^{-1}B$. The left interactor matrix $\xi_m(s)$ is defined in [11].

To meet the control objective, we need to assume the following

Design condition 1 All zeros of $G(s) = C(sI - A)^{-1}B$ are stable, and the system (16) is stabilizable and detectable.

Design condition 2 $G(s)$ is strictly proper with full rank and has a known modified interactor matrix $\xi_m(s)$ such that $K_p = \lim_{s\to\infty} \xi_m(s)G(s)$ is finite and nonsingular.

Design condition 3 There exists a known matrix $S_p \in R^{M\times M}$ such that $\Gamma_p = K_p^T S_p^{-1} = \Gamma_p^T > 0$.

Design conditions 1 and 2 are needed to match the plant with the stable reference model $W_m(s)$. Note that stabilizable and detectable are weaker conditions compared to controllable and observable. They indicate that the non-controllable or non-observable states are stable. Design condition 3 is used for the adaptive parameter update law. To verify Design condition 2, the following lemma is introduced.

Lemma 1 *If the gain matrix*

$$K_s = \begin{bmatrix} C_1 A^{\rho_1 - 1} B \\ \vdots \\ C_M A^{\rho_M - 1} B \end{bmatrix} \in R^{M\times M}, \tag{18}$$

is nonsingular for some $\rho_i > 0,\ i = 1, 2, \ldots, M$, *then* $G(s)$ *has a diagonal interactor matrix* $\xi_m(s) = diag\{d_1(s), d_2(s), \ldots, d_M(s)\}$, *where* $d_i(s) = s^{\rho_i} + a^*_{i\rho_i - 1} s^{\rho_i - 1} + \cdots + a^*_{i1} s + a^*_{i0}$, $i = 1, 2, \ldots, M$, *are any stable monic polynomials.*

In this case, ρ_i, $i = 1, 2, \ldots, M$ are the control relative degrees, and $K_p = K_s$ is the high frequency gain matrix associated with $\xi_m(s)$.

Design Condition Verification

Controllability and observability For the PV inverter system, from the definition of (A,B,C) in (8)–(10), by using Matlab we can verify that the controllability matrix S_6 and observability matrix L_6

$$S_6 = \left[B\ AB\ A^2B \ldots A^5B\right], \tag{19}$$

$$L_6 = \left[C^T\ (CA)^T\ (CA^2)^T \ldots (CA^5)^T\right]^T, \tag{20}$$

both have rank 6. This means that (A,B,C) for the three phase PV inverter system is controllable and observable, signifying that (A,B,C) is stabilizable and detectable.

System zeros It is hard to obtain a generic solution to prove that all system zeros are stable due to the high-order matrices in this inverter system. As a result, a set of system parameters was chosen for the inverter system (A,B,C) in Table 1 from Sect. 4. By applying this set of parameters to the system, we can verify that for this system the zeros are $(-5.56+j0.0031)\times 10^5$, $(-5.56-j0.0031)\times 10^5$, and they are indeed stable. Thus, the system satisfies design condition 1.
High-frequency gain matrix We can verify that for the inverter system (A,B,C) which is stated in Sect. 2, it follows that

$$CB = \begin{bmatrix} 0 & 0 \\ 0 & 0 \end{bmatrix}, \tag{21}$$

$$CAB = \begin{bmatrix} \frac{\sqrt{3}*U_{PV}(R_c+R_g)}{3L_f L_g} & 0 \\ 0 & \frac{\sqrt{3}*U_{PV}(R_c+R_g)}{3L_f L_g} \end{bmatrix} = K_p. \tag{22}$$

Because all the system parameters are nonzero, CAB is positive definite. Thus the inverter systems has relative degrees $\rho_i = 2,\ i = 1, 2$, and $K_p = CAB$ is a diagonal matrix. Together with Lemma 2, we can verify that design condition 2 is satisfied, and G(s) has a diagonal interactor matrix $\xi_m(s) = \text{diag}\{d_1(s), d_2(s)\}$, where $d_i(s) = s^2 + a_{i1}^* s + a_{i0}^*$ are stable monic polynomials. Also, because the high frequency gain matrix is positive and diagonal, one can choose $S_p = \Gamma_1$ such that $\Gamma_1 = \Gamma_1^T > 0$ and satisfy design condition 3.

For our control problem, the state space equation with presence of disturbance is

$$\dot{x} = Ax + Bu + B_d\delta,\ y = Cx, \tag{23}$$

where $A \in R^{6\times 6}$, $B, B_d \in R^{6\times 2}$ and $C \in R^{2\times 6}$. In this case, the disturbance relative degrees are defined as $\upsilon_i,\ i = 1, 2,$ such that $C_i A^{\upsilon_i - 1} B_d \neq 0,\ i = 1, 2$. With the definition of B_d for the inverter system given in (9), the disturbance relative degrees for the inverter system are $\upsilon_i = 1,\ i = 1, 2$. For the reference output given in (17), we can define the reference input

$$r(t) = \xi_m(s)[y_m](t), \tag{24}$$

to be used for the control design.

3.2 Nominal Control Design

In this subsection, the nominal model reference control design is presented based on knowledge of the system parameters. This forms the basis for establishing the structure and parametrization of the adaptive controller to be developed in the next subsection for the case of unknown parameters.

Nominal Controller The nominal state feedback model reference controller structure is

$$u(t) = K_1^* x(t) + K_2^* r(t) + K_3^*(t), \tag{25}$$

where $K_1^* \in R^{2\times 6}$ and $K_2^* \in R^{2\times 2}$ are used for output tracking and $K_3^*(t) \in R^2$ is used for eliminating the effect of the disturbance $\delta(t)$ in (6). For the inverter system, we can establish the following theorem.

Theorem 1 *For the multi-input multi-output inverter system (6) with disturbance (11), there exist a state feedback control law (25) to achieve output tracking:* $\lim_{t\to\infty}(y(t) - y_m(t)) = 0$, *where* $y_m(t) = W_m(s)[r](t)$ *is defined in (17), and asymptotic disturbance rejection of the disturbance is achieved.*

Proof With the system Eq. (6), which is restated here

$$\dot{x}(t) = Ax(t) + Bu(t) + B_d\delta(t), \ y_i(t) = C_i x(t), \tag{26}$$

for $y(t) = [y_1(t), y_2(t)]^T$, $i = 1, 2$, we can obtain the second-order derivative of $y_i(t)$ as

$$y_i^{(2)}(t) = C_i A^2 x(t) + C_i ABu(t) + C_i AB_d\delta(t) + C_i B_d\dot{\delta}(t), \tag{27}$$

in which $u(t)$ appears. For K_p defined in (22), we have $K_{pi} = C_i A^{\rho_i - 1} B, \ i = 1, 2.$ We can choose the following control law

$$u(t) = K_P^{-1} v(t), \ v(t) = \left[v_1(t) \ v_2(t)\right]^T, \tag{28}$$

to make (27) become

$$y^{(2)}(t) = C_i A^2 x(t) + v_i(t) + C_i AB_d\delta(t) + C_i B_d\dot{\delta}(t). \tag{29}$$

Then, we can choose the feedback laws as

$$v_i(t) = -\, CA^2 x(t) - a_{i1}^* \dot{y}_i - a_{i0}^* y(t) + r_i(t) \tag{30}$$

$$- \, C_i AB_d\delta(t) - C_i B_d\dot{\delta}(t) \tag{31}$$

to make (29) become

$$y_i^{(2)}(t) + a_{i1}^* \dot{y}_i + a_{i0}^* y(t) = r_i(t) \Rightarrow y_i(s) = \frac{1}{d_i(s)} r_i(s), \tag{32}$$

Then, we have

$$y(s) = \text{diag}\left\{\frac{1}{d_1(s)}, \frac{1}{d_2(s)}\right\} = W_m(s) r(s), \tag{33}$$

with $W_m(s) = \xi_m^{-1}(s)$. Similarly, signal $v(t)$ can be expressed as $v(t) = K_0 x(t) + r(t) + K_d\delta(t) + K_{d1}\dot{\delta}(t)$, where

$$K_0 = [k_{01}^T, k_{02}^T]^T,\ k_{0i}^T = -C_i A^2 - a_{i1}^* C_i A - a_{i0}^* C_i \tag{34}$$

$$K_d = [k_{d1}^T, k_{d2}^T]^T,\ k_{di} = -C_i A B_d \tag{35}$$

$$K_d 1 = [k_{d11}^T, k_{d12}^T]^T,\ k_{d1i} = -C_i B_d. \tag{36}$$

From the definition of $u(t)$ in (28), we have

$$u(t) = K_p^{-1}(K_0 x(t) + r(t) + K_d\delta(t) + K_{d1}\dot{\delta}(t)) \tag{37}$$

$$= K_1^* x(t) + K_2^* r(t) + K_3^*(t), \tag{38}$$

where,

$$K_1^* = K_p^{-1} K_0,\ K_2^* = K_p^{-1}, \tag{39}$$

$$K_3^* = K_p^{-1}(K_d\delta(t) + K_{d1}\dot{\delta}(t)). \tag{40}$$

Note that for the inverter system, the disturbance relative degree is lower than the system relative degree. Thus, the knowledge of $\dot{\delta}(t)$ is needed. By observing (11), we can find that the disturbance is the summation of a series of sinusoidal signals. As a result, $\dot{\delta}(t)$ can be easily obtained by taking derivatives of sinusoidal signals.

Applying $u(t)$ in (38) to the system (6), we have,

$$y(s) = W_m(s) r(s) \Rightarrow y(t) = W_m(s)[r](t). \tag{41}$$

From (41) and design condition 2, we have established that the controller (37) can ensure that the output $y(t)$ and the states $x(t)$ are bounded, and $y(t)$ tracks $y_m(t)$ asymptotically with the presence of disturbance. ∇

Matching Properties for $K_1^*, K_2^*, K_3^*(t)$ From (25), (26), (41), it can be verified that there exist $K_1^* \in R^{2\times 6}$, $K_2^* \in R^{2\times 2}$ and a nonsingular matrix function $K_3^*(t) \in R^{2\times 3}$ satisfying the matching properties

$$C(sI - A - BK_1^{*T})^{-1} BK_2^{*T} = W_m(s) \tag{42}$$

$$W_m(s) K_2^{*-1} K_3^*(s) + C(sI - A - BK_1^{*T})^{-1} B_d d(s) = 0. \tag{43}$$

Parameterization of $K_3^*(t)$ We need to construct parameterization for $K_3^*(t)$ for the adaptive control design. We can parameterize $K_3^*(t)$ in to the following form

$$K_3^*(t) = K_p^{-1}(K_d\delta(t) + K_{d1}\dot{\delta}(t)) \tag{44}$$

$$= K_p^{-1}(K_d\Phi^* f(t) + K_{d1}\bar{\Phi}^* f(t)) = K_{3f}^* f(t), \tag{45}$$

$$K_{3f}^* = K_p^{-1}(K_d\Phi^* + K_{d1}\bar{\Phi}^*), \tag{46}$$

where K_d and K_{d1} are control parameters from (35) and (36), Φ^* and $\bar{\Phi}^*$ are the parametrization matrices for $\delta(t)$ and $\dot{\delta}(t)$, respectively; a detailed parameterization process can be found in [18].

In summary, we have the following proposition.

Proposition 1 *For the disturbance signal $\delta(t)$ in (11) being the summation of sinusoidal signals, the disturbance related control parameter $K_3^*(t) = K_p^{-1}(K_d\delta(t) + K_{d1}\dot{\delta}(t))$ can be parameterized to the form $K_{3f}^* f(t)$, where $f(t)$ contains all the fundamental sinusoidal signals.*

Next, the adaptive control scheme is derived when system parameters are unknown.

3.3 Adaptive Control Design

With the nominal controller obtained, this subsection presents the adaptive version of the model reference controller.

Controller Structure Similar to the nominal controller structure (25), the adaptive controller structure is written as

$$u(t) = K_1(t)x(t) + K_2(t)r(t) + K_{3f}f(t), \tag{47}$$

where $K_1 \in R^{2\times 6}$, $K_2 \in R^{2\times 2}$ and $K_{3f} \in R^{2\times 3}$ are the estimates of K_1^*, K_2^* and K_{3f}^*, respectively.

Tracking Error The closed-loop system with the adaptive controller is given by applying the adaptive controller to the plant (6):

$$\begin{aligned} \dot{x}(t) &= (A + BK_1^{*T})x(t) + BK_2^* r(t) + BK_3^*(t) + B_d\delta(t) \\ &\quad + B(K_1^T - K_1^{*T})x(t) + B(K_2 - K_2^*)r(t) + B[K_3(t) - K_3^*(t)] \quad (48) \\ y(t) &= Cx(t). \quad (49) \end{aligned}$$

Based on the reference signal definition in (17), the parameter matching properties (42)–(43), and this closed-loop system, we can obtain the output tracking error equation as follows:

$$\begin{aligned} e(t) &= y(t) - y_m(t) \quad (50) \\ &= W_m(s)K_2^{*-1}[\tilde{\Theta}^T\omega](t) + Ce^{(A+BK_1^{*T})t}x(t), \quad (51) \end{aligned}$$

where

$$\begin{aligned} \Theta^{*T} &= [K_1^*, K_2^*, K_{3f}^*] \in R^{2\times 11} \quad (52) \\ \omega(t) &= [x^T(t), r^T(t), f^T(t)]^T \in R^{11} \quad (53) \\ \Theta^T(t) &= [K_1(t), K_2(t), K_{3f}(t)] \in R^{2\times 11} \quad (54) \end{aligned}$$

$$\tilde{\Theta}(t) = \Theta(t) - \Theta^*. \tag{55}$$

Due to design condition 1, term $Ce^{(A+BK_1^{*T})t}x(t)$ is an exponential decay term. As a result, we can write the output tracking error as

$$e(t) = W_m(s)K_2^{*-1}[\tilde{\Theta}^T\omega](t),\ W_m(s) = \xi_m^{-1}(s). \tag{56}$$

Estimation Error To derive the adaptive parameter update laws for the controller parameter $\Theta(t)$, further definition of the estimation error is needed. First we define a stable filter $h(s) = \frac{1}{f(s)}$ with $f(s)$ being a stable polynomial $f(s) = s^2 + f_1 s + f_0$ of degree 2. The estimation error is defined as

$$\varepsilon(t) = \xi_m(s)h(s)[e](t) + \Psi(t)\xi(t), \tag{57}$$

where $\Psi(t)$ estimates the high-frequency gain matrix $\Psi^* = K_p$ defined in design condition 2, $\xi_m(s)$ is the diagonal interactor matrix, $\xi(t)$ and $\zeta(t)$ are defined as

$$\xi(t) = \Theta^T(t)\zeta(t) - h(s)[u](t) \tag{58}$$

$$\zeta(t) = h(s)[\omega](t). \tag{59}$$

Applying (56) to (57), we have

$$\varepsilon(t) = \Psi^*\tilde{\Theta}(t)\zeta(t) + \tilde{\Psi}(t)\xi(t),\ \tilde{\Psi}(t) = \Psi(t) - \Psi^*, \tag{60}$$

to be the parametrized estimation error expression.
Adaptive Laws With the estimation error defined in (60), the adaptive parameter update law is chosen to be

$$\dot{\Theta}^T(t) = -\frac{\Gamma_1\varepsilon(t)\zeta^T(t)}{m^2(t)} \tag{61}$$

$$\dot{\Psi}(t) = -\frac{\Gamma_2\varepsilon(t)\xi^T(t)}{m^2(t)}, \tag{62}$$

where $\Gamma_1 = \Gamma_1^T > 0, \Gamma_2 = \Gamma_2^T > 0$ are design parameters, $m^2(t) = 1 + \zeta^T(t)\zeta(t) + \xi^T(t)\xi(t)$.
Stability Analysis With the above adaptive parameter update laws, we can present the following lemma

Lemma 2 *The adaptive laws (61), and (62) guarantee that*

(i) $\Theta(t) \in L^\infty$, $\Psi(t) \in L^\infty$, $\frac{\varepsilon(t)}{m(t)} \in L^2 \cap L^\infty$;

(ii) $\dot{\Theta}(t) \in L^\infty \cap L^2$, $\dot{\Psi}(t) \in L^\infty \cap L^2$.

Proof Choose the positive definite function

$$V = \frac{1}{2}\left(\mathrm{tr}[K_p\tilde{\Theta}^T\Gamma_1^{-1}\tilde{\Theta}] + \mathrm{tr}[\tilde{\Psi}^T\Gamma_2^{-1}\tilde{\Psi}]\right), \tag{63}$$

where $\mathrm{tr}(A)$ means the *trace* of A. Recalling that $K_p = k_pI, k_p > 0$ is diagonal and positive definite, with the adaptive laws (61), (62), the time derivative of V can be derived as follows:

$$\dot{V} = -\frac{\varepsilon^T(t)\varepsilon(t)}{m^2(t)} \leq 0. \tag{64}$$

Thus, we can conclude that $\Theta(t) \in L^\infty$, $\Psi(t) \in L^\infty$, $\frac{\varepsilon(t)}{m(t)} \in L^2 \cap L^\infty$, $\dot{\Theta}(t) \in L^\infty \cap L^2$, $\dot{\Psi}(t) \in L^\infty \cap L^2$. ∇

With Lemma 2, we can estabilish the following theorem:

Theorem 2 *For the plant (6) with uncertain parameters and the reference model (17), the adaptive controller (47) with the adaptive parameter update laws (61)–(62) guarantees uniform boundedness for all closed loop signals and asymptotic tracking with respect to the reference output signal.*

The proof of this theorem is based on multivariable model reference adaptive control theory [11], and its main ideas can be found in [26, 27].

Remark 1 Note that the algorithm in this work has used state feedback to realize a portion of the state components (i.e., the outputs of the system) to perfectly track their desired reference signals generated from the reference model, where we have assumed that the state information is obtainable. In comparison to an output feedback tracking control scheme, the advantage of such a state feedback control scheme is that it will result in a simpler adaptive control structure with fewer parameters to estimate than that of output feedback control, where the number of calculations required can thus be significantly reduced when it is implemented in practical systems.

With Theorem 2, the adaptive controller design presented in this chapter achieves its objectives, that is the asymptotic output tracking and rejection of a disturbance signal, with the presence of system uncertainties.

4 Simulation Study

This section first introduces the PV inverter system for simulation study, then some simulation results are given to verify that our adaptive control design meets the desired control objective.

Simulation System

The three-phase grid-connected PV inverter system is represented by the following equation:

Table 1 Simulation study parameters [13, 28]

Parameter	Unit	Value
Inductor L_f	(mH)	0.9
Capacitor C	(μF)	30
Inductor L_g	(mH)	0.54
Resistance R_f	(mΩ)	0.05
Resistance R_g	(mΩ)	0.03
Resistance R_c	(mΩ)	0.05
Line-to-line grid voltage $U_{a,b,c}$	(V)	380
Grid frequency ω	(rad/s)	314.15
Input voltage U_{PV}	(V)	750–820
Control parameter Γ_1, Γ_2		0.1I

$$\dot{x}(t) = Ax(t) + Bu(t) + B_d\delta(t), \; y(t) = Cx(t), \tag{65}$$

where the signals $x(t)$, $u(t)$, $\delta(t)$, $y(t)$ and the matrices A, B, B_d and C are defined in (7)–(10). The nominal values of the components in system matrices (A, B, C) and B_d can be found in Table 1, where we can obtain the PV inverter system parameter matrices as follows:

$$A = \begin{bmatrix} -88.89 & 314.16 & 33.33 & 0 & -1111.11 & 0 \\ -314.16 & -88.89 & 0 & 33.33 & 0 & -1111.11 \\ 111.11 & 0 & -55.56 & 314.16 & 1851.85 & 0 \\ 0 & 111.11 & -314.16 & -55.56 & 0 & 1851.85 \\ 33333.33 & 0 & -33333.33 & 0 & 0 & 314.16 \\ 0 & 33333.33 & 0 & -33333.33 & -314.16 & 0 \end{bmatrix}, \tag{66}$$

$$B = \begin{bmatrix} 5.13 \times 10^5 & 0 & 0\,0\,0\,0 \\ 0 & 5.13 \times 10^5 & 0\,0\,0\,0 \end{bmatrix}, \tag{67}$$

$$B_d = \begin{bmatrix} 0\,0 & -1.85 \times 10^3 & 0 & 0\,0 \\ 0\,0 & 0 & -1.85 \times 10^3 & 0\,0 \end{bmatrix}, \; C = \begin{bmatrix} 0\,0\,1\,0\,0\,0 \\ 0\,0\,0\,1\,0\,0 \end{bmatrix}. \tag{68}$$

Simulations are conducted on this linear model to verify the adaptive control scheme performance.

Simulation Results

Four different scenarios were simulated. We have chosen the reference model to be $r(t) = [17a_0, 0]^T$, $a_0 = 1$ and $y_m(0) = [0, 0]^T$ (for case 1, 2 and 3), and the initial error to be $e(0) = [-3, 1]$. Control parameters $K_1(t), K_2(t), K_3(t)$ were obtained from the adaptive parameter update laws.

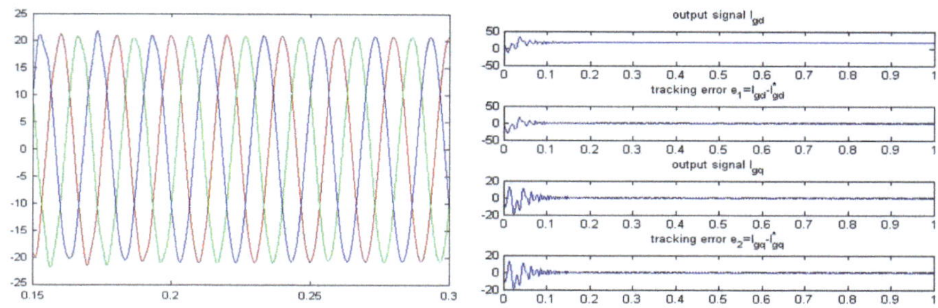

(a) Three-phase output current. (b) I_{gd}, I_{gq} and their tracking error.

Fig. 4 System response for Case 1, recovery from initial error

Case 1: Choose $\delta = [310, 0]^T$. Case 1 shows the inverter system works in a standard grid ($\delta(t) = [310, 0]^T$) with unknown system uncertainties (i.e., A, B, C, B_d unknown). This case tests the adaptive controller under system uncertainties. From Fig. 4 we can see that the system output meets the desired performance.

Case 2: Case 2 shows the inverter works in a polluted grid (i.e., there exist some unknown high order harmonics from the grid side). This will test the disturbance rejection ability of the adaptive controller. The polluted grid is represented by

$$\delta(t) = \begin{bmatrix} \delta_1(t) \\ \delta_2(t) \end{bmatrix} = \begin{bmatrix} 310 + 15.5\cos(6\omega t) + 7.75\cos(12\omega t) \\ 15.5\cos(6\omega t) + 7.75\cos(12\omega t) \end{bmatrix}. \tag{69}$$

Figure 5 shows the simulation results. The grid-side Total Harmonic Distortion (THD) is greater than 5% and the output current THD is around 1.2%. Thus, we can conclude that the inverter output distortion can remain at a sufficiently low level even when connected to a polluted grid.

Case 3: Choose $\delta = [310, 0]^T$. Case 3 is a comparison between the adaptive controller and the nominal controller under the scenario where a parameter in the inverter system is going through sudden change. We consider U_{PV} to change from 800 V to 750 V at $t = 1$. Figure 6 shows the simulation results, from which we can see that as parameter changes apply, the adaptive controller can handle the changes while the nominal controller becomes unstable immediately.

Case 4: Choose $\delta = [310, 0]^T$. Case 4 shows the inverter works in the standard grid with the reference signal containing sinusoidal components (simulating the variation of sunlight). This case tests the adaptive controller's ability to track a time-varying signal. For this purpose, we selected the reference input to be $r(t) = [17 + 2sin(5t), 0]$ and simulated the change of sunlight intensity. Simulation results are shown in Fig. 7. We can see that the output current I_{gd} performs a periodic trajectory (which asymptotically tracks the periodic MPPT output), indicating that the output power is always at the maximum power point as the reference signal changes. As shown in Fig. 7a, the magnitude of the output three-phase current is slightly increasing. Because the

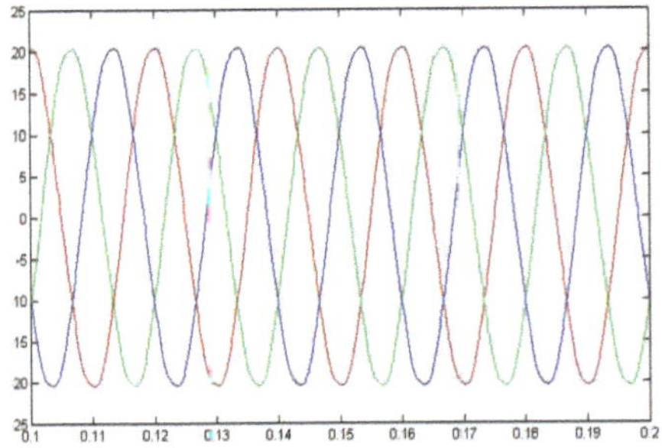

(a) Three-phase output current.

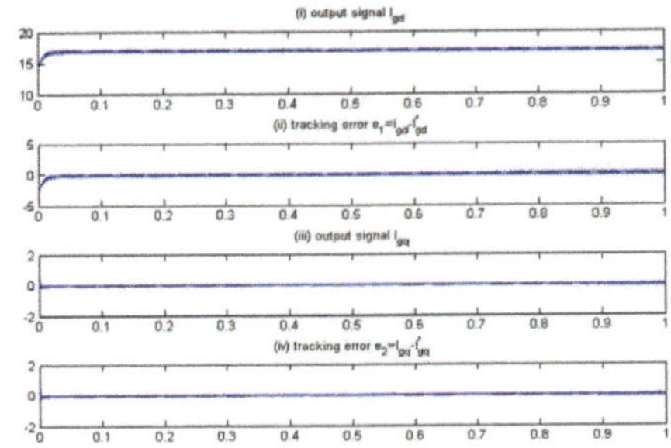

(b) I_{gd}, I_{gq} and their tracking error.

Fig. 5 System response for Case 2, steady-state grid voltage distortion

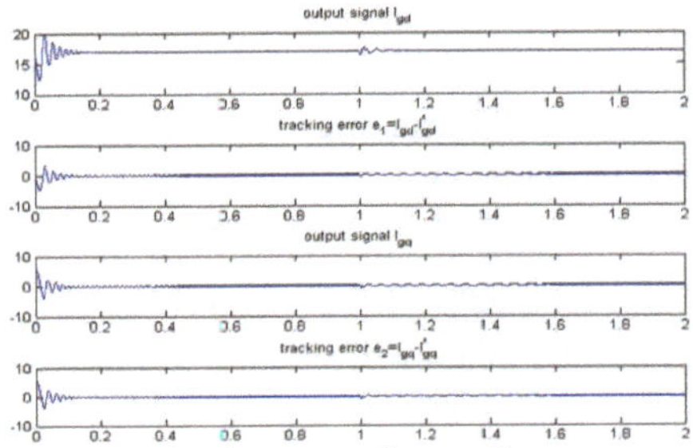

(a) Adaptive controller performance.

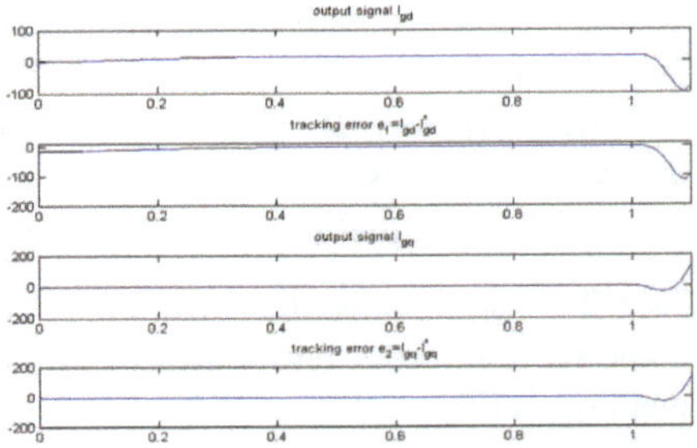

(b) Nominal controller performance.

Fig. 6 Adaptive and nominal response for Case 3, sudden change in U_{PV}

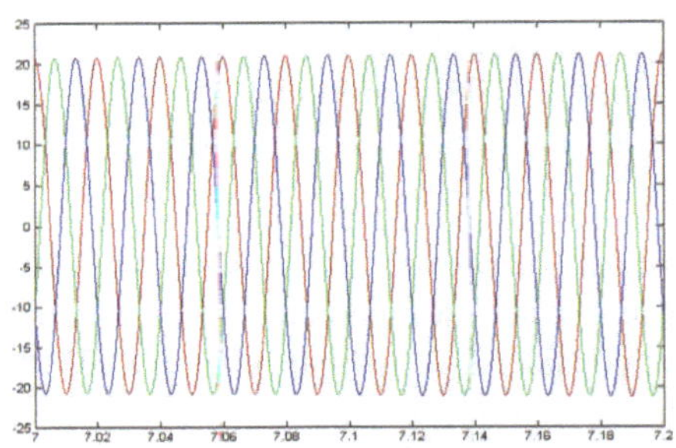

(a) Three-phase output current.

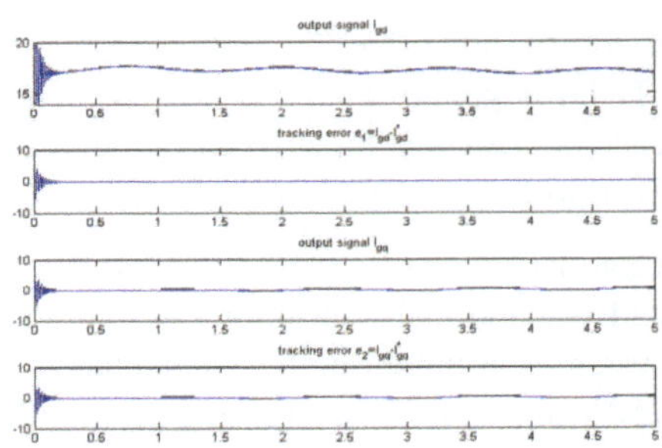

(b) I_{gd}, I_{gq} and their tracking error.

Fig. 7 System response for Case 4, variation in sunlight

change in sunlight intensity is slow with respect to the grid frequency, the magnitude change in Fig. 7a is also slow.

The simulation results indicate that the adaptive controller we designed can make the grid-connected PV inverter system work normally under system uncertainties and grid side harmonics disturbance, and the asymptotic tracking is achieved. This indicates that the PV inverter system can continuously work at the maximum power point P*.

5 Conclusions and Future Topics

This chapter has described the development of some technical foundations for applying adaptive control techniques to three-phase grid-connected PV inverter systems, including system modeling, design condition verification, adaptive controller development, control system analysis, and simulation validation. It shows that adaptive control techniques are suitable for dealing with the parameter uncertainties of PV inverter systems and have advantages in assuring desired system performance in the presence of large uncertainties.

Based on this work, further research using adaptive control techniques for renewable energy-based power generation systems can be carried out as listed below

Adaptive Control for Distributed PV System Distributed solar power generation systems are often subjected to load variations, weather uncertainties (variations of sunlight intensity), and uncertain faults such as actuator failures and component damage. By using effective adaptive fault detection and fault-tolerant control techniques, one can design controllers capable of recovering the desired distributed PV inverter system with parameter fault uncertainties.

Adaptive Frequency Control For the study presented in this chapter, we considered the grid fundamental frequency to be fixed. For more general cases when the grid side frequency may vary, further study for phase tracking should be considered.

Extension to Other Renewable Power Systems Many other types of renewable energy sources systems share some of the solar energy system characteristics of system uncertainty. The developed adaptive control technique for PV systems can be extended to other renewable energy based power generation systems such as wind power and marine power systems.

References

1. Munsell M (2017) US Solar Market Grows 95% in 2016, Smashes Records, Greentech Media, 15 February 2017
2. Fraas L, Partain L (2010) Solar Cells and Their Applications (Sect. 10.2), 2nd edn. Wiley, New Jersey
3. Vasquez JC, Guerrero JM, Savaghebi M, Eloy-Garcia J, Teodorescu R (2013) Modeling, analysis, and design of stationary-reference-frame droop-controlled parallel three-phase voltage source inverters. IEEE Trans Ind Electron 60(4):1271–1280
4. Teodorescu R, Blaabjerg F, Borup U, Liserre M (2004) A new control structure for grid-connected LCL PV inverters with zero steady-state error and selective harmonic compensation. In: Nineteenth annual IEEE applied power electronics conference and exposition, APEC'04, February 2004, pp 22–26

5. Rahim NA, Saidur R, Solangi KH, Othman M, Amin N (2012) Survey of grid-connected photovoltaic inverters and related systems. Clean Technol Environ Policy 14:521–533
6. Bernstein DS (1997) A student's guide to classical control. IEEE Control Syst Mag 17(4):96–100
7. Nise NS (2011) Control systems engineering, 6th edn. Wiley, Hoboken
8. Lavretsky E, Wise KA (2013) Robust and adaptive control with aerospace application. Springer, London
9. Chen C (1995) Linear system theory and design. Oxford University Press Inc., Oxford
10. Slotine JJE, Li W (1991) Applied nonlinear control. Prentice Hall Inc., Upper Saddle River
11. Tao G (2003) Adaptive control design and analysis. Wiley, New York
12. Wang H, Liu GP, Harris CJ, Brown M (1995) Advanced adaptive control. Elsevier Science (formerly Pergamon Press Ltd.), Amsterdam
13. Pedro RS, Carlos RC, Enrique RC, Husev O, Makovenko E (2013) Control scheme of a three-phase three level NPC qz-source inverter with LCL filter for RES applications. J Electr Eng Technol 8(3):544–558
14. Rigatos G, Siano P, Zervos N, Cecati C (2016) Control of three-phase voltage source converters with the derivative-free nonlinear kalman filter. Intell Ind Syst 2(1):21–33
15. Vasquez JC, Guerrero JM, Luna A, Rodríguez P, Teodorescu R (2009) Adaptive droop control applied to voltage-source inverters operating in grid-connected and islanded modes. IEEE Trans Ind Electron 56(10):4088–4096
16. Jung J, Vu N, Dang D, Do T, Choi Y, Choi H (2014) A three-phase inverter for a standalone distributed generation system: adaptive voltage control design and stability analysis. IEEE Trans Energy Convers 29(1):46–56
17. Gonzalez-Espin F, Gargera G, Patrao I, Figueres E (2012) An adaptive control system for three-phase photovoltaic inverters working in a polluted and variable frequency electric grid. IEEE Trans Power Electron 27(10):4248–4261
18. Hong W, Tao G (2018) An adaptive control scheme for three-phase grid-connected inverters in photovoltaic power generation system. In: Proceedings of the 2018 American control conference, Milwaukee, WI, 27–29 June 2018, pp 899–904
19. XC164 different PWM waveforms generation for 3-phase AC induction motor with XC164CS. Infineon Technologies AG 81726 Minchen, Germany, 2006
20. Wu Y, Lin J, Lin H (2017) Standards and guidelines for grid-connected photovoltaic generation systems: a review and comparison. IEEE Trans Ind Appl 53(4):3205–3216
21. Chung S (2000) A phase tracking system for three phase utility interface inverters. IEEE Trans Ind Electron 15(3):431–438
22. Timbus A, Teodoresecu R, Blaabjerg F (2005) Synchronization methods for three phase distributed power generation systems. In: Proceedings of power electronics specialists conference, pp 2474–2481
23. Figueres E, Garcerá G, Sandia J (2009) Sensitivity study of the dynamics of three-phase photovoltaic incerters with an LCL grid filter. IEEE Trans Ind Electron 56(3):706–717
24. Kamalakannan C, Suresh LP (2015) Design and analysis of three phase four wire. Power electronics and renewable energy systems. Springer, New York, p 1029
25. Park RH (1929) Two reaction theory of synchronous machines. AIEE Trans 48:716–730
26. Wen L, Tao G, Liu Y (2016) Multivariable adaptive output rejection of unmatched input disturbances. Int J Adapt Control Signal Process 30:1203–1227
27. Wen L, Tao G, Liu Y (2015) Aircraft turbulence compensation using adaptive multivariable disturbance rejection techniques. J Guid Control Dyn 28(5):954–962
28. Chen L, Amirahmadi A, Zhang Q, Kutkut N, Batarseh I (2014) Design and implementation of three-phase two-stage grid-connected module integrated converter. IEEE Trans Power Electron 28(9)

Application of Sliding-Mode Control for Maximum Power Point Tracking of PV Systems

M. R. Mojallizadeh and M. A. Badamchizadeh

Abstract Sliding-mode controllers are widely utilized for increasing the output power of photovoltaic sources. These sliding-mode controllers are primarily based on one-loop or two-loop schemes. The two-loop scheme is composed of two loops, i.e., searching and tracking loops. A maximum power point searching unit is utilized in the searching loop, and a tracking controller is utilized in the other loop to extract the maximum photovoltaic power. Compared to this scheme, the one-loop scheme can extract the maximum photovoltaic power without any searching algorithm. In this study, dynamic equations of a typical photovoltaic power source are derived using the state-space averaging method. Afterwards, one-loop and two-loop sliding-mode control schemes are used for extracting the maximum power. Stability of both schemes is guaranteed using Lyapunov theory. Conditions of the robust stability are derived analytically for both schemes. A deterministic cuckoo search algorithm is used for maximum power point searching in the two-loop scheme. Performances of the schemes are evaluated by some experiments and numerical simulations. Results are compared in uniform and partially shaded conditions.

Nomenclature

I_P	Current of the PV array
V_P	Nominal PV voltage
ΔV_P	Uncertainty of PV voltage
T_p	Temperature
T_{ref}	Reference temperature
Γ	Irradiance level
Γ_m	Maximum irradiance level
N_s	No. of series cells
N_p	No. of parallel cells

M. R. Mojallizadeh · M. A. Badamchizadeh (✉)
University of Tabriz, Tabriz, Iran
e-mail: mbadamchi@tabrizu.ac.ir

M. R. Mojallizadeh
e-mail: mojalli@tabrizu.ac.ir

R.-E. Precup et al. (eds.), *Solar Photovoltaic Power Plants*, Power Systems,
https://doi.org/10.1007/978-981-13-6151-7_2

q	Electron charge
A	Ideality coefficient
i_r	Saturation current at T_{ref}
I_{rs}	Saturation current
I_{ph}	Light generated current
K_0	Boltzmann's constant
r_p	Equivalent resistance
I_{sc}	Short circuit current
E_{go}	Semiconductor band-gap energy
C	Nominal capacitance
L	Nominal inductance
R	Nominal impedance of the load
ΔC	Uncertainty of the capacitance
ΔL	Uncertainty of the inductance
ΔR	Uncertainty of the impedance
V_O	Output voltage
W_1 and W_2	Lumped uncertainties
s	Sliding surface
u	Control signal
u_{eq}	Equivalent control
u_n	Curbing control
η	Control gain of the one-loop scheme
c	Control gain of the two-loop scheme
V_{MPP}	Optimal PV voltage
I_{MPP}	Optimal PV current
P_{MPP}	Maximum power
T	Dwell-time of irradiance levels

1 Introduction

Maximum power point (MPP) tracking units should be utilized in photovoltaic (PV) sources to increase their efficiency. Figure 1 shows that DC converters are usually utilized between the load and PV cells for MPP tracking. However, a tracking controller is required to control this converter. In this case, many MPP tracking con-

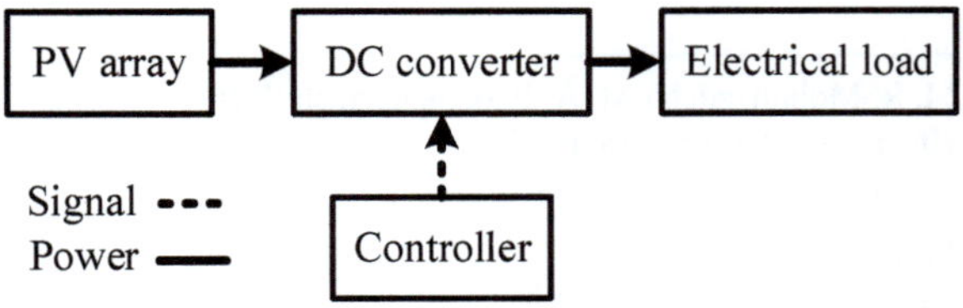

Fig. 1 A typical photovoltaic power source

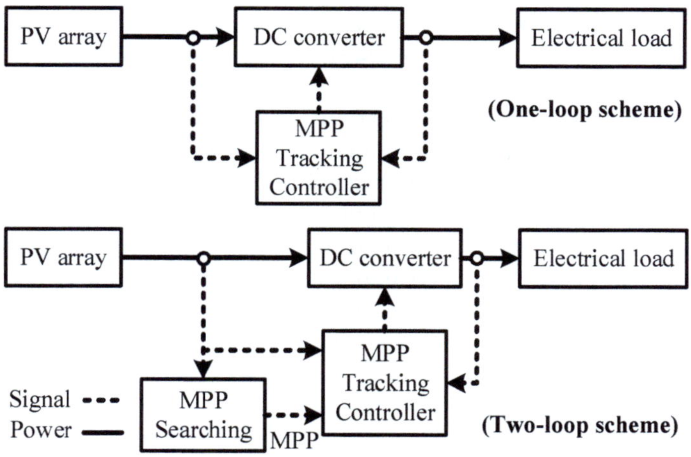

Fig. 2 One-loop and two-loop control schemes

trollers have been introduced, e.g., PID controller [1, 2], passivity-based control [3, 4], input-output linearization [5], feedback linearization [6] and sliding-mode control (SMC) [7–16]. Among these controllers, SMC is more popular due to its global stability and robustness. However, the primary problem of the SMC is its chattering problem. The discontinuous nature of the control law usually leads to the chattering problem. Chattering deteriorates the performance of the control systems [17].

SMC is a type of variable structure controller [17]. Each SMC is composed of a switching surface and a control law. One-loop and two-loop SMCs are shown in Fig. 2. In the one-loop scheme, the MPP tracking controller does not require MPP. On the other hand, the two-loop scheme requires an additional algorithm for MPP searching. These schemes are analyzed in Sects. 3 and 4, respectively.

The one-loop SMC was introduced in [7]. In this scheme, the switching surface is selected as dP/dI [7] or dP/dV [18], where P, I and V are power, current and voltage, respectively. By converging the switching surface to zero, the maximum power will be extracted. The number of required voltage sensors for this scheme is decreased in [8]. The one-loop control scheme can also be used for grid-connected PV systems [15]. Robust version of one-loop SMC is proposed in [12].

In two-loop scheme, the first and second loops are used for MPP searching and MPP tracking, respectively. The interaction between the loops must be considered in the designing procedure, i.e., the tracking loop should be faster than the searching loop. In this context, a terminal SMC was introduced in [10], which ensures the finite time stability. Thus, it can be shown that the tracking loop is always faster than the searching loop.

In order to calculate the MPP in the searching loop, many algorithms have been proposed in the literature [19]. Perturbation and observation (P&O) [19] is a well-known algorithm. This algorithm changes the current/voltage of PV cells and calculates the MPP by measuring the resulting power. It has a straightforward implementation. However, its performance is affected by the size of perturbations. Incremental conductance (IC) is another common MPP searching algorithm [20]. In this algorithm, the MPP is calculated using the sign of dP/dV. The key problem of the aforementioned methods is that they cannot be used in partially shaded conditions. Deterministic cuckoo search (CS) algorithm has been recently developed in [21]. By comparing the performance of this algorithm, it is shown that it presents better responses than conventional methods during partially shaded conditions [21].

The dynamic equations of a PV source are derived in Sect. 2. To provide a fair comparison, typical one-loop and two-loop control schemes are designed in Sects. 3 and 4, respectively. The characteristics of these schemes have been compared using numerical simulations and experiments in Sects. 5 and 6. Results are presented in Sect. 7.

2 Equations of the PV Source

Figure 3 shows a typical PV source. The system has a PV panel and a converter which are modeled in Sects. 2.1 and 2.2.

2.1 Equations of the PV Array

The following equation describes the characteristic of the PV array [22]:

$$I_P = \left(I_{ph} - I_{rs} \left(-1 + e^{\dfrac{qV_P}{T_P N_s K_0 A}} \right) \right) N_p \tag{1}$$

where I_P and V_P are current and voltage, respectively. I_{ph} is the light generated current, ideality coefficient is denoted by A, K_0 denotes the Boltzmann's constant, I_{rs} denotes current of the saturation, N_p and N_s indicate no. of parallel and series cells, respectively, and T_p denotes the temperature. I_{rs} and I_{ph} can be described by Eqs. (2) and (3) [22].

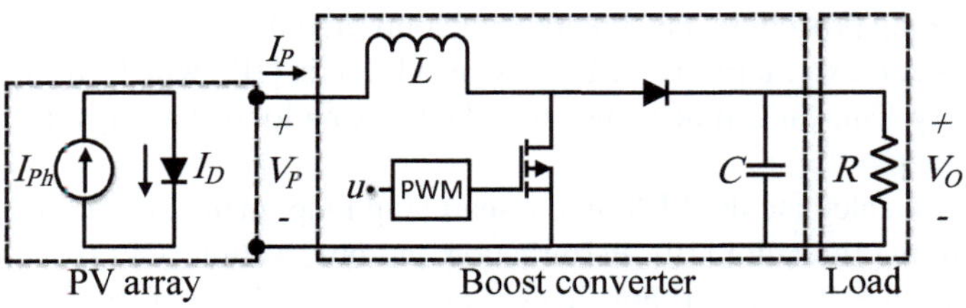

Fig. 3 Photovoltaic source

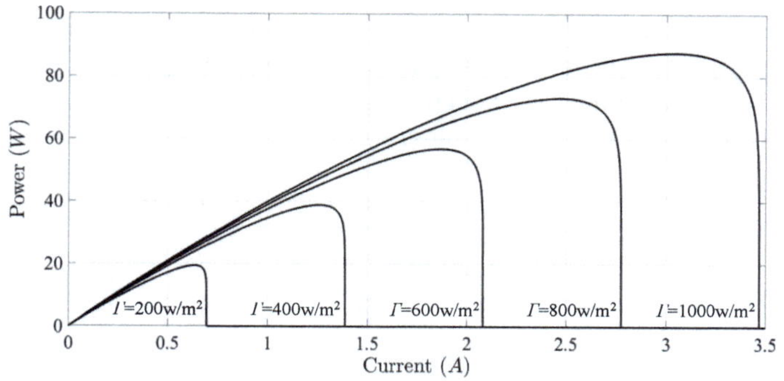

Fig. 4 The power-current characteristic

$$I_{rs} = i_r(T_p/T_{ref})^3 e^{\frac{qE_{go}}{K_0 A}\left[1/T_{ref} - 1/T_p\right]} \quad (2)$$

$$I_{ph} = \Gamma\big(I_{sc} + (T_p - T_{ref})K_l\big)/1000 \quad (3)$$

where i_r denotes the saturation current at T_{ref}, $K_l(A/K)$ denotes the temperature effect. Γ denotes the irradiance. A series resistance r_p is typically considered for modeling the total Ohmic loss. Figure 4 shows the characteristic of a typical PV panel.

2.2 Equations of the Converter

The converter converges the array to its MPP. Using averaging method [23], the equations of the PV source is derived as follows [24]:

$$\begin{cases} \dot{I}_P = -\frac{(1-u)}{\Delta L + L} V_O + \dfrac{V_P + \Delta V_P}{\Delta L + L} \\ \dot{V}_O = \frac{(1-u)}{\Delta C + C} I_P - \dfrac{V_O}{(C + \Delta C)(\Delta R + R)} \end{cases} \quad (4)$$

where C, L and R are nominal capacitance, inductance and load, respectively. u denotes the PWM control signal, V_O and V_P are output and input voltages of the converter. ΔR, ΔL, ΔC, and ΔV_P are uncertainties of R, L, C and V_P, respectively. Equation (4) can be rewritten as

$$\begin{cases} \dot{I}_P = -\dfrac{1-u}{L} V_O + \dfrac{V_P}{L} + \dfrac{W_1}{L} \\ \dot{V}_O = \dfrac{1-u}{C} I_P - \dfrac{V_O}{CR} + \dfrac{W_2}{C} \end{cases} \quad (5)$$

where W_1 and W_2 are the lumped uncertainties. W_1 and W_2 are calculated in Eq. (6). From Eqs. (5) and (6), it can be seen that the real model of a PV system is uncertain. Note that these uncertainties are not considered in many control design procedures. This usually leads to a lack of robustness.

$$\begin{cases} W_1 = L\left(-\dfrac{V_P}{L} - \dfrac{1-u}{L+\Delta L}V_O + \dfrac{V_P+\Delta V_P}{L+\Delta L} + \dfrac{1-u}{L}V_O\right) \\ W_2 = C\left(\dfrac{V_O}{RC} - \dfrac{I_P(1-u)}{C} - \dfrac{V_O}{(\Delta R+R)(C+\Delta C)} + \dfrac{I_P(1-u)}{\Delta C + C}\right) \end{cases} \quad (6)$$

3 One-Loop Sliding-Mode Control System

Each SMC has a switching surface and a control law. Based on Fig. 4, the switching surface is considered as follow to converge the PV array to its MPP.

$$\frac{\partial P_P}{\partial I_P} = \frac{\partial R_P I_P^2}{\partial I_P} = I_P\left(2R_P + \frac{\partial R_P}{\partial I_P} I_P\right) = 0 \quad (7)$$

where $P_P = I_P V_P$ and $R_P = V_P/I_P$. Since $I_P \neq 0$, the switching surface is

$$s \triangleq 2R_P + I_P\frac{\partial R_P}{\partial I_P} \quad (8)$$

The control signal is proposed as

$$u = u_{eq} + u_n \quad (9)$$

where u_n is the curbing control, and u_{eq} denotes the equivalent control. u_n drives the PV panel to the MPP. By setting $W_1 = 0$, u_{eq} is determined as follows [12]:

$$\dot{s} = (\partial s/\partial I_P)\dot{I}_P = 0 \rightarrow (-(1-u_{eq})V_O + V_P)/L = 0 \rightarrow u_{eq} = 1 - V_P/V_O \quad (10)$$

Moreover, u_n is selected as Eq. (11) [12]:

$$\begin{cases} u_n = k \cdot s + \cdot sgn(s)\eta \\ \eta = \eta_0(\mid s \mid + \delta) \end{cases} \quad (11)$$

where δ, k, and η_0 are some positive constants and sgn is the sign function. Equation (11) shows η depends on the switching surface (s). Therefore, the chattering will be suppressed as $s \rightarrow 0$.

From Eq. (3), $\partial s/\partial I_P$ can be calculated as follows:

$$\frac{\partial s}{\partial I_P} = 3\frac{\partial R_P}{\partial I_P} + I_P\frac{\partial^2 R_P}{\partial {I_P}^2} \tag{12}$$

Considering $R_P = V_P/I_P$, the following equations are achieved:

$$\begin{aligned} \frac{\partial R_P}{\partial I_P} &= \frac{\partial}{\partial I_P}\left(\frac{V_P}{I_P}\right) = \frac{1}{I_P}\frac{\partial V_P}{\partial I_P} - \frac{V_P}{I_P^2} \\ \frac{\partial^2 R_P}{\partial {I_P}^2} &= 2\frac{V_P}{{I_P}^3} + \frac{1}{I_P}\frac{\partial^2 V_P}{\partial {I_P}^2} - \frac{2}{{I_P}^2}\frac{\partial V_P}{\partial I_P} \end{aligned} \tag{13}$$

From Eq. (1), V_P can be written as Eq. (14).

$$V_P = \frac{N_s \cdot A \cdot K_0 \cdot T}{q} ln\left(\frac{N_P \cdot I_{ph} + N_P \cdot I_{rs} - I_P}{N_P \cdot I_{rs}}\right) \tag{14}$$

Derivatives of the V_P are as follows:

$$\frac{\partial V_P}{\partial I_P} = -\frac{N_s \cdot A \cdot K_0 \cdot T}{q}\frac{N_P \cdot I_{rs}}{N_P \cdot I_{ph} + N_P \cdot I_{rs} - I_P} < 0 \tag{15}$$

$$\frac{\partial {V_P}^2}{\partial^2 I_P} = -\frac{N_s \cdot A \cdot K_0 \cdot T}{q}\frac{N_P \cdot I_{rs}}{(N_P \cdot I_{ph} + N_P \cdot I_{rs} - I_P)^2} < 0 \tag{16}$$

Based on Eqs. (13)–(16), Eq. (12) is rewritten:

$$\frac{\partial s}{\partial I_P} = \frac{1}{I_P}\frac{\partial V_P}{\partial I_P} + \frac{\partial^2 V_P}{\partial {I_P}^2} - \frac{V_P}{{I_P}^2} < 0 \tag{17}$$

From Eqs. (5), (9) and (10):

$$\begin{aligned} \dot{I}_P &= -\frac{1 - u_{eq} - u_n}{L}V_O + \frac{V_P}{L} + W_1 = -\frac{1 - 1 + V_P/V_O - u_n}{L}V_O + \frac{V_P}{L} + \frac{W_1}{L} = \\ &(V_O/L)u_n + \frac{W_1}{L} \end{aligned} \tag{18}$$

Considering $V = (1/2)s^2$ as the Lyapunov function and Eqs. (10), (11), (18), $\dot{V}$ is

$$\begin{aligned} \dot{V} &= s\dot{s} = s\frac{\partial s}{\partial I_P}\left(\frac{V_O}{L}u_n + \frac{W_1}{L}\right) = s\frac{\partial s}{\partial I_P}\left(\frac{V_O}{L}(\eta_0(\mid s \mid + \delta) \cdot sgn(s) + k \cdot s) + \frac{W_1}{L}\right) = \\ &\frac{\partial s}{\partial I_P}\left(\frac{V_O}{L}\eta_0 s^2 + \frac{V_O}{L}\eta_0\delta \mid s \mid + \frac{V_O}{L}ks^2 + s\frac{W_1}{L}\right) \le \\ &\frac{\partial s}{\partial I_P}\left(\frac{V_O}{L}\eta_0 s^2 + \frac{V_O}{L}\eta_0\delta \mid s \mid + \frac{V_O}{L}ks^2 - \mid s \mid\mid \frac{W_1}{L} \mid\right) = \\ &\frac{\partial s}{\partial I_P}\left(\frac{V_O}{L}(\eta_0 s^2 + ks^2) + \frac{V_O}{L}\eta_0\delta \mid s \mid - \mid s \mid\mid \frac{W_1}{L} \mid\right) < 0, s \neq 0 \end{aligned} \tag{19}$$

Thus, the control system is stable for:

$$\eta_0 \delta > \frac{| W_1 |}{V_O}, \qquad k > 0 \tag{20}$$

Remark 1 To decrease the chattering amplitude, the parameters are designed such that $\eta_0 >> \delta$.

Remark 2 At $t = 0$, $V_O = 0$ may be achieved. However, with any arbitrary control signal, $V_P < V_O$ will be held and Eq. (20) can be satisfied with reasonable amounts of $\eta_0 \delta$.

4 Two-Loop Sliding-Mode Control Scheme

A typical two-loop SMC is as follows:

$$\begin{aligned} u &= u_{eq} + u_n \\ u_{eq} &= 1 - \frac{V_P}{V_O} \\ u_n &= -csgn(s) \end{aligned} \tag{21}$$

where $c > 0$. Sliding surface is defined as follows:

$$s = I_P - I_{MPP} \tag{22}$$

where I_{MPP} is current of the MPP. Considering $V = \frac{1}{2}s^2$ as the Lyapunov function, $\dot{V}$ is:

$$\dot{V} = s\dot{s} \tag{23}$$

Using Eq. (5),

$$\dot{V} = \big(- (1 - u)V_O + V_P + W_1 \big) \frac{1}{L} s \tag{24}$$

Based on Eqs. (21) and (24):

$$\begin{aligned} \dot{V} &= \frac{1}{L}\big(- c|s|V_O + W_1 s \big) \\ \dot{V} &< \frac{1}{L}\big(- c|s|V_O + |W_1||s| \big) \\ \dot{V} &< \frac{1}{L}|s|\big(- cV_O + |W_1| \big) \end{aligned} \tag{25}$$

Equation (25) shows that the two-loop control system is stable for:

$$c > \frac{|W_1|}{V_O} \tag{26}$$

Remark 3 At the initial time ($t = 0$), V_O may be equal to zero. In this case, the Eq. (26) does not hold. However, since a boost converter is used, $V_O > V_P$ is always achieved for any arbitrary condition. Therefore, Eq. (26) holds after a transient time.

Remark 4 To suppress the chattering amplitude, a continues functions, i.e., the saturation function (*sat*) is usually replaced with the *sign* function in Eq. (21).

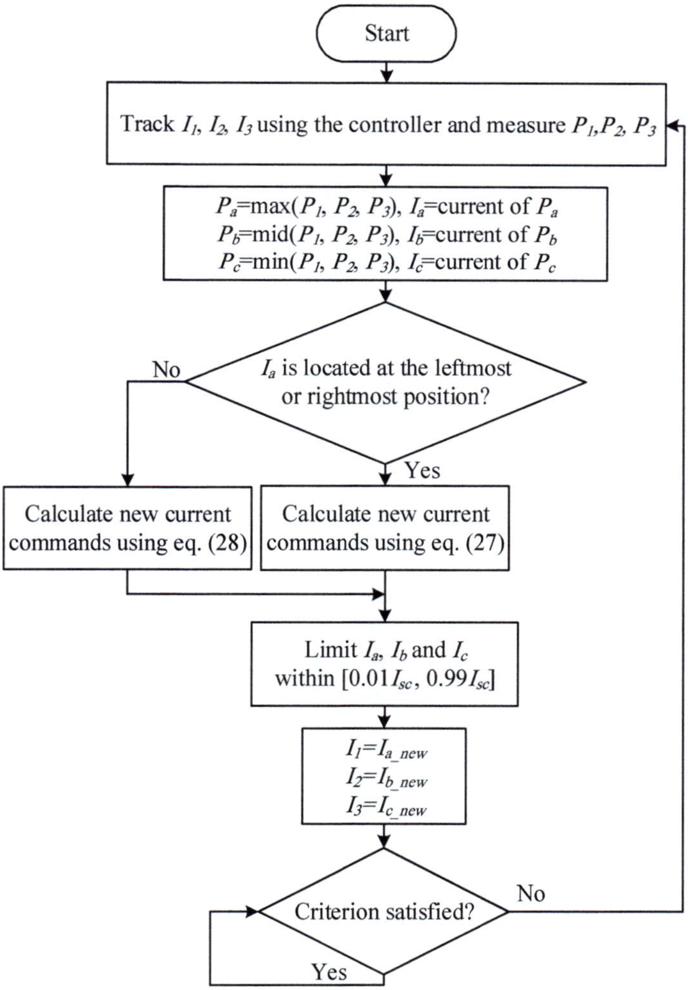

Fig. 5 Flowchart of the deterministic CS algorithm [21]

Remark 5 From Eqs. (21) and (22), it can be seen that the two-loop MPP scheme requires an extra MPP searching for calculating the I_{MPP}. It should be noted that the one-loop scheme does not need MPP reference or MPP searching loop.

Deterministic cuckoo search (CS) algorithm has been used for MPP searching. Flowchart of this algorithm is shown in Fig. 5. This algorithm is as follows [21]:

1. $I_a = 0.05I_{sc}$, $I_b = 0.5I_{sc}$ and $I_c = 0.95I_{sc}$, where I_a, I_b and I_c are currents of the PV panel at three different instants.
2. Sort the selected currents such that $P(I_a) \geq P(I_b) \geq P(I_c)$, where P is the power.
3. If I_a is located at the rightmost or leftmost positions, modify the currents as follows Eq. (27).

$$\begin{aligned} I_{a,new} &= I_a \\ I_{b,new} &= I_b + \varepsilon(I_a - I_b) \\ I_{c,new} &= I_a + \varepsilon(I_a - I_{b,new}). \end{aligned} \tag{27}$$

   Else, update the currents using Eq. (28)

$$\begin{aligned} I_{a,new} &= I_a \\ I_{b,new} &= I_b + \varepsilon(I_a - I_b) \\ I_{c,new} &= I_c + \varepsilon(I_a - I_c) \end{aligned} \tag{28}$$

   where $\varepsilon > 0$ is the multiplication constant.
4. If the following objective is met, stop the search.

$$max\{|I_a - I_b|, |I_a - I_c|\} < 0.01\,\text{A}. \tag{29}$$

5 Numerical Simulations

Responses of the one-loop and two-loop control schemes are evaluated in a number of simulations. Tables 1 and 2 show the coefficients of the PV panel and related MPPs. Parameters of the converter are designed for continuous conduction mode (CCM) and shown in Table 3 [25]. Thus, currents of the inductors will never reach zero.

Table 1 Parameters of the PV array

$K_0 = 1.3805 \times 10^{-23}$ J/K	$K_l = 12 \times 10^{-4}$ A/K
$N_p = 1$	$N_s = 92$
$i_r = 5.98 \times 10^{-8}$ A	$I_{sc} = 3.5$ A
$A = 1.13$	$q = 1.6 \times 10^{-19}$ C
$E_{go} = 1.21$ ev	$r_p = 3\,\Omega$
$T_{ref} = 298$ K	

Table 2 MPPs of the PV array in uniform insolation conditions

Γ (w/m^2)	V_{MPP} (V)	I_{MPP} (A)	$P_{MPP} = V_P I_P$ (W)
200	30.71	0.631	19.38
400	30.8	1.26	38.81
600	30.38	1.87	56.8
800	29.72	2.46	73.11
1000	28.84	3.04	87.67

Table 3 Parameters of the boost converter

Load R	100 Ω
Inductors L	1 mH
Capacitor C	100 μF

Parameters of the one-loop controller are designed based on Eq. (20) as $k = 0.01$, $\eta_0 = 1$ and $\delta = 0.01$. Moreover, the only parameter of the two-loop scheme $c = 0.01$ is designed based on Eq. (26).

The performances of the controllers are evaluated in uniform insolation and partially shaded conditions in Sects. 5.1 and 5.2, respectively.

5.1 *Uniform Insolation Condition*

Responses of the controllers are examined under the irradiance profile of the international standard EN50530:2010 which is shown in Fig. 6. This profile evaluates the performances in five different irradiance levels with uniform insolations. In this case, the maximum irradiance level and dwell-time of the irradiance are $\Gamma_m = 1000$ w/m^2 and $T = 10$ s, respectively.

Responses of the one-loop and two-loop schemes under test profile EN50530:2010 are shown in Figs. 7 and 8, respectively. Based on the numerical simulations, the tracking controller of the two-loop scheme can track the reference MPP of the MPP searching loop within 20 ms. In order to ensure the perfect tracking and avoid the

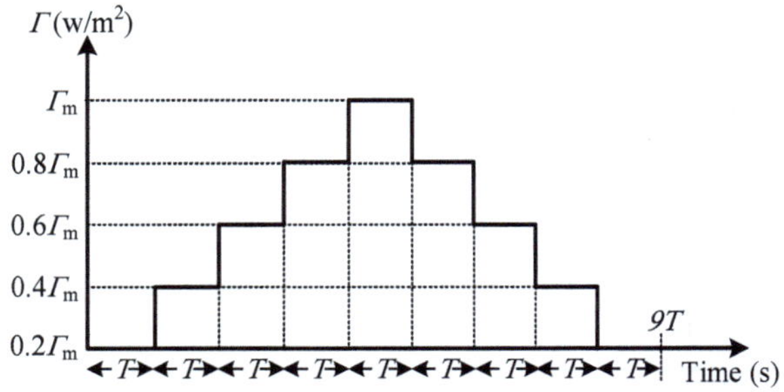

Fig. 6 Insolation profile of the standard EN50530:2010

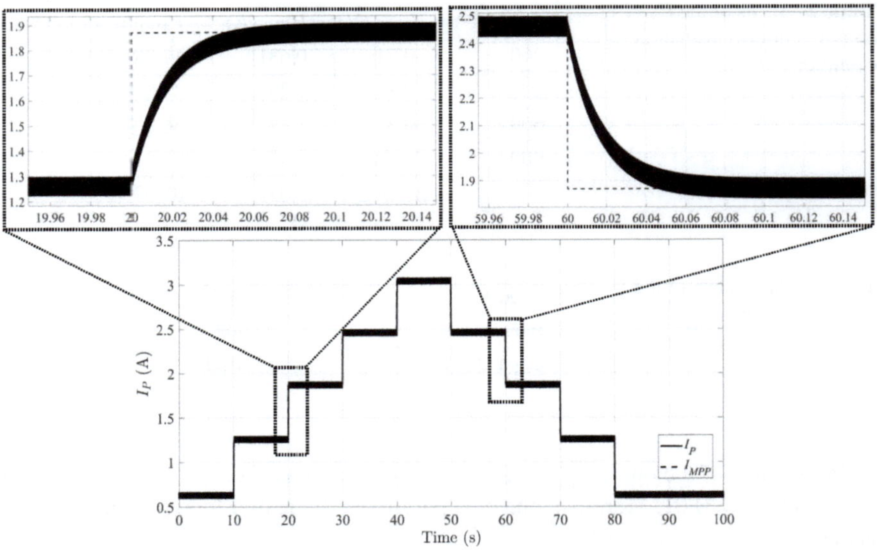

Fig. 7 Response of the one-loop scheme under test profile EN50530:2010

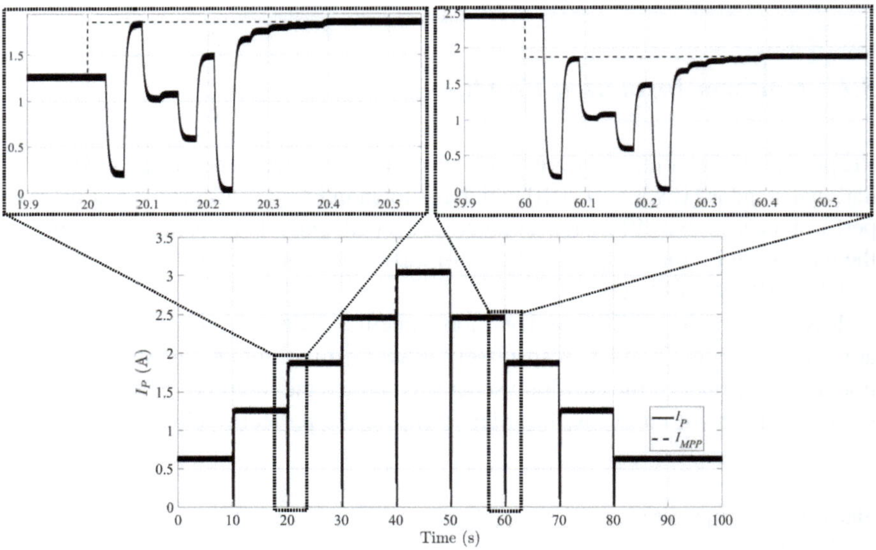

Fig. 8 Response of the two-loop scheme under test profile EN50530:2010

transient fluctuations, the tracking loop should be faster than the searching loop. Therefore, the period of the searching loop is selected as 30 ms which is higher than the settling time of the tracking loop. Figure 8 shows that the two-loop scheme converges to the MPP in 400 ms with zero steady-state error.

From Fig. 7, it can be seen that the one-loop scheme eliminates the disturbances of the irradiance change, completely. In this case, the one-loop scheme tracks its MPP in 80 ms. Therefore, compared to the two-loop scheme, it converges to its MPP 400 ms/80 ms = 5 times faster.

Conversion efficiencies ($\frac{\int_0^{90s} I_P V_P dt}{\int_0^{90s} I_{MPP} V_{MPP} dt}$) of the one-loop and two-loop schemes are 99.99 and 98.96%, respectively. Thus, the one-loop scheme presents better efficiency under uniform insolation conditions. Note that power loss of the converter is not considered in the definition of the efficiency.

5.2 Partially Shaded Condition

In order to examine the controllers more precisely, the controllers are tested under a partially shaded condition. Figure 9 shows the power-current curve under this partially shaded condition. In this case, the power-current curve presents a local maximum which is not the MPP. It can be seen that the current of the MPP is ($I_{MPP} = 2.9$ A) which corresponds to $P_{MPP} = 42.1$ W.

Figures 10, 11 and 12 show the output current, voltage, and power in the one-loop scheme, respectively. Since the one-loop scheme is mainly designed for uniform insolation conditions, it is trapped in local maxima ($I_P = 0.83$ A). It can be seen from Figs. 10 and 11 that the voltage and current do not converge to the MPP. As a result, based on Fig. 12, the output power is 27.56 W in the steady-states. In this case, the steady-state efficiency ($\frac{I_P V_P}{I_{MPP} V_{MPP}}$) of the one-loop scheme is 65%.

Figures 13, 14 and 15 show the output current, voltage and power in the two-loop scheme, respectively. Note that the two-loop scheme tracks the MPP within 270 ms. Thus, the two-loop scheme converges to the MPP even in partially shaded conditions. Figure 15 shows that the MPP is achieved. Therefore, the efficiency is 99.99% for the two-loop scheme. As mentioned before, the power loss of the converter is not taken into account.

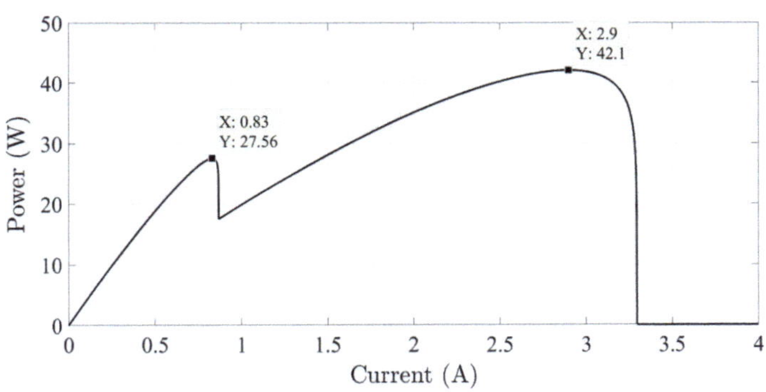

Fig. 9 Power-current curve under a partially shaded insolation condition

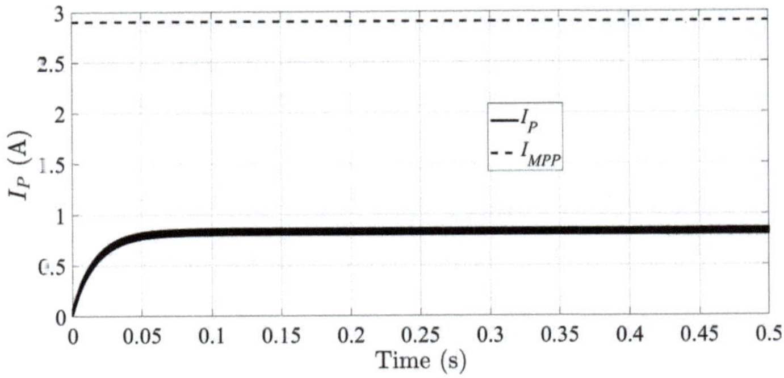

Fig. 10 One-loop control scheme under the partially shaded insolation: output current

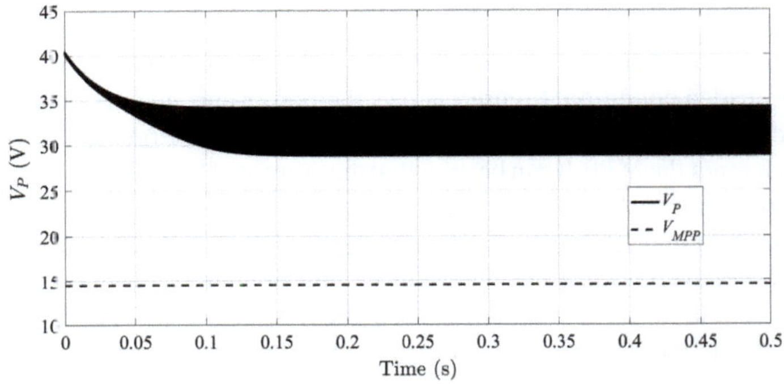

Fig. 11 One-loop control scheme under the partially shaded insolation: output voltage

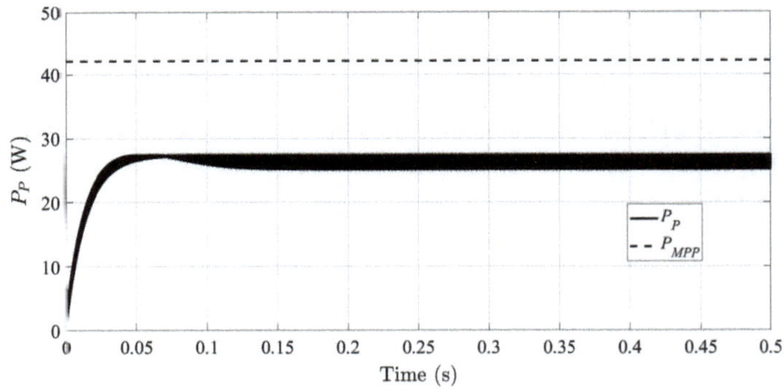

Fig. 12 One-loop control scheme under the partially shaded insolation: output power

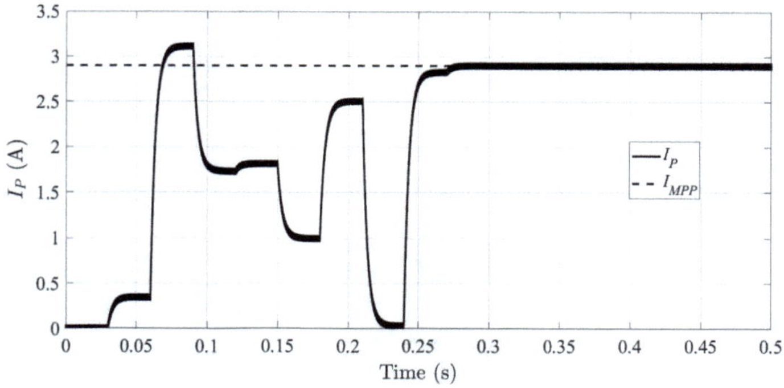

Fig. 13 Two-loop control scheme under the partially shaded insolation: output current

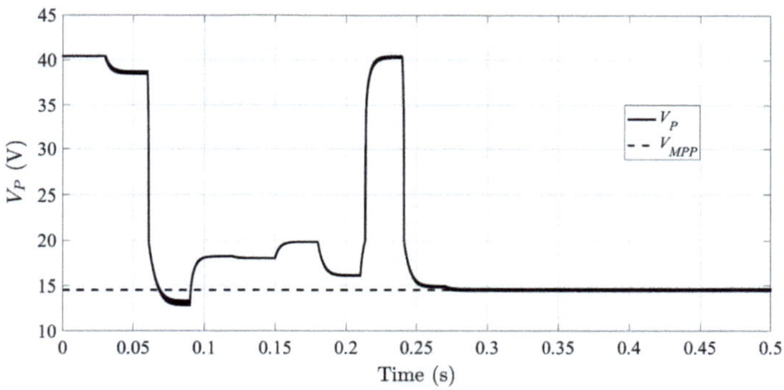

Fig. 14 Two-loop control scheme under the partially shaded insolation: output voltage

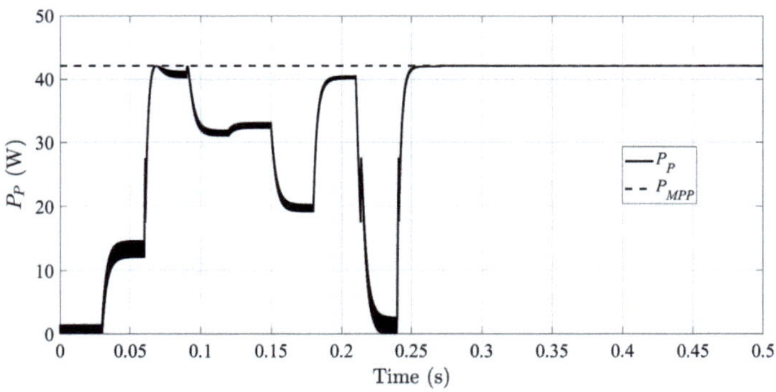

Fig. 15 Two-loop control scheme under the partially shaded insolation: output power

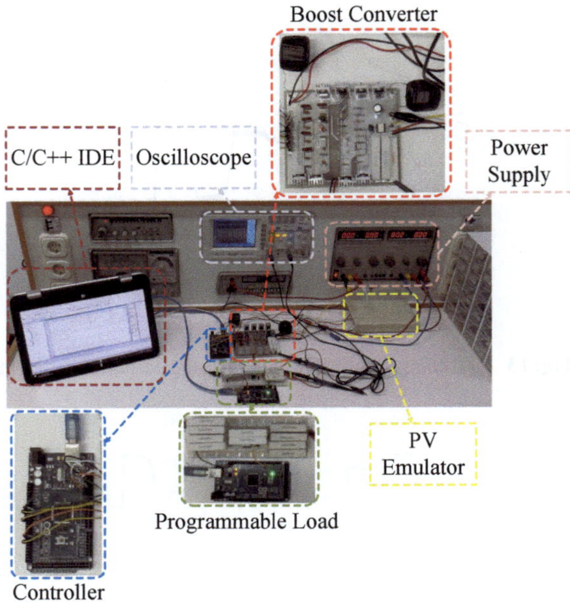

Fig. 16 Test bench

6 Experimental Results

Digital implementations of the schemes have been written in *C* language. ATmega2560 microcontroller is used to implement the control schemes as shown in Fig. 16. The arduino software package is utilized for compiling the programs. TLP250 gate driver is used to drive a IRLZ44n MOSFET switch. The PV current is measured using a current module. The PV array is emulated by a emulator [26].

Conditions of the experiments are similar to the partially shaded case of the numerical simulations (Sect. 5.2). Waveforms are recorded in the partially shaded profile (Fig. 9). Figures 17 and 18 show the current for the one-loop and two-loop schemes, respectively. Similar to the simulations, the one-loop scheme is trapped in the local maximum ($I_P = 0.83$ A), while the two-loop scheme tracks the MPP ($I_{MPP} = 2.9$ A) perfectly. In this case, the steady-state efficiencies of the one-loop and two-loop schemes are 65 and 99.9%, respectively. Results of the simulations and experiment are summarized in Table 4. It can be seen that the one-loop scheme can increase the efficiency by 1% in uniform insolation conditions. However, it presents poor performances in partially shaded conditions.

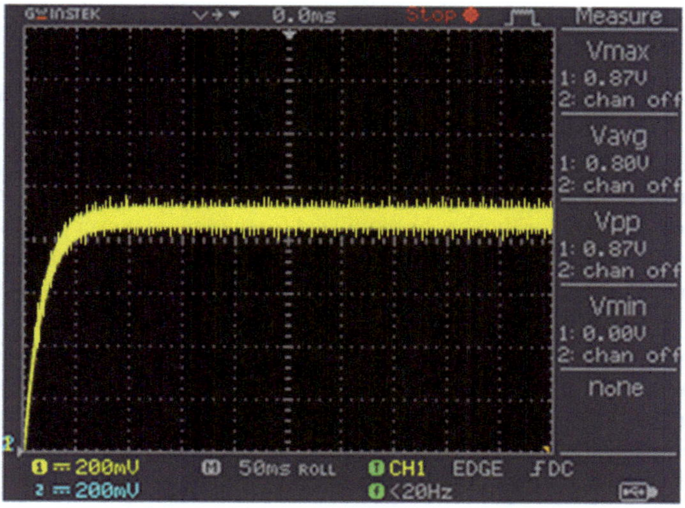

Fig. 17 Experimental result of the one-loop scheme under the partially shaded condition: I_P (1V/A)

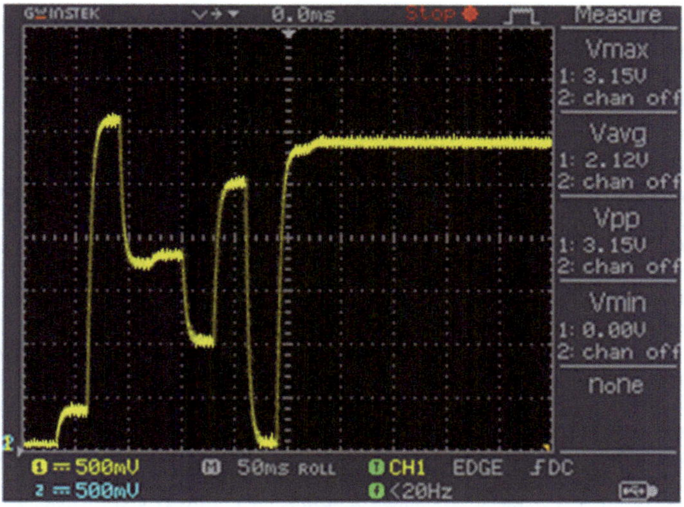

Fig. 18 Experimental result of the two-loop scheme under the partially shaded condition: I_P (1V/A)

Table 4 Summarized results

Parameters	One-loop scheme (%)	Two-loop scheme (%)
Average efficiency under test profile EN50530:2010[a]	99.9	98.9
Steady-state efficiency in the partially shaded condition[a]	65	99.9

[a]Note that power loss of the converter is not taken into account

7 Conclusion

Typical one-loop and two-loop sliding-mode control schemes are designed for a photovoltaic power source. Robust stability of the schemes is guaranteed using Lyapunov stability theory. Compared to the one-loop controller, the two-loop scheme requires an additional maximum power point searching algorithm. Thus, deterministic cuckoo search algorithm has been utilized for maximum power point searching. Characteristics of the control schemes are evaluated in a number of simulations and experiments. Results of the control schemes are compared in uniform and partially shaded conditions. It is shown that the one-loop scheme presents faster responses than the two-loop scheme. As a result, it can increase the efficiency in uniform insolation condition. However, since the one-loop scheme is mainly designed for uniform insolation conditions, it will be trapped in local maxima. On the other hand, two-loop scheme converges to the MPP and shows better responses in partially shaded conditions compared to the one-loops scheme.

Acknowledgements The project was financially supported by Iran National Science Foundation (INSF) and University of Tabriz.

References

1. Liu Y-H, Liu C-L, Huang J-W, Chen J-H (2013) Neural-network-based maximum power point tracking methods for photovoltaic systems operating under fast changing environments. Sol Energy 89:42–53
2. Mahammad AK, Saon S, Chee WS (2013) Development of optimum controller based on MPPT for photovoltaic system during shading condition. Procedia Eng 53:337–346
3. Tofighi A, Kalantar M (2011) Power management of PV/battery hybrid power source via passivity-based control. Renew Energy 36(9):2440–2450
4. Mojallizadeh MR, Badamchizadeh MA (2016) Adaptive passivity-based control of a photovoltaic/battery hybrid power source via algebraic parameter identification. IEEE J Photovolt 6(2):532–539
5. Espinoza-Trejo DR, Brcenas-Brcenas E, Campos-Delgado DU, Angelo CHD (2015) Voltage-oriented input-output linearization controller as maximum power point tracking technique for photovoltaic systems. IEEE Trans Ind Electron 62(6):3499–3507
6. Lalili D, Mellit A, Lourci N, Medjahed B, Boubakir C (2013) State feedback control and variable step size MPPT algorithm of three-level grid-connected photovoltaic inverter. Sol Energy 98(Part C):561–571
7. Chu C-C, Chen C-L (2009) Robust maximum power point tracking method for photovoltaic cells: a sliding mode control approach. Sol Energy 83(8):1370–1378
8. Zhang F, Maddy J, Premier G, Guwy A (2015) Novel current sensing photovoltaic maximum power point tracking based on sliding mode control strategy. Sol Energy 118:80–86
9. Ghazanfari J, Farsangi MM (2013) Maximum power point tracking using sliding mode control for photovoltaic array. Iran J Electr Electron Eng 9:189–196
10. Chiu C-S, Ouyang Y-L, Ku C-Y (2012) Terminal sliding mode control for maximum power point tracking of photovoltaic power generation systems. Sol Energy 86(10):2986–2995
11. Mojallizadeh MR, Badamchizadeh MA (2016) Chattering free full-order terminal sliding-mode control for maximum power point tracking of photovoltaic cells. IET Renew Power Gener 532–539

12. Mojallizadeh MR, Badamchizadeh M, Khanmohammadi S, Sabahi M (2016) Designing a new robust sliding mode controller for maximum power point tracking of photovoltaic cells. Sol Energy 132:538–546
13. Dahech K, Allouche M, Damak T, Tadeo F (2017) Backstepping sliding mode control for maximum power point tracking of a photovoltaic system. Electr Power Syst Res 143:182–188
14. Pradhan R, Subudhi B (2016) Double integral sliding mode mppt control of a photovoltaic system. IEEE Trans Control Syst Technol 24(1):285–292
15. Valencia PAO, Ramos-Paja CA (2015) Sliding-mode controller for maximum power point tracking in grid-connected photovoltaic systems. Energies 8(11):12 363–12 387
16. Mojallizadeh MR, Badamchizadeh M (2017) Second-order fuzzy sliding-mode control of photovoltaic power generation systems. Sol Energy 149:332–340
17. Utkin V (2013) Sliding mode control of dc/dc converters. J Frankl Inst 350(8):2146–2165
18. Yau H-T, Chen C-L (2012) Fuzzy sliding mode controller design for maximum power point tracking control of a solar energy system. Trans Inst Meas Control 34(5):557–565
19. Enany MA, Farahat MA, Nasr A (2016) Modeling and evaluation of main maximum power point tracking algorithms for photovoltaics systems. Renew Sustain Energy Rev 58:1578–1586
20. Kish GJ, Lee JJ, Lehn PW (2012) Modelling and control of photovoltaic panels utilising the incremental conductance method for maximum power point tracking. IET Renew Power Gener 6(4):259–266
21. Peng BR, Ho KC, Liu YH (2018) A novel and fast MPPT method suitable for both fast changing and partially shaded conditions. IEEE Trans Ind Electron 65(4):3240–3251
22. Park J-Y, Choi S-J (2015) A novel datasheet-based parameter extraction method for a single-diode photovoltaic array model. Sol Energy 122:1235–1244
23. Ngamkong P, Kochcha P, Areerak K, Sujitjorn S, Areerak K (2012) Applications of the generalized state-space averaging method to modelling of dc-dc power converters. Math Comput Model Dyn Syst 18(3):243–260
24. Wai R-J, Shih L-C (2011) Design of voltage tracking control for dc-dc boost converter via total sliding-mode technique. IEEE Trans Ind Electron 58(6):2502–2511
25. Ayop R, Tan CW (2018) Design of boost converter based on maximum power point resistance for photovoltaic applications. Sol Energy 160:322–335
26. Lu DD, Nguyen QN (2012) A photovoltaic panel emulator using a buck-boost dc/dc converter and a low cost micro-controller. Sol Energy 86(5):1477–1484

Predictive Control of Four-Leg Converters for Photovoltaic Energy Systems

Venkata Yaramasu, Marco Rivera, Apparao Dekka and Jose Rodriguez

Abstract Photovoltaic energy systems are one of the most widely adopted distributed generation facilities. This book chapter presents predictive based current and voltage control strategies for four-leg converters employed in grid-connected and standalone photovoltaic energy systems, respectively. The proposed approach employs the novel stationary frame sampled-data models of the four-leg converters with inductive (L) and inductive-capacitive (LC) filters on the output side to predict the control variables such as output currents and load voltages. These predictions are performed using all the possible switching states of four-leg converters. The objective of minimizing the error between reference and predicted variables (load currents or voltages) is fulfilled through a cost function in the predictive control schemes. In addition, the voltage balancing of DC-bus capacitors is considered with the four-leg neutral-point clamped converters. The optimal switching states corresponding to the minimal cost function value are chosen and directly applied to the converter. The predictive control strategies fulfil the control requirements such as load current/voltage control, DC-bus voltage balancing, and neutral-leg switching frequency minimization. The simulation and experimental studies conducted using unbalanced and nonlinear loads to validate the proposed predictive control strategies.

V. Yaramasu (✉)
Northern Arizona University, Flagstaff, AZ, USA
e-mail: Venkata.Yaramasu@nau.edu

M. Rivera
Universidad de Talca, Curico, Chile
e-mail: marcoriv@utalca.cl

A. Dekka
Ryerson University, Toronto, ON, Canada
e-mail: dapparao@ieee.org

J. Rodriguez
Universidad Andres Bello, Santiago, Chile
e-mail: jose.rodriguez@unab.cl

R.-E. Precup et al. (eds.), *Solar Photovoltaic Power Plants*, Power Systems,
https://doi.org/10.1007/978-981-13-6151-7_3

1 Introduction

Photovoltaic energy systems (PVES) are growing rapidly and worldwide installed capacity reached about 403 GW by the end of 2017 [1]. In terms of cumulative installed capacity, the PVES are the third most important renewable energy sources after hydro and wind energy. In 2017, the PVES continued their global expansion and reached almost 100 GW installed capacity. The cumulative installed capacity represents 500 billion kWh electricity produced, which is equivalent to 2% of electricity demand worldwide. In some countries, PVES have high contribution to the national electricity demand, for example, 13% in Honduras, 7.5% in Germany, and 7.3% in Greece. The cost of PVES has dropped by 59% over the last decade as the industry has escalated manufacturing and incrementally improved the technology with new materials and power electronics technologies. For PVES, power electronic converters and their digital control greatly contributes to minimize the energy cost, increase the energy yield from the sun and power density, improvement in reliability, smaller footprint, lower weight, and enhanced power quality and grid codes compliance [2].

The PVES are mainly used for grid-connected operations with few off-grid (standalone) applications. The grid-connected PVES performs active filter function to feed grid with sinusoidal currents, while supplying local loads [3]. The standalone distribution generation (DG) systems are economical and efficient in supplying electrical energy to the rural area customers where grid expansion is expensive and complicated [4]. In the standalone operation, local loads are supplied by PVES with constant voltage and frequency. The voltage source converters with output inductive (L) and inductive-capacitive (LC) filters are used for grid-connected and standalone operations, respectively. The output current is regulated in grid-connection operation, whereas load voltage is controlled in standalone application.

In three-phase four-wire applications, the three-leg converters offer simple structure for grid connected and standalone operations, where the load neutral is connected to the mid-point of the DC-bus capacitors. Due to the unbalanced and nonlinear loads, the load neutral current flows through the DC-bus capacitors and damages them if they are not overrated. Moreover, a voltage surge caused by the step-change in loading conditions can also destroy the DC-bus capacitors. On the other hand, the four-leg converters are an attractive alternative to three-leg converters and provide transformerless connection to the load neutral. In four-leg converters, the load neutral is connected to the mid-point of fourth-leg instead of DC-bus capacitors mid-point. Furthermore, the four-leg converter improves the DC-bus utilization by 15% and minimizes the DC-bus capacitance value and its voltage ripple compared with the three-leg converters [5]. The four-leg two-level and three-level neutral-point clamped (NPC) converters can handle the neutral currents effectively in low-power and high-power applications, respectively [6, 7]. For high-power applications, the NPC converter offers several advantages such as high quality output voltages and currents, higher voltage handling capability, and smaller common-mode voltages compared with the two-level converters [8].

The classical output current and voltage control techniques for four-leg converters use PI regulators to minimize steady-state tracking errors and a pulse width modulator to generate the switching signals. Particularly, the three-dimensional space vector modulation (3D-SVM) scheme is widely used to generate the switching signals for the four-leg two-level and NPC converters due to higher DC-bus utilization and less total harmonic distortion compared with the carrier-based pulse width modulation (C-PWM) scheme [9–11]. However, due to complex modeling and higher computational burden, the 3D-SVM is less attractive to control the four-leg converters [12]. Recently, the predictive control has emerged as a promising and power tool to control the four-leg converters. The predictive control eliminates the PI regulators and pulse width modulator, and thus improves the dynamic response and reduces the complexity of the control scheme [13]. Furthermore, it is possible to achieve multilevel control objectives using a cost function in the predictive control [14]. However, the predictive control operates with the variable switching frequency, similar to the hysteresis control.

Considering the discrete nature of the power converters, the predictive current control (PCC) and predictive voltage control (PVC) schemes for four-leg two-level and NPC converters are studied in this chapter. To simplify the implementation of PCC and PVC schemes in the digital control platforms, the number of predictions is optimized. Furthermore, a novel mathematical model of the entire system is developed in discrete-time domain for PCC and PVC schemes by incorporating the neutral-leg inductor. These sampled-data models are defined in terms of converter switching states and used to predict the future behavior of control variables (load currents or voltages). In each sampling interval, the predicted variables are evaluated against the reference variables with the help of a cost function. The reduction of switching frequency constraint is also incorporated in the cost function to improve the reliability of the fourth-leg (neutral-leg). The feasibility of the PCC and PVC schemes are verified by simulation and experimental results during different operating conditions such as unbalanced references and unbalanced/nonlinear loads.

This chapter is organized as follows:

- The continuous-time modeling of four-leg converters (two-level and NPC) with L and LC filters on the output side is discussed in Sect. 2.
- The PCC scheme is analyzed for four-leg two-level and NPC converters in Sects. 3 and 4, respectively. The performance is validated through simulation and experimental studies.
- The PVC scheme is studied for four-leg two-level and NPC converters in Sects. 5 and 6, respectively. The simulation and experimental results are presented to verify the control schemes.
- Conclusions of this chapter are presented in Sect. 7.

2 Modeling of Four-Leg Converters

To design the predictive control schemes for four-leg converters, the relationship between converter input and output variables is modeled in terms of converter switching states. This section deals with the system configurations, two-level and NPC converter output voltage models, DC-bus capacitors voltage model of NPC converter, and load models with *L* and *LC* filters.

2.1 *Four-Leg Power Converter Topologies*

The connection of four-leg two-level converter for grid-connected and standalone operation are shown in Figs. 1 and 2, respectively. The boost converter performs maximum power point tracking (MPPT) and the grid-connected two-level converter controls the DC-bus voltage and grid reactive power [15]. The grid-connected two-level converter supplies the unbalanced and nonlinear load currents and thereby maintains sinusoidal grid currents [16]. The standalone two-level converter controls the three-phase load voltages at their rated values [17], whereas the converter fourth-leg controls the zero sequence voltage of the load [18].

The converter and load neutrals are denoted with n and o, respectively. The load neutral o is connected to the mid-point n of the fourth-leg through a neutral inductor. L_n and r_n represent the neutral inductance and its internal resistance, respectively. The ripple in the converter fourth-leg current is minimized by the neutral inductor. i_{ni}, i_{no}, and i_{ng} are converter, load, and grid neutral currents, respectively. The converter voltages with respect to the fourth-leg mid-point n are represented by v_{an}, v_{bn}, and v_{cn}. The load (output) voltages for standalone operation are represented by v_{ao}, v_{bo}, and v_{co}. i_{ai}, i_{bi}, and i_{ci} are the converter currents. i_{ao}, i_{bo}, and i_{co} are the load (output) currents. i_{ag}, i_{bg}, and i_{cg} are the grid currents. The filter inductance, internal resistance of filter inductance, and filter capacitance are represented by L_f,

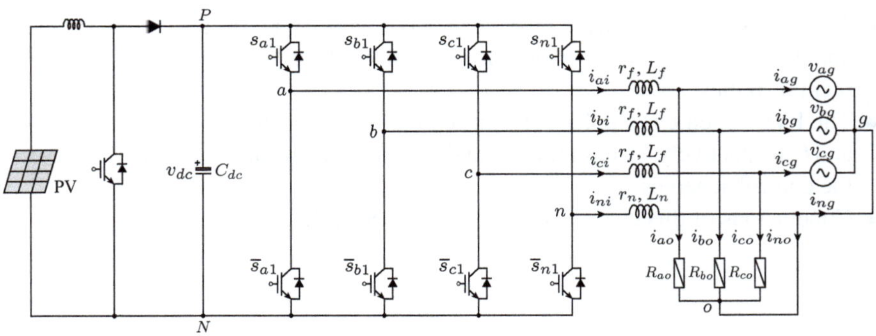

Fig. 1 Power circuit of a grid-connected four-leg two-level converter with arbitrary local loads

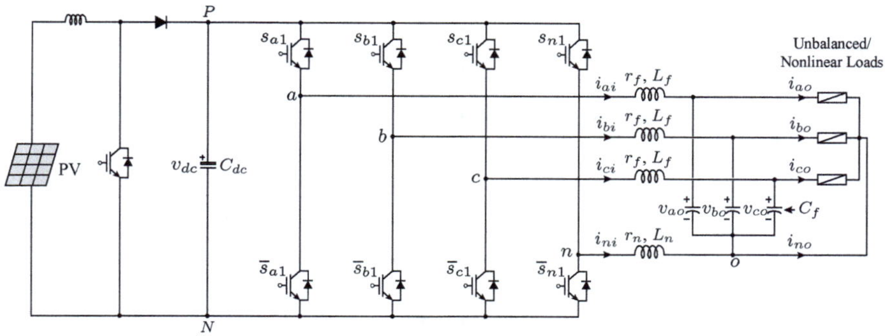

Fig. 2 Power circuit of a standalone four-leg two-level converter with arbitrary loads

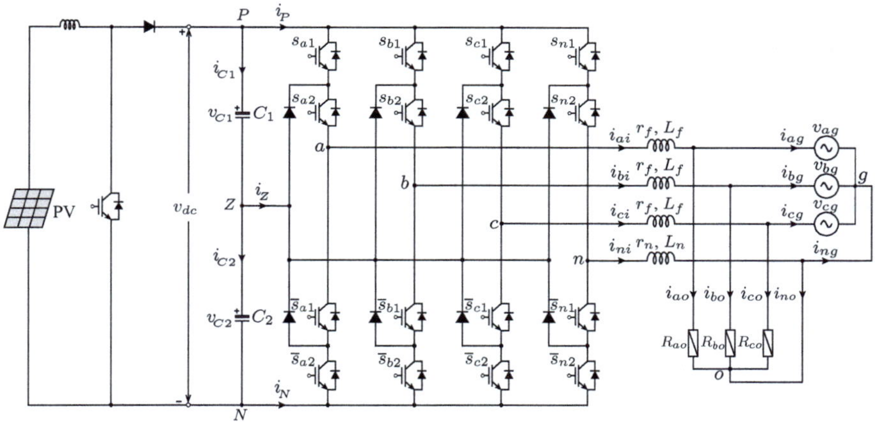

Fig. 3 Power circuit of a grid-connected four-leg NPC converter with arbitrary local loads

r_f, and C_f, respectively. The switching signals s_{a1}, s_{b1}, s_{c1}, and s_{n1} form a total of 16 (2^4) switching state combinations. The switching signals s_{x1} and $\bar{s}_{x1}$ operate in complementary manner for all $x \in \{a, b, c, n\}$. The increase in switching state combinations in comparison to the three-leg converters offers flexibility in control and improvement in the converter output voltage and current quality.

The power circuit of a four-leg NPC converter for grid-connected and standalone operation are shown in Figs. 3 and 4, respectively. Each leg of the converter is realized by four active switches and two clamping diodes, whereas the DC-bus is realized with two identical capacitors. In each leg, the mid-point (neutral-point) of DC-bus capacitors and clamping diodes are tied together. The net DC-bus voltage is equally distributed among the two DC-bus capacitors (i.e., $v_{C1} = v_{C2} = v_{dc}/2$). The neutral-point Z is assumed to be floating and it is not connected to the boost converter. Each phase of four-leg converter has three switching states and a total of 81 (3^4) switching combinations are available to control a four-leg converter.The

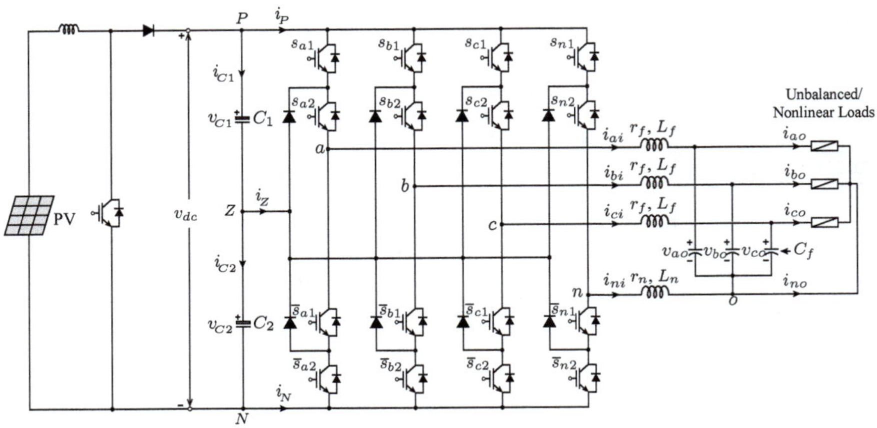

Fig. 4 Power circuit of a standalone four-leg NPC converter with arbitrary loads

Table 1 Switching states for four-leg NPC converter $\forall\ x \in \{a, b, c\}$

Switching Signals				Output Voltage	Input Currents		DC-bus Current Coefficients	
s_{x1}	s_{x2}	$\overline{s}_{x1}$	$\overline{s}_{x2}$	v_{xN}	i_P	i_Z	K_{x1}	K_{x2}
'1'	'1'	'0'	'0'	$v_{C1} + v_{C2}$	i_x	0	−1	−1
'0'	'1'	'1'	'0'	v_{C2}	0	i_x	−1	0
'0'	'0'	'1'	'1'	0	0	0	0	0

switching states and their respective output voltages, DC-bus capacitors current, and coefficients to estimate DC-bus capacitors current are summarized in Table 1 [19].

2.2 *Four-Leg Two-Level Converter Model*

With the help of semiconductor switches, the load terminal is connected to the positive or negative DC-bus points (P or N) to generate the positive or zero voltage levels at the output, respectively. The four-leg two-level converter output voltage with respect to the negative DC-bus (N) is expressed as [20]:

$$\begin{bmatrix} v_{aN} \\ v_{bN} \\ v_{cN} \\ v_{nN} \end{bmatrix} = v_{dc} \begin{bmatrix} s_{a1} \\ s_{b1} \\ s_{c1} \\ s_{n1} \end{bmatrix} \tag{1}$$

where v_{dc} and v_{nN} are DC-bus and load neutral voltages, respectively.

The four-leg two-level converter AC-side voltages in terms of switching signals and DC-bus voltage are given as follows:

$$\begin{bmatrix} v_{an} \\ v_{bn} \\ v_{cn} \end{bmatrix} = v_{dc} \begin{bmatrix} s_{a1} - s_{n1} \\ s_{b1} - s_{n1} \\ s_{c1} - s_{n1} \end{bmatrix}. \tag{2}$$

The stationary ($\alpha\beta\gamma$) reference frame uses less number of online calculations for the predictive model in comparison to the natural (*abc*) reference frame. For this reason, the four-leg two-level converter output voltages in (2) are transformed to the $\alpha\beta\gamma$-frame as follows:

$$\begin{bmatrix} v_{\alpha n} \\ v_{\beta n} \\ v_{\gamma n} \end{bmatrix} = v_{dc} \begin{bmatrix} s_{\alpha 1} \\ s_{\beta 1} \\ s_{\gamma 1} \end{bmatrix} \tag{3}$$

where $s_{\alpha 1}$, $s_{\beta 1}$, and $s_{\gamma 1}$ represent the switching signals in $\alpha\beta\gamma$-frame and they can be defined offline.

The *abc*-frame to $\alpha\beta\gamma$-frame transformation matrix is given as follows:

$$\begin{bmatrix} s_{\alpha 1} \\ s_{\beta 1} \\ s_{\gamma 1} \end{bmatrix} = \frac{2}{3} \begin{bmatrix} 1 & -\frac{1}{2} & -\frac{1}{2} \\ 0 & \frac{\sqrt{3}}{2} & -\frac{\sqrt{3}}{2} \\ \frac{1}{2} & \frac{1}{2} & \frac{1}{2} \end{bmatrix} \begin{bmatrix} s_{a1} - s_{n1} \\ s_{b1} - s_{n1} \\ s_{c1} - s_{n1} \end{bmatrix}. \tag{4}$$

2.3 Four-Leg NPC Converter Model

Depending on the switching signals, the converter terminals *a*, *b*, *c*, and *n* are connected to the positive, neutral-point, or negative (P, Z, or N) DC-bus. From Table 1, the four-leg NPC converter output voltage measured from N point is given in terms of switching states and the DC-bus capacitors voltage [7],

$$\begin{bmatrix} v_{aN} \\ v_{bN} \\ v_{cN} \\ v_{nN} \end{bmatrix} = v_{C1} \begin{bmatrix} s_{a1} \\ s_{b1} \\ s_{c1} \\ s_{n1} \end{bmatrix} + v_{C2} \begin{bmatrix} s_{a2} \\ s_{b2} \\ s_{c2} \\ s_{n2} \end{bmatrix}. \tag{5}$$

The above converter voltages are given with respect to the mid-point (n) of the fourth-leg as,

$$\begin{bmatrix} v_{an} \\ v_{bn} \\ v_{cn} \end{bmatrix} = v_{C1} \begin{bmatrix} s_{a1} - s_{n1} \\ s_{b1} - s_{n1} \\ s_{c1} - s_{n1} \end{bmatrix} + v_{C2} \begin{bmatrix} s_{a2} - s_{n2} \\ s_{b2} - s_{n2} \\ s_{c2} - s_{n2} \end{bmatrix}. \tag{6}$$

The converter output voltages are given in the $\alpha\beta\gamma$-frame as follows:

$$\begin{bmatrix} v_{\alpha n} \\ v_{\beta n} \\ v_{\gamma n} \end{bmatrix} = v_{C1} \begin{bmatrix} s_{\alpha 1} \\ s_{\beta 1} \\ s_{\gamma 1} \end{bmatrix} + v_{C2} \begin{bmatrix} s_{\alpha 2} \\ s_{\beta 2} \\ s_{\gamma 2} \end{bmatrix} \tag{7}$$

where $s_{\alpha 1}, s_{\beta 1}, s_{\gamma 1}, s_{\alpha 2}, s_{\beta 2}$, and $s_{\gamma 2}$ represent the switching signals in the $\alpha\beta\gamma$-frame.

Along with the load voltage and currents, the DC-bus capacitors voltage needs to be controlled in a four-leg NPC converter. The DC-bus capacitors voltage are given as follows [21]:

$$\begin{aligned} \frac{d}{dt} v_{C1} &= \frac{1}{C_1} i_{C1} \\ \frac{d}{dt} v_{C2} &= \frac{1}{C_2} i_{C2} \end{aligned} \tag{8}$$

where i_{C1} and i_{C2} represent the current flowing through the DC-bus capacitors C_1 and C_2, respectively.

The DC-bus capacitor currents are given in terms of converter output currents and coefficients as,

$$\begin{aligned} i_{C1} &= K_{a1}\, i_a + K_{b1}\, i_b + K_{c1}\, i_c \\ i_{C2} &= K_{a2}\, i_a + K_{b2}\, i_b + K_{c2}\, i_c. \end{aligned} \tag{9}$$

For a four-leg NPC with output L filter, i_a, i_b, and i_c in the above expression are replaced by i_{ao}, i_{bo}, and i_{co}, respectively, whereas for a four-leg NPC with output LC filter, i_a, i_b, and i_c in the above expression are replaced by i_{ai}, i_{bi}, and i_{ci}, respectively. The coefficients for the DC-bus capacitors current are defined in Table 1. The calculation of these coefficients is not necessarily required to be done in online; therefore the computational burden of overall control scheme becomes low.

2.4 Load Model with Output L filter

The converter's control design and its performance is independent of the grid voltages [15]. Hence, a simple R load is considered by eliminating grid connection in the present study. Figure 5 shows the equivalent model of four-leg converter with an output L filter and R load. Due to the elimination of the grid, the converter output currents in Fig. 1 are equal to the load currents, that is, $i_{ai} = i_{ao}$, $i_{bi} = i_{bo}$, $i_{ci} = i_{co}$, and $i_{ni} = i_{no}$.

By applying Kirchhoff's voltage law to Fig. 5, the converter AC-side voltages are obtained as follows:

$$\begin{aligned} \begin{bmatrix} v_{an} \\ v_{bn} \\ v_{cn} \end{bmatrix} = r_f \begin{bmatrix} i_{ao} \\ i_{bo} \\ i_{co} \end{bmatrix} + L_f \frac{d}{dt} \begin{bmatrix} i_{ao} \\ i_{bo} \\ i_{co} \end{bmatrix} + \begin{bmatrix} R_{ao} & 0 & 0 \\ 0 & R_{bo} & 0 \\ 0 & 0 & R_{co} \end{bmatrix} \begin{bmatrix} i_{ao} \\ i_{bo} \\ i_{co} \end{bmatrix} \\ - r_n i_{no} - L_n \frac{d}{dt} i_{no}. \end{aligned} \tag{10}$$

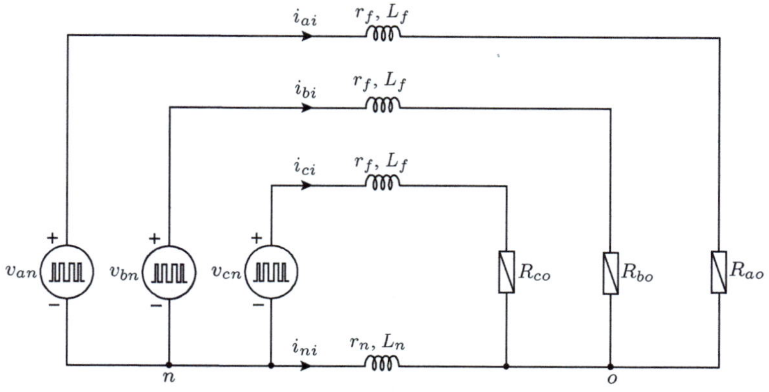

Fig. 5 Representation of an L filter and load connected to the four-leg converter

The load neutral current is the summation of three-phase load currents as expressed below:

$$i_{no} = -(i_{ao} + i_{bo} + i_{co}). \tag{11}$$

By combining (10) and (11), the converter's output voltages are obtained as,

$$\begin{bmatrix} v_{an} \\ v_{bn} \\ v_{cn} \end{bmatrix} = \mathbf{R_{eq}} \begin{bmatrix} i_{ao} \\ i_{bo} \\ i_{co} \end{bmatrix} + \mathbf{L_{eq}} \frac{d}{dt} \begin{bmatrix} i_{ao} \\ i_{bo} \\ i_{co} \end{bmatrix} \tag{12}$$

where,

$$\mathbf{R_{eq}} = \begin{bmatrix} r_f + r_n + R_{ao} & r_n & r_n \\ r_n & r_f + r_n + R_{bo} & r_n \\ r_n & r_n & r_f + r_n + R_{co} \end{bmatrix} \tag{13}$$

and,

$$\mathbf{L_{eq}} = \begin{bmatrix} L_f + L_n & L_n & L_n \\ L_n & L_f + L_n & L_n \\ L_n & L_n & L_f + L_n \end{bmatrix}. \tag{14}$$

The dynamic model of load currents is obtained from (12) as shown below:

$$\frac{d}{dt} \begin{bmatrix} i_{ao} \\ i_{bo} \\ i_{co} \end{bmatrix} = \mathbf{A} \begin{bmatrix} i_{ao} \\ i_{bo} \\ i_{co} \end{bmatrix} + \mathbf{B} \begin{bmatrix} v_{an} \\ v_{bn} \\ v_{cn} \end{bmatrix}. \tag{15}$$

where,

$$\mathbf{A} = -\mathbf{L}_{\mathbf{eq}}^{-1}\,\mathbf{R}_{\mathbf{eq}} = \begin{bmatrix} -\frac{L_f\,(r_f+r_n)+2\,L_n\,r_f}{L_f^2+3\,L_n\,L_f} & \frac{L_n\,r_f-L_f\,r_n}{L_f^2+3\,L_n\,L_f} & \frac{L_n\,r_f-L_f\,r_n}{L_f^2+3\,L_n\,L_f} \\ \frac{L_n\,r_f-L_f\,r_n}{L_f^2+3\,L_n\,L_f} & -\frac{L_f\,(r_f+r_n)+2\,L_n\,r_f}{L_f^2+3\,L_n\,L_f} & \frac{L_n\,r_f-L_f\,r_n}{L_f^2+3\,L_n\,L_f} \\ \frac{L_n\,r_f-L_f\,r_n}{L_f^2+3\,L_n\,L_f} & \frac{L_n\,r_f-L_f\,r_n}{L_f^2+3\,L_n\,L_f} & -\frac{L_f\,(r_f+r_n)+2\,L_n\,r_f}{L_f^2+3\,L_n\,L_f} \end{bmatrix} \tag{16}$$

and,

$$\mathbf{B} = \mathbf{L}_{\mathbf{eq}}^{-1} = \begin{bmatrix} \frac{L_f+2\,L_n}{L_f^2+3\,L_n\,L_f} & \frac{-L_n}{L_f^2+3\,L_n\,L_f} & \frac{-L_n}{L_f^2+3\,L_n\,L_f} \\ \frac{-L_n}{L_f^2+3\,L_n\,L_f} & \frac{L_f+2\,L_n}{L_f^2+3\,L_n\,L_f} & \frac{-L_n}{L_f^2+3\,L_n\,L_f} \\ \frac{-L_n}{L_f^2+3\,L_n\,L_f} & \frac{-L_n}{L_f^2+3\,L_n\,L_f} & \frac{L_f+2\,L_n}{L_f^2+3\,L_n\,L_f} \end{bmatrix}. \tag{17}$$

The state-space model in (15) is converted to the stationary-frame as follows:

$$\frac{d}{dt}\begin{bmatrix} i_{\alpha o} \\ i_{\beta o} \\ i_{\gamma o} \end{bmatrix} = \mathbf{A}\begin{bmatrix} i_{\alpha o} \\ i_{\beta o} \\ i_{\gamma o} \end{bmatrix} + \mathbf{B}\begin{bmatrix} v_{\alpha n} \\ v_{\beta n} \\ v_{\gamma n} \end{bmatrix}. \tag{18}$$

2.5 Load Model with Output LC filter

The equivalent model of the four-leg converter with an output *LC* filter is shown in Fig. 6. With the help of Kirchhoff's voltage law, the converter's AC-side voltages are given as follows:

$$\begin{bmatrix} v_{an} \\ v_{bn} \\ v_{cn} \end{bmatrix} = r_f\begin{bmatrix} i_{ai} \\ i_{bi} \\ i_{ci} \end{bmatrix} + L_f\frac{d}{dt}\begin{bmatrix} i_{ai} \\ i_{bi} \\ i_{ci} \end{bmatrix} + \begin{bmatrix} v_{ao} \\ v_{bo} \\ v_{co} \end{bmatrix} - r_n i_{ni} - L_n\frac{d}{dt}i_{ni}. \tag{19}$$

The converter neutral current is equal to the sum of three-phase converter currents, that is,

$$i_{ni} = -(i_{ai} + i_{bi} + i_{ci}). \tag{20}$$

Measurement of the converter's neutral current is unnecessary as it adds cost and complexity to the system. By combining (19) and (20), the converter's neutral current term is eliminated from the converter AC-side voltages, that is,

$$\begin{bmatrix} v_{an} \\ v_{bn} \\ v_{cn} \end{bmatrix} = \mathbf{R}_{\mathbf{eq}}\begin{bmatrix} i_{ai} \\ i_{bi} \\ i_{ci} \end{bmatrix} + \mathbf{L}_{\mathbf{eq}}\frac{d}{dt}\begin{bmatrix} i_{ai} \\ i_{bi} \\ i_{ci} \end{bmatrix} + \begin{bmatrix} v_{ao} \\ v_{bo} \\ v_{co} \end{bmatrix} \tag{21}$$

where,

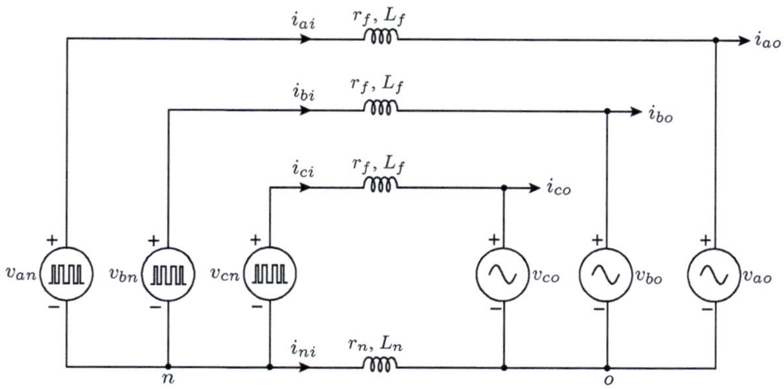

Fig. 6 Representation of an *LC* filter and load connected to the four-leg converter

$$\mathbf{R_{eq}} = \begin{bmatrix} r_f + r_n & r_n & r_n \\ r_n & r_f + r_n & r_n \\ r_n & r_n & r_f + r_n \end{bmatrix} \tag{22}$$

and,

$$\mathbf{L_{eq}} = \begin{bmatrix} L_f + L_n & L_n & L_n \\ L_n & L_f + L_n & L_n \\ L_n & L_n & L_f + L_n \end{bmatrix}. \tag{23}$$

The dynamic model of the converter's currents is obtained from (21) as,

$$\frac{d}{dt}\begin{bmatrix} i_{ai} \\ i_{bi} \\ i_{ci} \end{bmatrix} = -\mathbf{L_{eq}^{-1}} \begin{bmatrix} v_{ao} \\ v_{bo} \\ v_{co} \end{bmatrix} - \mathbf{L_{eq}^{-1}}\,\mathbf{R_{eq}} \begin{bmatrix} i_{ai} \\ i_{bi} \\ i_{ci} \end{bmatrix} + \mathbf{L_{eq}^{-1}} \begin{bmatrix} v_{an} \\ v_{bn} \\ v_{cn} \end{bmatrix}. \tag{24}$$

From Fig. 6, the relationship between the converter and the load currents is obtained as,

$$\begin{bmatrix} i_{ai} \\ i_{bi} \\ i_{ci} \end{bmatrix} = \begin{bmatrix} i_{ao} \\ i_{bo} \\ i_{co} \end{bmatrix} + C_f \frac{d}{dt} \begin{bmatrix} v_{ao} \\ v_{bo} \\ v_{co} \end{bmatrix}. \tag{25}$$

The dynamic model of output voltages is deducted from the above model as follows:

$$\frac{d}{dt}\begin{bmatrix} v_{ao} \\ v_{bo} \\ v_{co} \end{bmatrix} = \frac{1}{C_f}\begin{bmatrix} i_{ai} \\ i_{bi} \\ i_{ci} \end{bmatrix} - \frac{1}{C_f}\begin{bmatrix} i_{ao} \\ i_{bo} \\ i_{co} \end{bmatrix}. \tag{26}$$

The system models in continuous-time demoain are given in (24) and (26) and they are represented in state-space form as,

$$\frac{d}{dt}\begin{bmatrix} v_{ao} \\ v_{bo} \\ v_{co} \\ i_{ai} \\ i_{bi} \\ i_{ci} \end{bmatrix} = \mathbf{A}\begin{bmatrix} v_{ao} \\ v_{bo} \\ v_{co} \\ i_{ai} \\ i_{bi} \\ i_{ci} \end{bmatrix} + \mathbf{B}\begin{bmatrix} v_{an} \\ v_{bn} \\ v_{cn} \\ i_{ao} \\ i_{bo} \\ i_{co} \end{bmatrix} \tag{27}$$

where,

$$\mathbf{A} = \begin{bmatrix} 0 & 0 & 0 & \frac{1}{C_f} & 0 & 0 \\ 0 & 0 & 0 & 0 & \frac{1}{C_f} & 0 \\ 0 & 0 & 0 & 0 & 0 & \frac{1}{C_f} \\ a_1 & a_2 & a_2 & a_3 & a_4 & a_4 \\ a_2 & a_1 & a_2 & a_4 & a_3 & a_4 \\ a_2 & a_2 & a_1 & a_4 & a_4 & a_3 \end{bmatrix} \tag{28}$$

with,

$$\begin{aligned} a_1 &= -\frac{L_f + 2\,L_n}{L_f^2 + 3\,L_n\,L_f} \\ a_2 &= \frac{L_n}{L_f^2 + 3\,L_n\,L_f} \\ a_3 &= -\frac{L_f\,(r_f + r_n) + 2\,L_n\,r_f}{L_f^2 + 3\,L_n\,L_f} \\ a_4 &= \frac{L_n\,r_f - L_f\,r_n}{L_f^2 + 3\,L_n\,L_f} \end{aligned} \tag{29}$$

and,

$$\mathbf{B} = \begin{bmatrix} 0 & 0 & 0 & -\frac{1}{C_f} & 0 & 0 \\ 0 & 0 & 0 & 0 & -\frac{1}{C_f} & 0 \\ 0 & 0 & 0 & 0 & 0 & -\frac{1}{C_f} \\ \frac{L_f+2\,L_n}{L_f^2+3\,L_n\,L_f} & \frac{-L_n}{L_f^2+3\,L_n\,L_f} & \frac{-L_n}{L_f^2+3\,L_n\,L_f} & 0 & 0 & 0 \\ \frac{-L_n}{L_f^2+3\,L_n\,L_f} & \frac{L_f+2\,L_n}{L_f^2+3\,L_n\,L_f} & \frac{-L_n}{L_f^2+3\,L_n\,L_f} & 0 & 0 & 0 \\ \frac{-L_n}{L_f^2+3\,L_n\,L_f} & \frac{-L_n}{L_f^2+3\,L_n\,L_f} & \frac{L_f+2\,L_n}{L_f^2+3\,L_n\,L_f} & 0 & 0 & 0 \end{bmatrix}. \tag{30}$$

The state-space model in (27) is converted to the stationary-frame as follows:

$$\frac{d}{dt}\begin{bmatrix} v_{\alpha o} \\ v_{\beta o} \\ v_{\gamma o} \\ i_{\alpha i} \\ i_{\beta i} \\ i_{\gamma i} \end{bmatrix} = \mathbf{A}\begin{bmatrix} v_{\alpha o} \\ v_{\beta o} \\ v_{\gamma o} \\ i_{\alpha i} \\ i_{\beta i} \\ i_{\gamma i} \end{bmatrix} + \mathbf{B}\begin{bmatrix} v_{\alpha n} \\ v_{\beta n} \\ v_{\gamma n} \\ i_{\alpha o} \\ i_{\beta o} \\ i_{\gamma o} \end{bmatrix}. \tag{31}$$

3 Predictive Current Control of Four-Leg Two-Level Converter

This section deals with the PCC scheme for a four-leg two-level converter with output L filter. As mentioned earlier, the control scheme is not affected by the grid voltages. To simplify the analysis, unbalanced R loads are considered here. To validate the PCC scheme, simulation and experimental results are presented.

3.1 Control Scheme

Figure 7 shows the structure of PCC scheme for a four-leg two-level converter. The PCC scheme is simple and intuitive to understand. The measured DC-bus voltage and load currents at present (kth) sampling instant are used in the prediction process. In grid-connected systems, the reference currents are obtained from the DC-bus voltage and the grid reactive power control loops [15]. The measured and reference load currents are converted into stationary frame by using the transformation matrix in (4). These variables at $(k+1)$ sampling instant are evaluated in the cost function subsystem.To calculate the load current error accurately, the vector angle extrapolation is used to extrapolate the reference currents to the $(k+1)$ sampling instant [22]:

$$\begin{bmatrix} \widehat{i^*_{\alpha o}}(k+1) \\ \widehat{i^*_{\beta o}}(k+1) \\ \widehat{i^*_{\gamma o}}(k+1) \end{bmatrix} = e^{j2\pi f_s(k)\,Ts} \begin{bmatrix} i^*_{\alpha o}(k) \\ i^*_{\beta o}(k) \\ i^*_{\gamma o}(k) \end{bmatrix} \tag{32}$$

where, f_s is the grid/load frequency which is updated at each sampling interval.

The proposed PCC scheme requires a discrete-time model of load currents to implement in the digital control platforms. The continuous-time state-space model in (18) is represented in discrete-time with zero-order hold (ZOH) discretization for one-sample-ahead prediction as,

$$\begin{bmatrix} i^p_{\alpha o}(k+1) \\ i^p_{\beta o}(k+1) \\ i^p_{\gamma o}(k+1) \end{bmatrix} = \mathbf{\Phi} \begin{bmatrix} i_{\alpha o}(k) \\ i_{\beta o}(k) \\ i_{\gamma o}(k) \end{bmatrix} + \mathbf{\Gamma} \begin{bmatrix} v^p_{\alpha n}(k) \\ v^p_{\beta n}(k) \\ v^p_{\gamma n}(k) \end{bmatrix}. \tag{33}$$

where, the superscript p denotes the predicted variable.

The discrete-time parameters in (33) are calculated by,

$$\mathbf{\Phi} = e^{\mathbf{A}T_s}, \quad \mathbf{\Gamma} = \mathbf{A}^{-1}(\mathbf{\Phi} - \mathbf{I})\mathbf{B} \tag{34}$$

where T_s is the sampling time.

The load voltages are predicted in stationary frame with the help of 16 switching state combinations and the DC-bus voltage (reproduced from (3)):

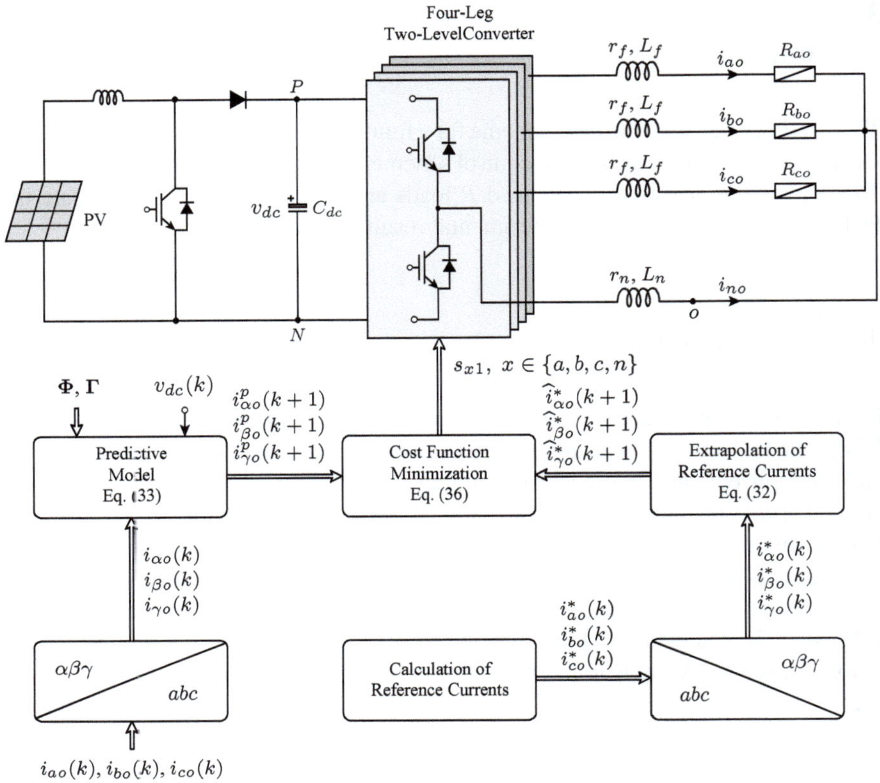

Fig. 7 Block diagram of the PCC scheme for four-leg two-level converter

$$\begin{bmatrix} v^p_{\alpha n}(k) \\ v^p_{\beta n}(k) \\ v^p_{\gamma n}(k) \end{bmatrix} = v_{dc}(k) \begin{bmatrix} s^p_{\alpha 1}(k) \\ s^p_{\beta 1}(k) \\ s^p_{\gamma 1}(k) \end{bmatrix} \tag{35}$$

The load-neutral voltage v_{nN} causes the fourth-leg to operate at a higher switching frequency compared with rest of the phases. Considering all these issues, the cost function is formulated to minimize the tracking error along with the neutral-leg's switching frequency, and is given as follows:

$$\begin{aligned} g_i(k) = & \left[\widehat{i}^*_{\alpha o}(k+1) - i^p_{\alpha o}(k+1)\right]^2 + \left[\widehat{i}^*_{\beta o}(k+1) - i^p_{\beta o}(k+1)\right]^2 \\ & + \left[\widehat{i}^*_{\gamma o}(k+1) - i^p_{\gamma o}(k+1)\right]^2 + \lambda_{sw}\left[s^p_{n1}(k) - s^{\text{op}}_{n1}(k)\right]^2 \end{aligned} \tag{36}$$

where λ_{sw} is the weighting factor and $s^{\text{op}}_{n1}(k)$ is the optimal gating signal at kth sampling instant. The λ_{sw} value is heuristically selected according to the desired switching frequency minimization.

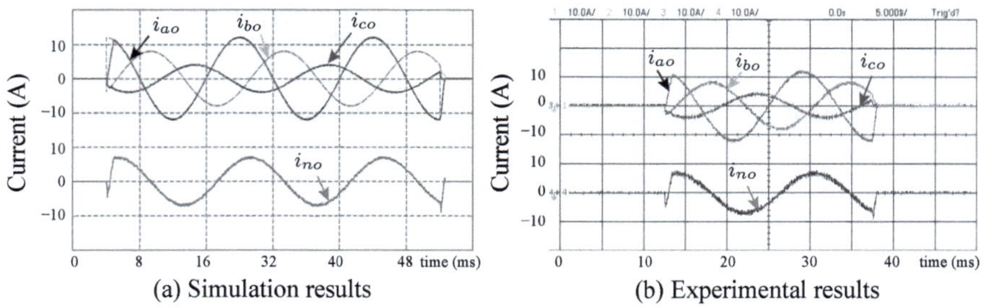

Fig. 8 Results for the PCC scheme of a four-leg two-level converter during a step-change in unbalanced reference currents

The minimum g_i value is accomplished by the optimization process. In each sampling instant, the switching combination which gives the minimum g_i value out of 16 functional values is selected and applied to the converter during the next sampling interval.

3.2 *Results and Analysis*

To validate the proposed PCC scheme, simulation and experimental studies are presented in Fig. 8a and b, respectively. The laboratory prototype is built using the dSPACE DS1103 controller, Semikron IGBT modules and gate drivers, LEM voltage and current sensors, and Xantrex DC power supply. The simulation and experimental parameters are as follows: $v_{dc} = 300\,\text{V}$, $C_{dc} = 1200\,\mu\text{F}$, $r_f = r_n = 0.07\,\Omega$, $L_f = L_n = 15\,\text{mH}$, $f_s = 60\,\text{Hz}$, and $T_s = 20\,\mu\text{s}$.

The results are presented with a step-change in unbalanced reference currents (i^*_{ao} from 0 to 12 A (*peak*) to 0 A, i^*_{bo} from 0 to 8 A (*peak*) to 0 A, and i^*_{co} from 0 to 4 A (*peak*) to 0 A) and unbalanced loads ($R_{ao} = 12\,\Omega$, $R_{bo} = 8\,\Omega$, $R_{co} = 6\,\Omega$). This is the typical scenario for three-phase four-wire systems, where each phase has different load demand. Each phase current is controlled independently by the PCC scheme, therefore the load currents follow their references with the least amount of steady-state error. The average current tracking error (e_i), average total harmonic distortion (THD) of load currents (THD$_i$), and average switching frequency (f_{sw}) are noted as 1.47%, 1.44%, and 11,000 Hz, respectively for simulation waveforms (Fig. 8a). With the experimental results (Fig. 8b), e_i, THD$_i$, and f_{sw} are noted as 3.12%, 3.23%, and 9200 Hz, respectively. The actual currents track their references without any overshoot, whereas the neutral current flows through the fourth-leg of the converter. The control scheme compensates the variations in load parameter changes and the actual currents continue to follow their references effectively.

4 Predictive Current Control of Four-Leg NPC Converter

The PCC scheme for a four-leg NPC converter with an output L filter and unbalanced R load is analyzed in this section through experimental results.

4.1 Control Scheme

The structure of PCC scheme for a four-leg NPC is shown in Fig. 9. The control objectives are reference tracking of load currents, voltage balancing of DC-bus capacitors, and neutral-leg switching frequency reduction. The PCC scheme predict the future values of the control variables by using the mathematical model of the system, load currents at (kth) sampling instant, DC-bus capacitors voltage, and 81 possible switching states. The calculation of reference currents and its extrapolation is similar to the PCC scheme discussed earlier with the four-leg two-level converter in Sect. 3. The load configuration is the same for two-level and NPC converters, therefore the two-level converter predictive model in (33) is used here for the NPC converter control.

The converter voltages in (33) are obtained by the following model (reproduced from (7)):

$$\begin{bmatrix} v^p_{\alpha n}(k) \\ v^p_{\beta n}(k) \\ v^p_{\gamma n}(k) \end{bmatrix} = v_{C1}(k) \begin{bmatrix} s^p_{\alpha 1}(k) \\ s^p_{\beta 1}(k) \\ s^p_{\gamma 1}(k) \end{bmatrix} + v_{C2}(k) \begin{bmatrix} s^p_{\alpha 2}(k) \\ s^p_{\beta 2}(k) \\ s^p_{\gamma 2}(k) \end{bmatrix} \tag{37}$$

The DC-bus capacitors voltage in discrete-time domain is calculated from (8) as shown below:

$$\begin{aligned} v_{C1}(k+1) &= v_{C1}(k) + \frac{T_s}{C_1}\, i^p_{C1}(k) \\ v_{C2}(k+1) &= v_{C2}(k) + \frac{T_s}{C_2}\, i^p_{C2}(k). \end{aligned} \tag{38}$$

The DC-bus capacitor currents are predicted by using the measured load currents and coefficients in Table 1 (reproduced from (9)) as:

$$\begin{aligned} i^p_{C1}(k) &= K_{a1}\, i_{ao}(k) + K_{b1}\, i_{bo}(k) + K_{c1}\, i_{co}(k) \\ i^p_{C2}(k) &= K_{a2}\, i_{ao}(k) + K_{b2}\, i_{bo}(k) + K_{c2}\, i_{co}(k). \end{aligned} \tag{39}$$

The cost function evaluates the predicted load currents and DC-bus capacitors voltage for all possible switching states. Finally, the switching state which minimizes the cost function is selected and applied to the converter.

The defined cost function has three objectives: reduce the tracking error between the reference and predicted load currents, voltage balancing of DC-bus capacitors, and neutral-leg switching frequency minimization, which are expressed as follows:

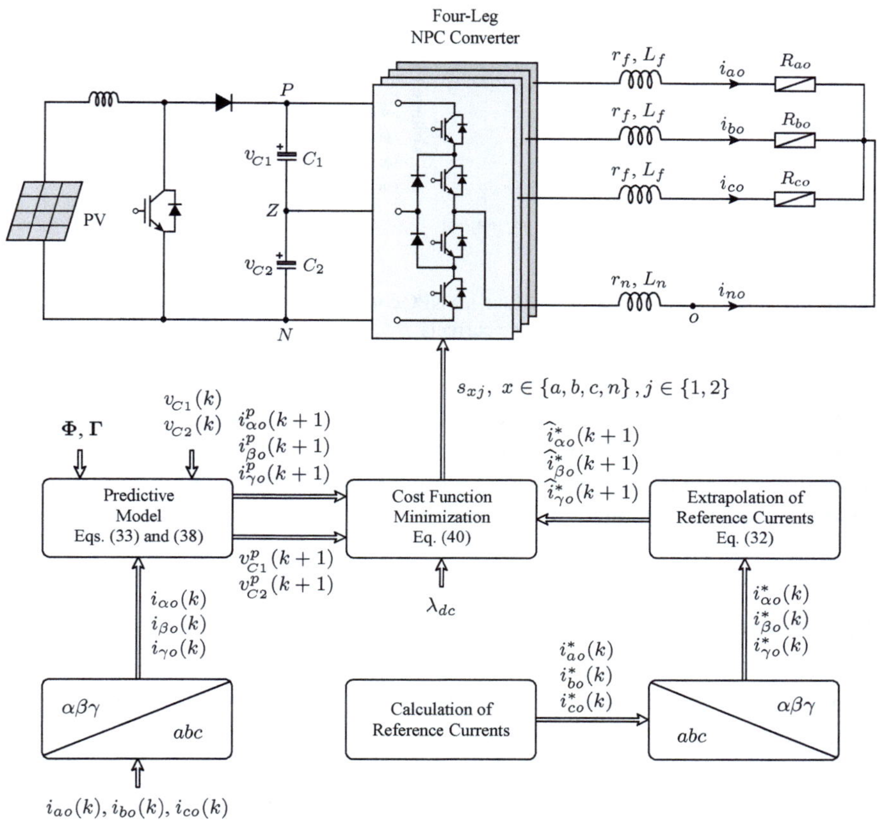

Fig. 9 Block diagram of the PCC scheme for a four-leg NPC converter

$$
\begin{aligned}
g_i(k) = & \left[\widehat{i^*_{\alpha o}}(k+1) - i^p_{\alpha o}(k+1)\right]^2 + \left[\widehat{i^*_{\beta o}}(k+1) - i^p_{\beta o}(k+1)\right]^2 \\
& + \left[\widehat{i^*_{\gamma o}}(k+1) - i^p_{\gamma o}(k+1)\right]^2 + \lambda_{dc}\left[v^p_{C1}(k+1) - v^p_{C2}(k+1)\right]^2 \\
& + \lambda_{sw}\left[s^p_{n1}(k) - s^{op}_{n1}(k)\right]^2 + \lambda_{sw}\left[s^p_{n2}(k) - s^{op}_{n2}(k)\right]^2
\end{aligned}
\tag{40}
$$

where λ_{dc} is the weighting factor for balancing the DC-bus capacitors voltage. $s^{op}_{n1}(k)$ and $s^{op}_{n2}(k)$ are the optimal gating signals at kth sampling instant.

The weighting factor λ_{dc} defines the importance of DC-bus capacitors voltage over the load current control and helps in accomplishing the desired control performance. The value of λ_{dc} is calculated as follows [14]:

$$
\lambda_{dc} = \frac{I_R}{V_{dc}} \tag{41}
$$

where I_R is the rated *rms* load current and V_{dc} is rated DC-bus voltage.

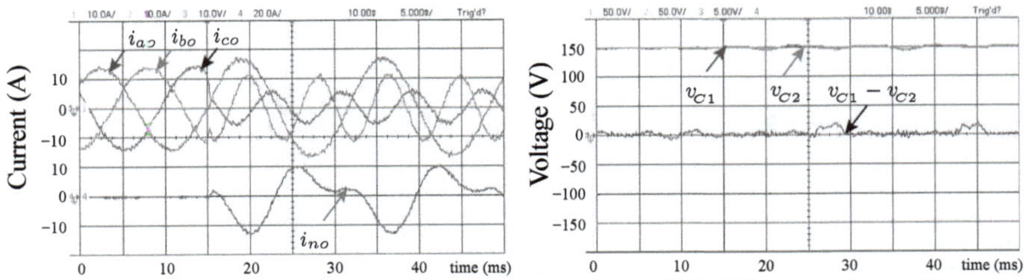

(a) Load currents (b) DC-bus capacitors voltage

Fig. 10 Results for the PCC scheme of a four-leg NPC converter during a step-change from balanced reference currents to unbalanced reference currents

When $\lambda_{sw} > 0$, the neutral-leg switching is reduced, however the load current tracking and neutral-point voltage errors are increased. In each sampling instant, the switching state combination (among 81) corresponding to the minimum cost function value is selected and applied to the four-leg NPC converter gating terminals.

4.2 Results and Analysis

To study the performance of the PCC scheme for a four-leg NPC converter, experimental results are presented in Fig. 10 with the following parameters: $v_{dc} = 300\,\text{V}$, $C_{dc} = 4700\,\mu\text{F}$, $r_f = r_n = 0.045\,\Omega$, $L_f = L_n = 10\,\text{mH}$, $f_s = 60\,\text{Hz}$, and $T_s = 100\,\mu\text{s}$. The experimental set-up is built similar to the four-leg two-level converter prototype discussed earlier.

The transient performance of the PCC scheme for a four-leg NPC converter is analyzed by applying a step-change from balanced reference currents ($i^*_{ao} = i^*_{bo} = i^*_{co} = 10\,\text{A}$ (rms)) to unbalanced reference currents ($i^*_{ao} = 12\,\text{A}$ (rms), $i^*_{bo} = 10\,\text{A}$ (rms), $i^*_{co} = 8\,\text{A}$ (rms)). Balanced loads ($R_{ao} = R_{bo} = R_{co} = 12\,\Omega$) are considered in this test. As previously mentioned in Sect. 3.2, this test represents a typical case for a four-leg converter connected to the unbalanced grid. Due to the decoupled control architecture, the actual and reference currents follow each other with a smaller steady-state error as shown in Fig. 10a. During the step-change in reference currents, the load currents exhibit fast dynamic response and no overshoots. These changes also effect the load neutral current accordingly. As depicted in Fig. 10b, the DC-bus capacitors voltage remain balanced despite the transient change in the reference currents. The average current tracking error, average THD of load currents, and average switching frequency are noted as 5%, 3.8%, and 2000 Hz, respectively. This PCC scheme is observed to be robust and it can handle the load and DC-bus parameter variations as well.

5 Predictive Voltage Control of Four-Leg Two-Level Converter

The PVC scheme for a four-leg two-level converter with an output *LC* filter and unbalanced/nonlinear arbitrary loads is discussed with respect to digital implementation, and simulation and experimental results.

5.1 Control Scheme

The block diagram of the proposed PVC scheme for a four-leg two-level converter with an output *LC* filter and arbitrary loads is shown in Fig. 11. The stationary-frame reference voltages are extrapolated to $(k+1)$ sampling instant using the extrapolation model in (32). The 16 switching states and the DC-bus voltage $v_{dc}(k)$ are used to predict the converter output voltages in stationary frame, that is, $v^p_{\alpha n}(k)$, $v^p_{\beta n}(k)$, and $v^p_{\gamma n}(k)$ (refer to Eq. (35)). The converter output voltage predictions, along with the measured load voltages, converter currents, and load currents are used to estimate the future values of load voltages as shown below (reproduced from (31)):

$$\begin{bmatrix} v^p_{\alpha o}(k+1) \\ v^p_{\beta o}(k+1) \\ v^p_{\gamma o}(k+1) \\ i^p_{\alpha i}(k+1) \\ i^p_{\beta i}(k+1) \\ i^p_{\gamma i}(k+1) \end{bmatrix} = \mathbf{\Phi} \begin{bmatrix} v_{\alpha o}(k) \\ v_{\beta o}(k) \\ v_{\gamma o}(k) \\ i_{\alpha i}(k) \\ i_{\beta i}(k) \\ i_{\gamma i}(k) \end{bmatrix} + \mathbf{\Gamma} \begin{bmatrix} v^p_{\alpha n}(k) \\ v^p_{\beta n}(k) \\ v^p_{\gamma n}(k) \\ i_{\alpha o}(k) \\ i_{\beta o}(k) \\ i_{\gamma o}(k) \end{bmatrix} \tag{42}$$

The below cost function is formulated to regulate the load voltages and to minimize the neutral-leg switching frequency as,

$$\begin{aligned} g_i(k) = {} & \left[\widehat{v}^*_{\alpha o}(k+1) - v^p_{\alpha o}(k+1)\right]^2 + \left[\widehat{v}^*_{\beta o}(k+1) - v^p_{\beta o}(k+1)\right]^2 \\ & + \left[\widehat{v}^*_{\gamma o}(k+1) - v^p_{\gamma o}(k+1)\right]^2 + \lambda_{sw}\left[s^p_{n1}(k) - s^{\text{op}}_{n1}(k)\right]^2. \end{aligned} \tag{43}$$

The optimal switching state combination is chosen on the basis of minimal cost function value and applied to the system during next sampling period.

5.2 Results and Analysis

To verify the PVC scheme, the simulation and experimental studies are conducted with the following system parameters: $v_{dc} = 350\,\text{V}$, $C_{dc} = 2350\,\mu\text{F}$, $r_f = r_n = 0.022\,\Omega$, $L_f = L_n = 2.5\,\text{mH}$, $C_f = 60\,\mu\text{F}$, $f_s = 60\,\text{Hz}$, and $T_s = 50\,\mu\text{s}$.

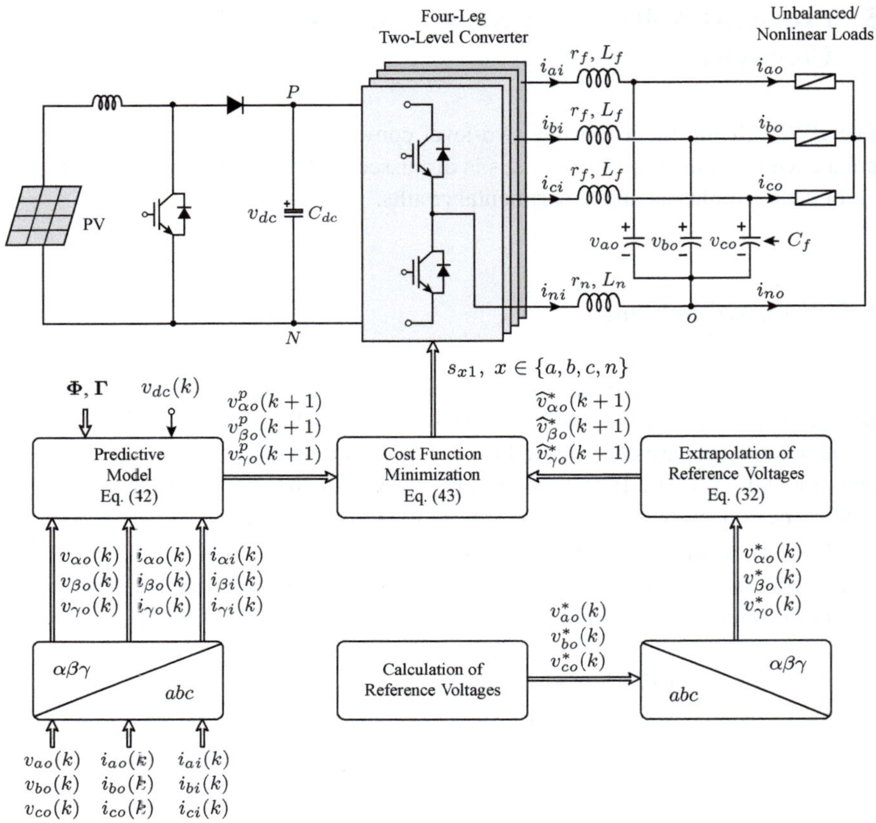

Fig. 11 Block diagram of PVC scheme for four-leg two-level converter

Figure 12a, b shows the simulation and experimental performance of PVC scheme with balanced reference voltages $v^*_{ao} = v^*_{bo} = v^*_{co} = 120\,\text{V}$ (rms). A step change from no load to unbalanced loads ($R_{ao} = 24\,\Omega, R_{bo} = 12\,\Omega, R_{co} = 8\,\Omega$) is applied as shown in Fig. 12c, d, where a fast transient response without any overshot is observed. From Fig. 12c, d, it is observed that the fourth-leg carries the neutral current due to the load unbalance. This case study replicates the standard four-wire DG application, where the consumers maintain different loads on each phase. The proposed PVC scheme maintains the load voltages at the given reference value irrespective of the loading condition. The average voltage tracking error, average THD of load voltages, and average switching frequency are noted as 3.1%, 2.3%, and 3400 Hz, respectively for simulation waveforms. With the experimental results, the average voltage tracking error, average THD of load voltages, and average switching frequency are noted as 5.91%, 3.19%, and 4050 Hz, respectively.

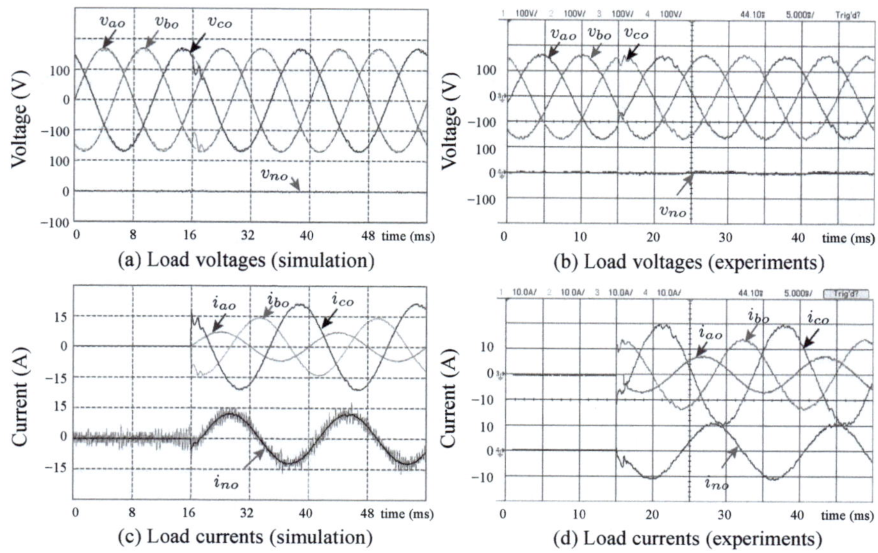

Fig. 12 Results for the PVC scheme of a four-leg two-level converter during a step-change from no-load to unbalanced load

6 Predictive Voltage Control of Four-Leg NPC Converter

The PVC scheme for a four-leg NPC converter with an output *LC* filter and arbitrary loads is analyzed in this section. The performance of the PVC scheme is validated via experimental results.

6.1 Control Scheme

Figure 13 shows the PVC scheme for a four-leg NPC converter. The reference voltages are defined and extrapolated similarly as in the PVC scheme of two-level converter discussed in Sect. 5.1. The proposed approach uses mathematical model of the converter and an *LC* filter in discrete-time domain to predict the future values of load voltage and DC-bus capacitors voltage. The discrete-time model in (37) is used to predict the future values of NPC converter output voltages $v^p_{\alpha n}(k)$, $v^p_{\beta n}(k)$, and $v^p_{\gamma n}(k)$. The present sampling instant converter currents, load voltages, and load currents, predicted converter output voltages, and discrete-time model in (42) are used to predict the future values of load voltages for all the switching state combinations. The discrete-time model in (38) is used to predict the DC-bus capacitors voltage.

The DC-bus capacitors are predicted by using the measured converter currents and coefficients in Table 1 (reproduced from (9)):

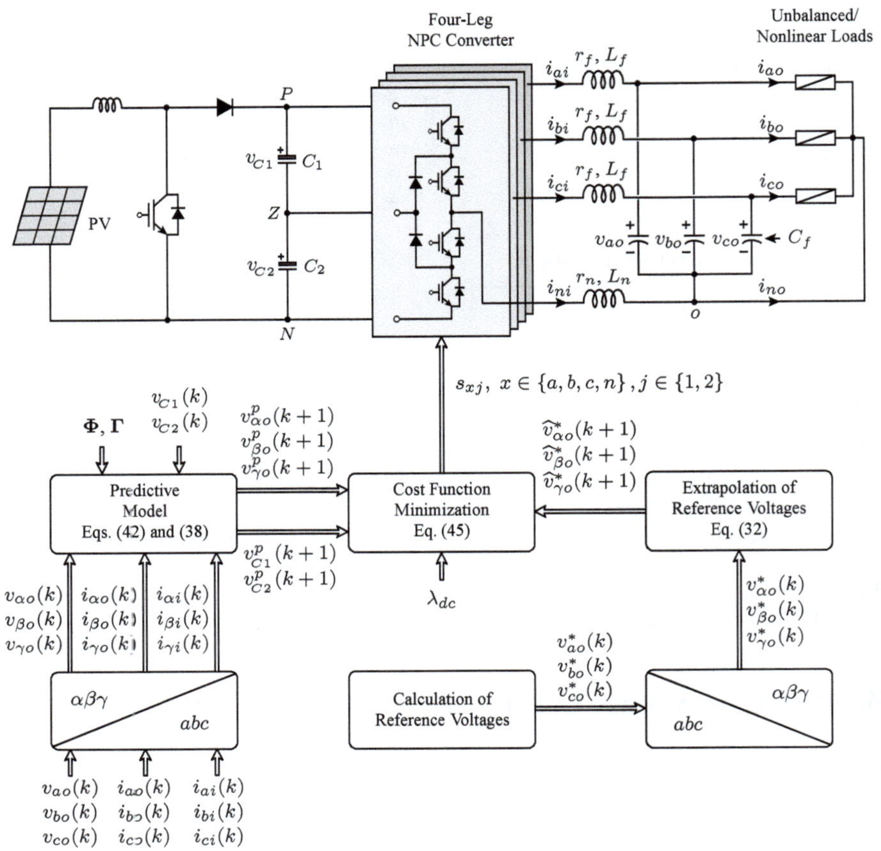

Fig. 13 Block diagram of the PVC scheme for a four-leg NPC converter

$$\begin{aligned} i^p_{C1}(k) &= K_{a1}\, i_{ai}(k) + K_{b1}\, i_{bi}(k) + K_{c1}\, i_{ci}(k) \\ i^p_{C2}(k) &= K_{a2}\, i_{ai}(k) + K_{b2}\, i_{bi}(k) + K_{c2}\, i_{ci}(k). \end{aligned} \tag{44}$$

The optimization process of cost function is carried out in each sampling interval such that all the control objectives are met. A cost function that takes into account the control objectives is defined as follows:

$$\begin{aligned} g_i(k) = &\left[\widehat{v}^*_{\alpha o}(k+1) - v^p_{\alpha o}(k+1)\right]^2 + \left[\widehat{v}^*_{\beta o}(k+1) - v^p_{\beta o}(k+1)\right]^2 \\ &+ \left[\widehat{v}^*_{\gamma o}(k+1) - v^p_{\gamma o}(k+1)\right]^2 + \lambda_{dc}\left[v^p_{C1}(k+1) - v^p_{C2}(k+1)\right]^2 \\ &+ \lambda_{sw}\left[s^p_{n1}(k) - s^{\text{op}}_{n1}(k)\right]^2 + \lambda_{sw}\left[s^p_{n2}(k) - s^{\text{op}}_{n2}(k)\right]^2. \end{aligned} \tag{45}$$

The first three terms in the above expression correspond to the tracking of load voltages. To achieve this objective, the 81 load voltage predictions are compared with their references in stationary reference frame. The fourth term is corresponding to the voltage balancing of DC-bus capacitors voltage. The fifth and sixth terms deal with the reduction of fourth-leg switching frequency in the NPC converter. At the kth instant, the algorithm applies the optimal switching state combination that minimizes the cost function at the $(k+1)$ instant.

6.2 Results and Analysis

The performance of PVC scheme is verified through the experimental studies conducted on a DS1103 based laboratory prototype. The parameters of the laboratory prototype converter are as follows: $v_{dc} = 350\,\text{V}$, $C_{dc} = 4700\,\mu\text{F}$, $r_f = r_n = 0.022\,\Omega$, $L_f = L_n = 2.5\,\text{mH}$, $C_f = 60\,\mu\text{F}$, $f_s = 60\,\text{Hz}$, and $T_s = 100\,\mu\text{s}$. To compensate the computational delay caused by the digital controller, the cost function is calculated with the modified one-sample-ahead prediction [23].

In order to demonstrate the dynamic response of the PVC scheme, a transient analysis has been performed using nonlinear loads and the results are presented in Fig. 14. As depicted in Fig. 14a, the reference load voltages are defined as balanced ($v_{ao}^* = v_{bo}^* = v_{co}^* = 120\,\text{V}\,(rms)$); therefore, load neutral voltage v_{no} is observed to be zero. A step change from no-load to nonlinear load is applied, and the load voltages remain sinusoidal and without any significant transients due to the connection of nonlinear load as shown in Fig. 14b. The nonlinear current flows through the load

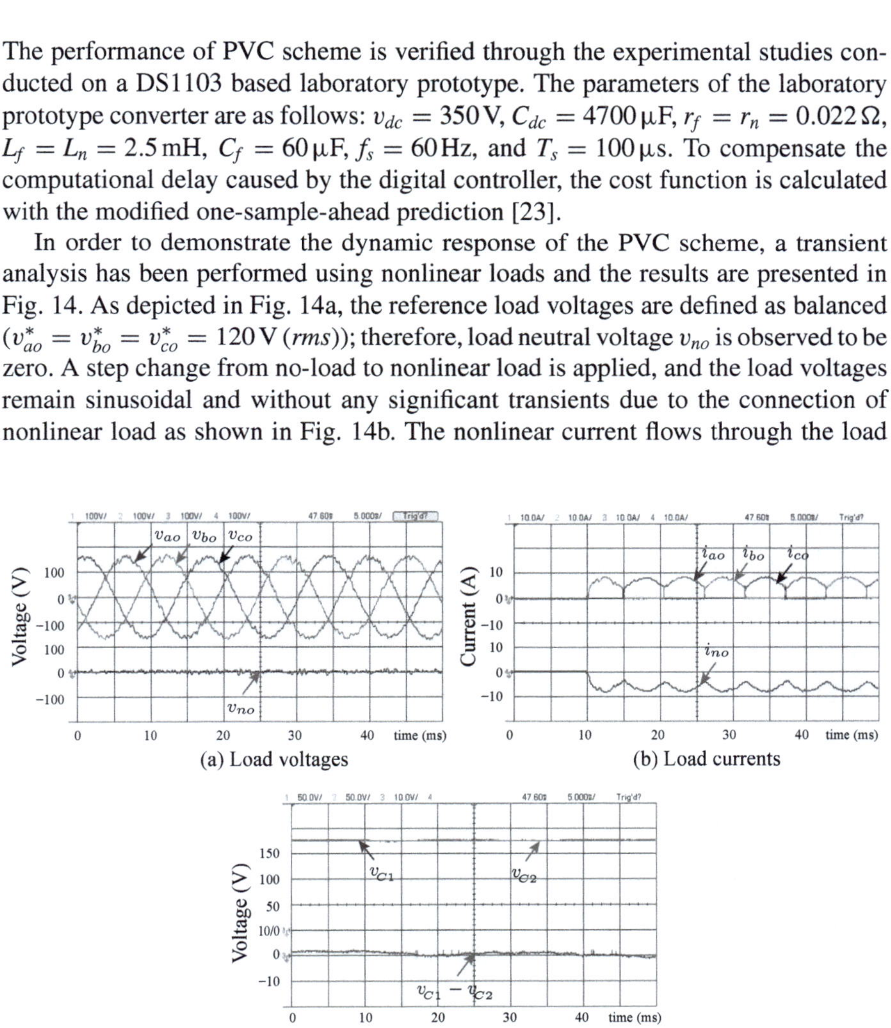

Fig. 14 Results for the PVC scheme of a four-leg NPC converter during a step-change from no-load to nonlinear load

during the positive fundamental cycle of the load voltage. The load neutral current can be observed circulating through the fourth-leg. Fig. 14c demonstrates perfect balancing of the DC-bus capacitors voltage, where the difference between capacitors voltage is negligible.

7 Summary

This book chapter provides a comprehensive analysis on the predictive control strategies for the four-leg two-level and NPC converters with output L and LC filters for grid-connected and standalone photovoltaic energy systems. To implement the predictive control schemes, a detailed stationary frame system models including power converters, output filters, and arbitrary loads are developed in terms of the converter switching states. The PCC scheme effectively reduces the load current tracking error, whereas the PVC scheme reduces the load voltage tracking error. Also, these schemes have an ability to achieve secondary objectives such as DC-bus capacitors voltage control in neutral-point clamped converter and neutral-leg switching frequency minimization. The control algorithm evaluates the cost function for all the possible switching states, and a switching state which gives the minimum cost function value is selected and applied to the converter. The PCC and PVC schemes ensure that the load currents/voltages are tracked to their references with minimal error, and the balancing of DC-bus capacitors during all the operating conditions. These control schemes can compensate the variations in load and filter parameters, while the output currents/voltages continue to track their references. This compensation has been achieved without sacrificing the performance of the converter. The PI controllers and modulation stage are eliminated in the control loop and as a result, a fast dynamic response has been achieved for output currents and load voltages. The analysis and results presented in this chapter demonstrate that predictive control is distinctly simple yet it is a promising tool to control four-leg converters in modern photovoltaic energy systems with excellent performance indices.

References

1. International Energy Agency. Snapshot of global photovoltaic markets. Technical report, 2018. Rep. IEA PVPS T1-33:2018
2. Kouro S, Leon JI, Vinnikov D, Franquelo LG (2015) Grid-connected photovoltaic systems: an overview of recent research and emerging PV converter technology. IEEE Ind Electron Mag 9(1):47–61
3. Jain C, Singh B (2017) Solar energy used for grid connection: a detailed assessment including frequency response and algorithm comparisons for an energy conversion system. IEEE Ind Appl Mag 23(2):37–50
4. Singh B, Sharma S (2012) Design and implementation of four-leg voltage-source-converter-based VFC for autonomous wind energy conversion system. IEEE Trans Ind Electron 59(12):4694–4703

5. Liang J, Green TC, Feng C, Weiss G (2009) Increasing voltage utilization in split-link, four-wire inverters. IEEE Trans. Power Electron. 24(6):1562–1569
6. Rivera M, Yaramasu V, Llor A, Rodriguez J, Wu B, Fadel M (2013) Digital predictive current control of a three-phase four-leg inverter. IEEE Trans Ind Electron 60(11):4903–4912
7. Yaramasu V, Rivera M, Narimani M, Wu B, Rodriguez J (2014) Finite state model-based predictive current control with two-step horizon for four-leg NPC converters. J Power Electron 14(6):1178–1188
8. Rodríguez J, Bernet S, Steimer PK, Lizama IE (2010) A survey on neutral-point-clamped inverters. IEEE Trans Ind Electron 57(7):2219–2230
9. Kim Jang-Hwan, Sul Seung-Ki (2004) A carrier-based PWM method for three-phase four-leg voltage source converters. IEEE Trans Power Electron 19(1):66–75
10. Li Xiangsheng, Deng Zhiquan, Chen Zhida, Fei Qingzhao (2011) Analysis and simplification of three-dimensional space vector PWM for three-phase four-leg inverters. IEEE Trans Ind Electron 58(2):450–464
11. Ceballos S, Pou J, Zaragoza J, Martin JL, Robles E, Gabiola I, Ibanez P (2008) Efficient modulation technique for a four-leg fault-tolerant neutral-point-clamped inverter. IEEE Trans Ind Electron 55(3):1067–1074
12. Yaramasu V, Rivera M, Wu B, Rodriguez J (2013) Model predictive current control of two-level four-leg inverters - Part I: concept, algorithm and simulation analysis. IEEE Trans Power Electron 28(7):3459–3468
13. Vazquez S, Rodriguez J, Rivera M, Franquelo LG, Norambuena M (2017) Model predictive control for power converters and drives: advances and trends. IEEE Trans Ind Electron 64(2):935–947
14. Yaramasu V, Wu B (2016) Model predictive control of wind energy conversion systems, 1st edn. Wiley-IEEE Press, Hoboken
15. Yaramasu V, Wu B, Chen J (2014) Model-predictive control of grid-tied four-level diode-clamped inverters for high-power wind energy conversion systems. IEEE Trans Power Electron 29(6):2861–2873
16. Sawant RR, Chandorkar MC (2009) A multifunctional four-leg grid-connected compensator. IEEE Trans Ind Appl 45(1):249–259
17. Yaramasu V, Rivera M, Narimani M, Wu B, Rodriguez J (2014) Model predictive approach for a simple and effective load voltage control of four-leg inverter with an output *LC* filter. IEEE Trans Ind Electron 61(10):5259–5270
18. Liu Z, Liu J, Li J (2013) Modeling, analysis and mitigation of load neutral point voltage for three-phase four-leg inverter. IEEE Trans Ind Electron 60(5):2010–2021
19. Yaramasu V, Rivera M, Narimani M, Wu B, Rodriguez J (2015) High performance operation for a four-leg NPC inverter with two-sample-ahead predictive control strategy. Int J Electric Power Syst Res 123:31–39
20. Yaramasu V, Rivera M, Wu B, Rodriguez J (2015) Predictive control of four-leg power converters. pp 121–125
21. Yaramasu V, Wu B, Rivera M, Rodriguez J (2013) Predictive current control and DC-link capacitor voltages balancing for four-leg NPC inverters. In: IEEE international symposium on industrial electronics (ISIE), Taipei, Taiwan. pp 1–6
22. Rodriguez J, Coŕtes P (2012) Predictive control of power converters and electrical drives, 1st edn. IEEE Wiley press, Chichester
23. Coŕtes P, Rodríguez J, Silva C, Flores A (2012) Delay compensation in model predictive current control of a three-phase inverter. IEEE Trans Ind Electron 59(2):1323–1325

A Novel Maximum Power Point Tracking Method for Photovoltaic Application Using Secant Incremental Gradient Based on Newton Raphson

Saber Arabi Nowdeh, Mohammad Jafar Hadidian Moghaddam, Manoochehr Babanezhad, Iraj Faraji Davoodkhani, Akhtar Kalam, Abdollah Ahmadi and Almoataz Y. Abdelaziz

Abstract In this chapter, some common methods of maximum power point tracking (MPPT) of the photovoltaic system such as perturb and observe, particle swarm optimization and grey wolf optimizer are described to solve the MPPT problem. Also, a novel method is proposed for MPPT of PV system titled secant incremental gradient based on Newton Raphson (SIGBNR) method. SIGBNR uses the chord slope passing through two points of the function instead of using the explicit derivative of the function, which is equal to tangent line tilt of the function. In addition to high

S. A. Nowdeh
Electrical Department, Payambarazam Student Research Center,
Aqqala, Golestan, Iran
e-mail: saber.arabi17@gmail.com

M. J. H. Moghaddam · A. Kalam
College of Engineering and Science, Victoria University,
Melbourne, Australia
e-mail: mohammad.hadidianmoghaddam@live.vu.edu.au

A. Kalam
e-mail: akhtar.kalam@vu.edu.au

M. Babanezhad
Department of Statistics, Faculty of Science, Golestan,
Golestan University, Gorgan, Iran
e-mail: mbaba22@yahoo.com

I. F. Davoodkhani
Department of Electrical Engineering, Islamic Azad University,
Khalkhal Branch, Khalkhal, Iran
e-mail: faraji.iraj@gmail.com

A. Ahmadi
School of Electrical Engineering and Telecommunications,
University of New South Wales, Sydney, NSW, Australia
e-mail: ahmadi.abdollah.janah@gmail.com

A. Y. Abdelaziz (✉)
Faculty of Engineering, Electric Power and Machine Department,
Ain Shams University, Cairo, Egypt
e-mail: almoatazabdelaziz@hotmail.com

R.-E. Precup et al. (eds.), *Solar Photovoltaic Power Plants*, Power Systems,
https://doi.org/10.1007/978-981-13-6151-7_4

convergence speed, the proposed method requires less computation and also has a higher accuracy in the number of repetitions when it solves MPPT problems. The results arising from the proposed method are compared and analyzed with the other methods to evaluate its performance for solving MPPT problem. The proficiency of the methods is investigated in different scenarios of partial shading condition and compared in view of various features especially efficiency and convergence velocity. The results showed that the proposed method has better performance in achieving to global maximum power point with more tracking efficiency and convergence speed than the other methods. Also, superior capabilities of the proposed method are demonstrated.

Nomenclature

AI	Artificial intelligence
d	Duty cycle
FL	Fuzzy logic
GA	Genetic algorithm
GMPP	Global maximum power point
GWO	Grey wolf optimizer
H&C	Hill-climbing
IncCond	Incremental conductance
LMPP	Local maximum power point
MPPT	Maximum power point tracking
OCV	Open circuit voltage
P&O	Perturb and observe
PSC	Partial shading condition
PSO	Particle swarm optimization
RCC	Ripple correlation control
SCC	Short circuit current
SIGBNR	Secant incremental gradient based on Newton Raphson
URC	Uniform radiation conditions
2S2P	2 parallel strings each with 3 PV cells
6S	6 series PV cells

1 Introduction

Photovoltaic (PV) energy is introduced as a clean and free energy source for residential and industrial applications. The PV cell is an essential part of the PV system. In this system, solar radiation energy is converted into the electricity by PV effects. PV cells often connect electrically to each other and are considered as a module. Today, electricity generation is possible from small to large scales by PV power technology [1, 2]. On the other hand, in addition to the availability of sunlight throughout the year, PV systems can be easily installed. However, PV systems have some problems such as very low efficiency that need to be considered. Receiving the maximum power point (MPP) from PV systems has a very important role due to increasing the efficiency. This can be done by connecting a controller to the MPP due to regulating the work cycle for the load. On the other hand, due to changing the output characteristic of PV systems under various factors such as radiation and temperature variations, changing the load size and partial shaded conditions (PSC), these systems work at the MPP rarely [3, 4]. The effect of the radiation change on the current is greater than the voltage so that reducing the amount of radiation reduces the current more than the voltage and ultimately decreases the power. Also, temperature reduction affects the voltage so that by increasing the temperature, the voltage is decreased more than the current and this problem reduces the power. Therefore, in the operation of PV cells, an algorithm that guarantees global MPP should be used when the operating point changes.

Much work has been done to improve the performance of PV systems through the development of new algorithms to access to maximum power point tracking (MPPT). For example, the perturbation and observation method (P&O) [5] and hill-climbing method (H&C) [6] are widely used for MPPT .The basis of the H&C algorithm is the same as the P&O method since it adjusts the PV voltage to follow the maximum point of the voltage. The main disadvantage of P&O and H&C methods are that if the work point of the system changes rapidly, the algorithm will encounter tracking errors and And not capable to accurately track the Maximum PV power. The incremental conductance (IncCond) algorithm [7] that compares incremental and momentary conductance of PV arrays can track the MPP of a PV system and deliver a high PV energy to the load. The basis of the IncCond method is the zero-derivative of the power per the voltage or the current in the MPP and is found by conducting a momentary conductance of the MPP. The major disadvantage of IncCond is that it requires some involved control circuits. Ripple correlation control (RCC) [8] with the help of the converter due to controlling MPPT causes the ripple in the control strategy. This method works very well in high sunlight, but tracking efficiency decreases in some conditions such as the low radiation. By removing the load of the PV array, the current and voltage in the MPP of the PV system can be determined by short circuit current (SCC) [9] and open circuit voltage (OCV) methods [10]. Although, the VOC and SCC methods have low cost and easy implementations, may be they cannot capable to tracking the PV power point accurately. Therefore, these two categories cannot be considered as the correct search for MPP due to the periodic removal of the load. However, the implementation simplicity of these algorithms causes that they can be used as parts of a new combination technique.

To solve the complex issues such as MPP, artificial intelligence (AI) methods are used instead of traditional methods [11]. For this purpose, the weight coefficients in the objective function must be proportional to the PV input and output. Nevertheless, in order to achieve a suitable input and output pattern of the neural network, testing of the array PV will be time consuming and requires a lot of information. One of the advantages of using the artificial neural network is to show a comparatively correct MPPT without needing to extensive data of PV parameters. The fuzzy logic (FL) method has also been successfully used to track the maximum general power in PV systems under PSC. Although the use of these methods has favorable outcomes, one of their biggest drawbacks is having difficult and complex calculations [12]. In recent years, several literature works have been done due to solving the MPPT of the PV system and the impact of PSC has been evaluated. Due to having the low ability of traditional algorithms to detect MPP, researchers developed other methods such as accidental algorithms based on the artificial intelligence that are inspired by the nature behavior and have ability to increase the efficiency of generated power by PV. These methods are particle swarm optimization (PSO) [13] and genetic algorithm (GA) [14]. PSO and GA optimization methods are easy to calculate and can be easily executed on a low-price digital controller. Also, these methods are very successful due to obtaining the global peak under PSC. Recently, a meta-heuristic algorithm, inspired from the hunting behavior of the gray wolves, called the grey wolf optimizer (GWO) has been developed [15] to solve the MPPT problem. Also, this method requires less parameters than other the evolutionary methods and has few operators, which is a clear feature when the design process is quick. The GWO method is used to solve the MPPT problem under PSC and has a high convergence rate. There are many literature works that discuss and compare the performance of each of the MPPT methods. However, the reviews of these papers are not yet update and do not cover all MPPT methods under uniform radiation conditions (URC) and PSC.

In this chapter, some common methods of MPPT problem such as P&O and intelligent optimization methods like PSO and GWO are described. Also, a new method is proposed for MPPT based on the secant incremental gradient based on Newton Raphson (SIGBNR) method in the PV systems. Secant gradient method uses the chord slope passing through two points of the function instead of using the explicit derivative of the function, which is equal to tangent line tilt of the function. In addition to high convergence speed of SGNR, the proposed method requires less computation and also has a higher accuracy in the number of repetitions when it solves MPPT problems. In this chapter, the results arising from the proposed method are compared and analyzed with other methods to evaluate its performance for solving MPPT problem. The proficiency of cited procedures and proposed method in different scenarios of partial shading condition is investigated and compared to various features especially efficiency and convergence speed. In Sect. 2, PV system modeling is presented. PV system in partial shading condition and partial shading patterns are described in Sect. 3. MPPT application and proposed method are presented in Sect. 4. The results are presented in Sect. 5, and in Sect. 6 the results are concluded.

2 PV System Modelling

One of the most commonly used models for the photovoltaic modules is the single-diode model. In this model, a current source with a diode and the resistor are in parallel. This set, is connected to a second resistance. According to Fig. 1, the value of energy generated by the PV cell is declared using current I_{ph} [16]. The mathematical

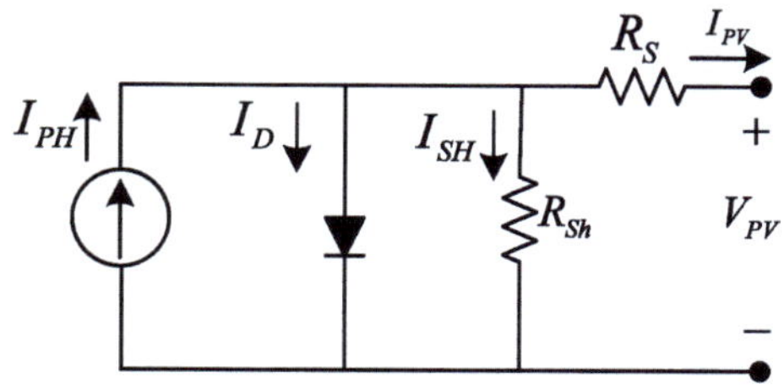

Fig. 1 The single-diode model of a PV cell

equation which defines the PV cell is written as Eq. (1);

$$I_{PV} = I_{Ph} - I_D - \frac{V_D}{R_{sh}} \tag{1}$$

The characteristics equation of the diode is expressed in Eq. (2);

$$I_D = I_0 \left[e^{\frac{V_D}{V_{VT}} A} - 1 \right] \tag{2}$$

The voltage of the diode is obtained [16] according to Eq. (3);

$$V_D = (V_{PV} + I_{PV} R_S) \tag{3}$$

The photocurrent (photic current), Iph, is defined as Eq. (4);

$$I_{Ph} = (I_{SC} + k_1(T - T_{Ref}))\lambda \tag{4}$$

In the above equations, I_0 is the saturation current of the cell, V_T is the thermal voltage of PV cells and is equal to kT/q, q is electrical charge (1.6×10^{-19} coulomb), k is the Boltzmann constant equal to $1.38 \times 10^{-23} J/C$, T is the temperature of p-n joint in K, A shows the ideality factors of diode dependent on the PV technology, I_{SC} represents the short circuit current of the cell under standard conditions (1000 W/m^2 in 25 °C, K_1 expresses the coefficient of cells short circuit current, R_S and R_{sh} are series and shunt resistors in Ω, T_{Ref} is the cells reference temperature, and λ is the solar radiation in W/m^2 [16].

Figures 2 and 3 depict the P-V and I-V characteristics curves of a PV system. Both of the curves show that the array output is nonlinear. Under full solar radiation, there is only one peak in the P-V characteristics curve. However, under partial shading conditions, the PV characteristics curve is varied, and there are several peak points.

3 PV System in Partial Shading Condition

The PV voltage and current depend on the temperature and solar radiation, respectively, which evaluate the PV maximum power point [5]. In other words, when the PV cell obtains more solar radiation, it gives greater current and vice versa. According to

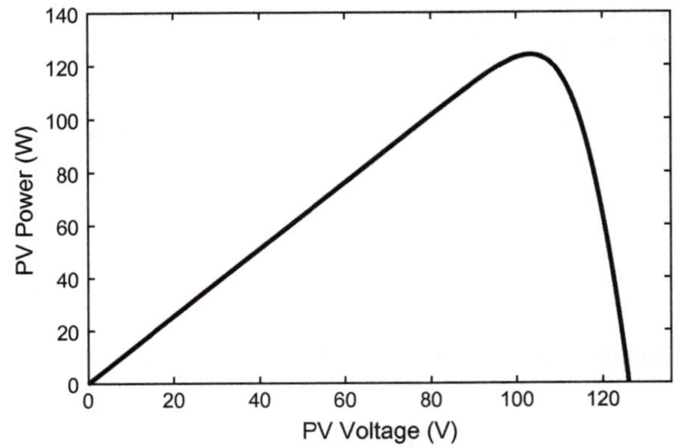

Fig. 2 P-V characteristics curve for a PV system

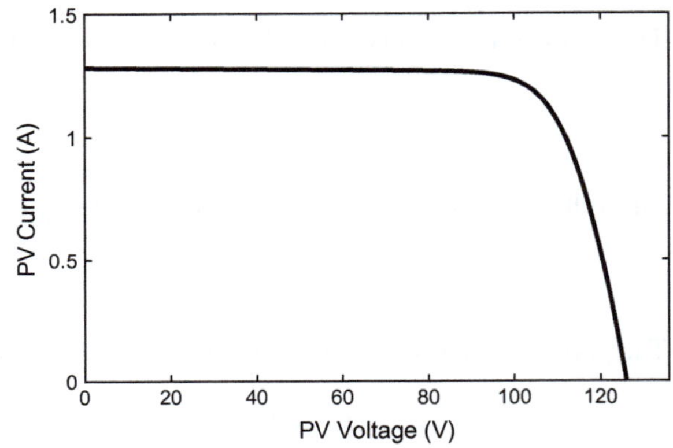

Fig. 3 I-V characteristics curve for a PV system

the I-V and P-V characteristics curves, there is only one optimal point that delivered the maximum power to the system.

Large PV systems usually consist of a series-parallel arrangement of PV modules that each module contains a chain of PV cells with the series connection. Some components of these modules may get poor radiation from the sun due to the movement of the clouds and the shadow arising from trees, buildings and so on. This phenomenon is called PSC [13, 15]. The voltage of the two modules in series mode and their current in parallel mode is different. This subject generates peak points in the intrinsic curve of the P-V. Among these courier points, the largest of them is called global MPP (GMPP) and another pints are Local MPP (LMPP) that their position and amplitude depend on the composition of PV modules and shadow pattern variations. Because of the common MPPT methods are not suitable for the convergence of the GMPP, it is necessary to develop the smart optimization algorithms that can get GMPP under PSC.

Technical parameters of the PV cell used in this study are listed in Table 1.

Two partial shading pattern configurations for PV modules have been considered in this study, i.e., 6S (6 series PV cells) and 2S2P (2 parallel strings each with 3 PV

Table 1 Parameters of the single PV cell

Parameter	Value
Maximum power	20.64 W
Open-circuit rated voltage (V_{ocn})	21 V
The voltage of maximum power point (V_{mp})	17.2 V
Short-circuit rated current (I_{scn})	1.28 A
Current of maximum power point (I_{mp})	1.2 A

cells). For each one of these configurations, two arrangements were assumed from the standpoint of partial shading with non-uniform radiation. Different arrangements of 6S and 3S2P along with I-V and P-V characteristics curves are shown in Figs. 4 and 5.

4 MPPT Application and Proposed Method

The employed MPPT system includes PV modules, a DC/DC boost converter, and a load. The system is shown in Fig. 6. The PV voltage and current are calculated, and the PV power is obtained via multiplying voltage by current and then is applied to the MPPT algorithm. Using the MPPT algorithm a duty cycle, d is generated, and DC/DC converter is activated. The amount of duty cycle (d) is considered as a decision-making variable in the MPPT algorithm, and its corresponding output power is assumed as the fitness value. Accordingly, the objective of the MPPT algorithm is to determine the optimal duty cycle with the aim of extracting the highest power from the PV system based on the MPPT method.

In this study, P&O, PSO, GWO and the proposed SIGBNR method are utilized to solve the MPPT problem. Following describes each of the methods mentioned above.

4.1 P&O Method

The basis of this algorithm is power comparison and voltage variations, so that the voltage and current of the module are first sampled and power and voltage changes are calculated [17, 18]. The flowchart of this method is depicted in Fig. 7, where X is the reference signal. In this algorithm, if the reference voltage is chosen as voltage (i.e., $X = V$), then the objective will be guiding the reference voltage signal towards V_{MPP}. The result is that the instantaneous voltage tracks V_{MPP}, and the output power reaches the maximum power point. For this purpose, a small yet constant disturbance is applied to the voltage of the PV cell. A series of such disturbances, shown by

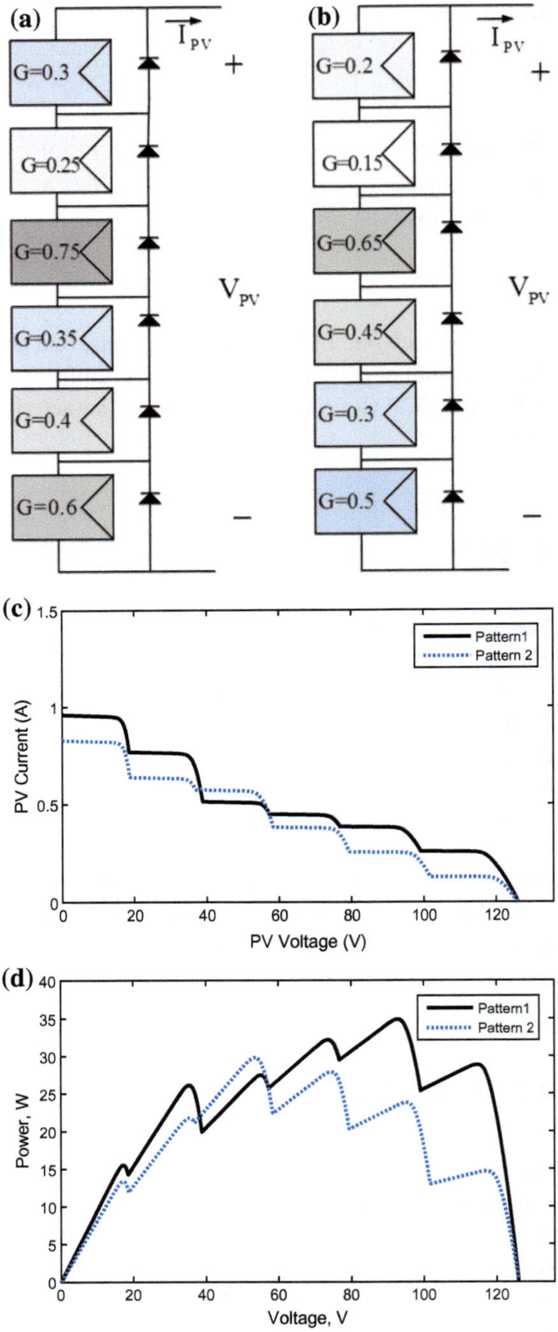

Fig. 4 6S configuration with different partial shading arrangements, **a** pattern 1, **b** pattern 2, **c** I-V characteristics curve, and **d** P-V characteristics curve under partial shading conditions

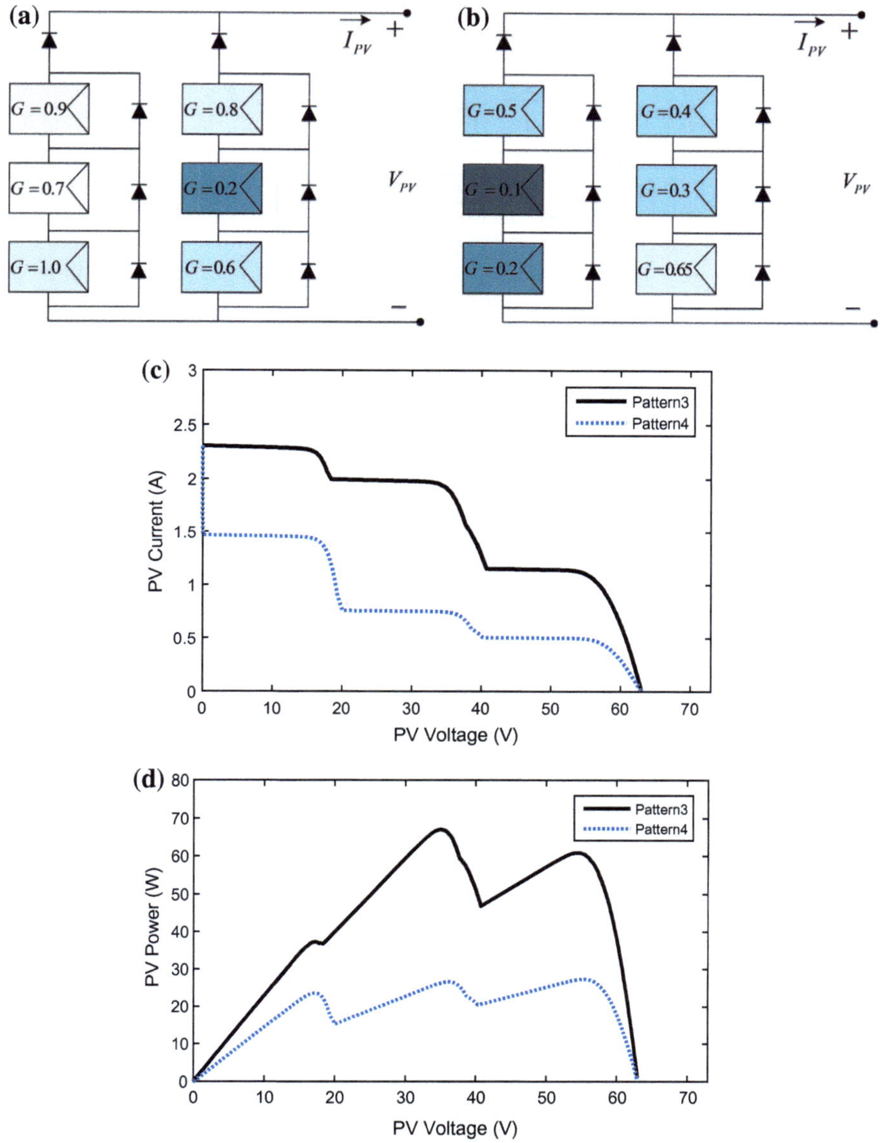

Fig. 5 3S2P configuration with different partial shading arrangements, **a** pattern 1, **b** pattern 2, **c** I-V characteristics curve, and **d** P-V characteristics curve under partial shading conditions

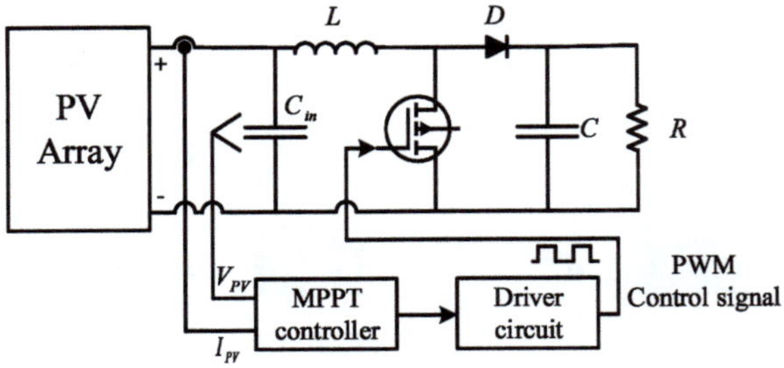

Fig. 6 A general block diagram of the MPPT algorithm [15]

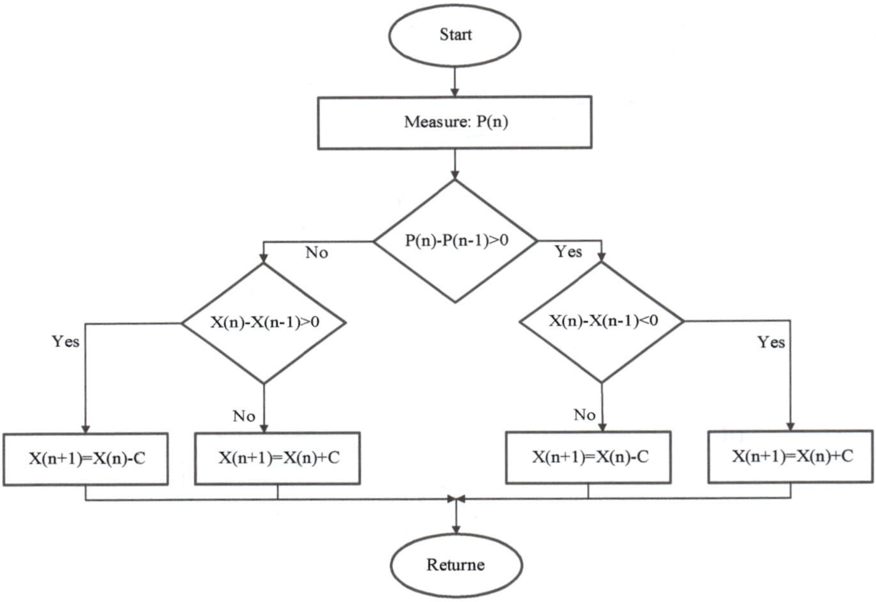

Fig. 7 Flowchart of P&O algorithm [17]

$C = \Delta V$, is applied step by step to change the operating point of the system. After each disturbance, a variation of the output power (ΔP) is measured. If ΔP is positive, then the power reaches the MPP. Therefore, a voltage disturbance with the positive sign should be applied. On the other hand, a negative ΔP means that the power moves away from the MPP and a disturbance with the positive sign should be applied. This process is continued until it approaches the MPP.

The P & O algorithm does not has a proper performance for the rapid changes if the radiation, and cannot follow the MPP. For example, in the case of a fast incensement of the radiation, the problem arises from the fact that the algorithm cannot detect that the power incensement is due to the radiation variation or based on the performance on the radiation curve and therefore, an incorrect command is issued.

4.2 Particle Swarm Optimization (PSO) Method

The PSO algorithm was first introduced by Kennedy and Eberhart. N, 1995 [19]. The algorithm initiates with a random group of solutions (i.e., particles) and then searches for an optimal solution in the problem space by updating the generations. The population of particles (duty cycle) is initiated by constant positions that Surround the search space, as such that d_{min} and d_{max} refer to the lowest and highest values of the dc-dc converters duty cycle. Using a controller similar to the situation of each particle, the dc-dc converter is activated, and its ability amount (power) is calculated after the acceptable settling time of the converter. In this step, a situation with the maximum ability amount obtained so far by the ith particle is known as the best personal value (pbesti). The fitness value of the ith particle is compared with pbesti each moment. If the current situation has a greater fitness amount, then it replaces pbesti. The situation of a particle with the maximum ability amount on the population is called the best global (gbest). There is only one *gbest* for the whole population, and the particles move towards it. The fitness value of the ith particle is compared with *gbest*. If its ability amount is bigger than *gbest*, then replaces it. The speed and situation of the ith particle in the kth iteration are calculated using Eqs. (5) and (6), respectively [13, 19, 20].

$$V_i^{k+1} = w_i V_i^k + r_1 c_1 (pbest_i - d_i^k) + r_2 c_2 (gbest - d_i^k) \tag{5}$$

where,

$$w_i = w_{\max} - \frac{k}{k_{\max}}(w_{\max} - w_{\min});\ c_1 = c_{1\max} - \frac{k}{k_{\max}}(c_{1\max} - c_{1\min});$$
$$c_2 = c_{2\max} - \frac{k}{k_{\max}}(c_{2\max} - c_{2\min}).$$

Equation (5) shows the difference speed of the particle arising from the positions of *pbest* and *gbest*. w_i scales the new speed that is increased along *pbest* and *gbest*. The coefficients c_1 and c_2 display the cognitive and social rates, respectively. In fact, c_1 represents the amount of permeability of the particle arising from *pbest* and c_2 represents the magnitude of the particle impact arising from the rest of the population. r_1 and r_2 are random numbers in the range of (0, 1). Equation (6) indicates the updated speed so that the particle moves to a new position [19, 20].

$$d_i^{k+1} = d_i^k + V_i^{k+1} \tag{6}$$

Flowchart of PSO algorithm for solving MPPT problem is given in Fig. 8.

4.3 Grey Wolf Optimizer

One of the population-based intelligent and evolutionary algorithms is the Gray Wolf algorithm (GWO) which was first announced by Mir-Jalali in 2014 [15, 21, 22]. In this algorithm, the performance and behavior of gray wolfs for hunting is simulated. The parameters such as α, β, δ, and ω represent the leaders of the group so that α directs the group as leader of the group and has important decisions about hunting, resting place and so on. The second group of leadership belongs to β. In addition to being able to help α due to having a good decision, β members are also the best substitutes for the α wolfs when they are old or dead. ω but is at the bottom of this group. The other group members that are not α, β and ω are called δ. The principles of GWO function are as follows:

1. Investigating, pursuing and following the hunt;
2. Pursuing, sieging and harassing the hunt until it stops; and
3. Attacking to the hunt.

In the simulation of the GWO, α is considered as a top answer. After that, the next two responses after α are considered β and δ. Finally, the rest answers are regarded ω. Modeling a suitable situation that gray wolves encircle their hunt during hunting are shown as Eqs. (7) and (8);

$$\vec{D} = \left| \vec{C} \cdot \vec{X_p}(t) - \vec{X_p}(t) \right| \tag{7}$$

$$\vec{X}(t+1) = \vec{X_p}(t) - \vec{A} \cdot \vec{D} \tag{8}$$

where, t represents the repetition, $\vec{A}$, $\vec{C}$, and $\vec{X_p}$ indicate the coefficients vectors, the position of the bait and the gray wolf respectively.

The coefficients vector is obtained from Eqs. (9) and (10);

$$\vec{A} = 2\vec{a} \cdot \vec{r_1} - \vec{a} \tag{9}$$

$$\vec{C} = 2\vec{r_2} \tag{10}$$

where vectors of random numbers at a distance are shown by $\vec{r_1}$ and $\vec{r_2}$ that are selected from [0, 1] area and vector $\vec{a}$ decreases from 0 to 2 during repetitions. To implement the MPPT based on GWO, the duty cycle, D, is defined as a grey wolf. Hence, Eq. (9) can be modified as [15] according to Eq. (11);

$$D_i(k+1) = D_i(k) - A \cdot D \tag{11}$$

Consequently, the GWO fitness is calculated according to Eq. (12).

$$P(d_i^k) > P(d_i^{k-1}) \tag{12}$$

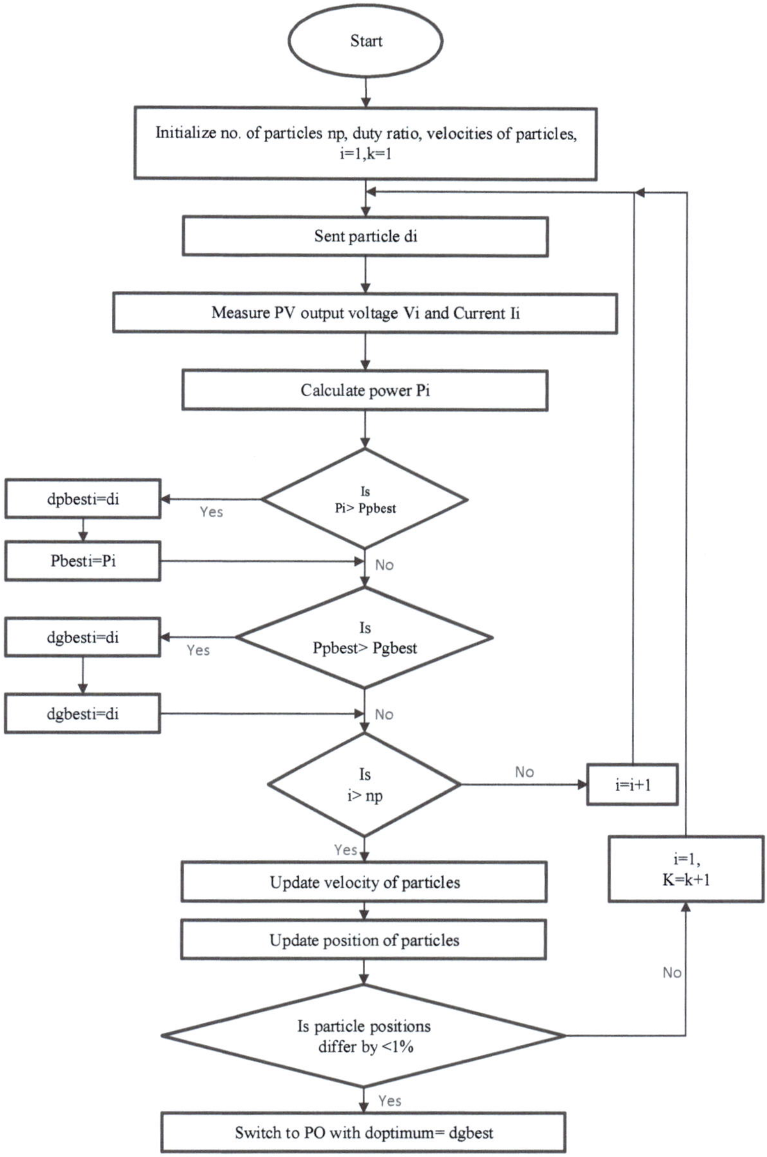

Fig. 8 Flowchart of PSO algorithm [20]

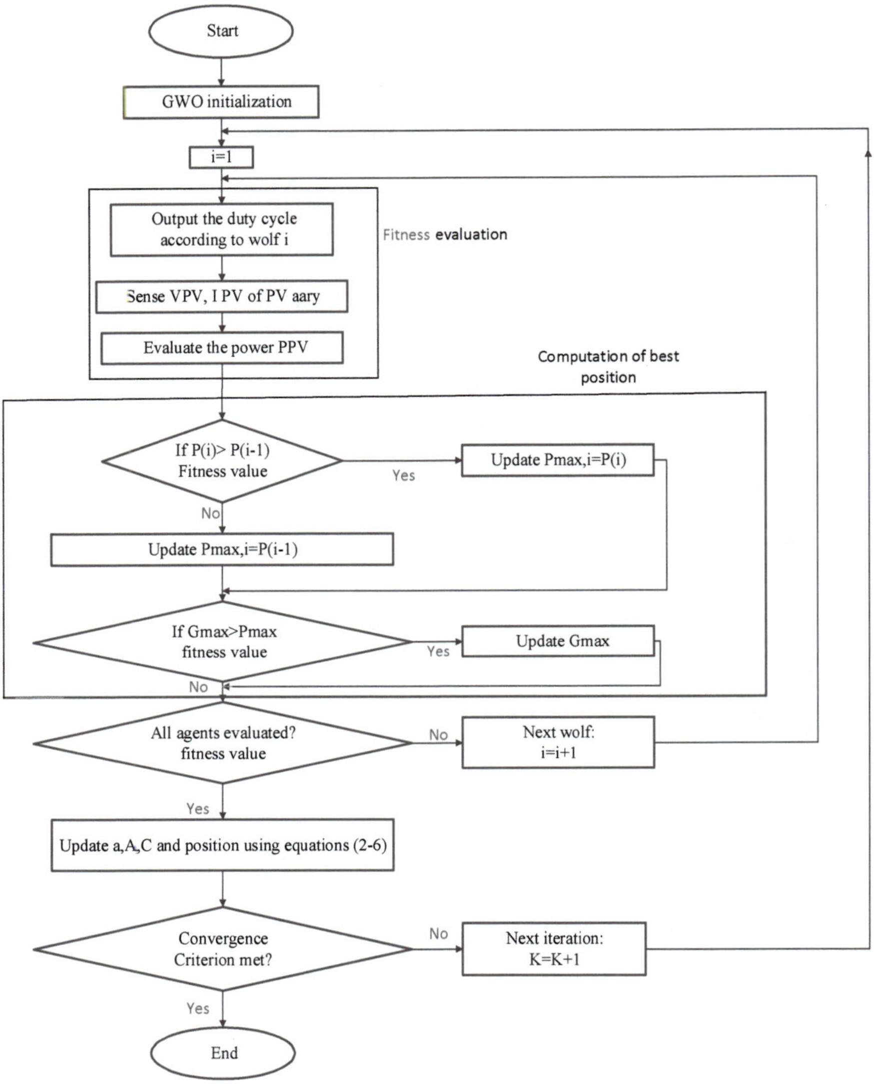

Fig. 9 Flowchart of GWO algorithm [15]

where *P* refers to PV power, *d* refer to duty factor, *i* is the grey wolves number, and *k* is GWO iterations. GWO algorithm flowchart in MPPT solution is illustrated in Fig. 9.

4.4 Secant Gradient Based on Newton Raphson

In this section, at first the proposed method is described, and then the solution of the MPPT problem using the proposed method is presented.

4.4.1 Equation Solution Using Newton Raphson (NR)

To find the roots of a function, i.e., to find a point where the value of function $f(x)$ is zero, many different methods have been suggested up to date, where they can be classified as analytic and numerical methods. In analytic methods, the term for function $f(x)$ is exactly specified, and it is possible to calculate its derivation up to different higher orders. However, numerical methods deal with only the values of the function or its zero-order derivatives, yet the solution method for $f(x) = 0$ is carried out in an iterative and recursive manner [23, 24]. One of the methods is Newton-Raphson (NR), which demands less amount of calculation in addition to having a higher convergence speed. This method finds the solution in a low number of iterations and results in a higher accuracy because it exploits the derivatives of the functions in addition to the functions values for determination of the future points. In NR method, Eq. (13) is used for numerical solving of $f(x) = 0$ [23, 24]:

$$x_{n+1} = x_n - \frac{f(x_n)}{df(x_n)} \tag{13}$$

where the initial value for x_0 is selected based on randomness and guess, and the numerical value of the initial value not impact the final solution of the problem, and the recursive algorithm converges to a specified point regardless of the initial value.

4.4.2 Incremental Gradient Method

In Calculus, the gradient of a scalar field is a vector field where its components show the variation rate of the first field in different directions. The direction of the gradient vector field is in the same direction of the maximum changes. In other words, for a vector that its amplitude and direction show the maximum space rate of change of a scalar vector, the gradient is defined in Eq. (14) as a numerical quantity, as follows [25, 26]:

$$\nabla f = \frac{\partial f}{\partial x} i + \frac{\partial f}{\partial y} j + \frac{\partial f}{\partial z} k \tag{14}$$

Gradient descent or incremental gradient method is a recursive and iteration-based first-order optimization algorithm that exploits the gradient of a function to find its minimum or maximum values. If the function under study is a single-variable and scalar function, its gradient is equal to the first-order derivative of the function concerning the variable itself.

As mentioned earlier, the direction of the functions gradient is aligned with the direction of the maximum changes. Hence, to find the minimum value of the function with the help of gradient, it suffices for the variable changes to be proportional to the minus of the gradient value. This method is called gradient descent. Similarly, if the objective is to determine the maximum fitness, the direction of changes for the functions variable should be in the same direction of the function gradient and proportional to the gradient value at each point. This latter method is called incremental gradient.

The considered objective function for optimization purpose, $F(x)$ is given. Now, the necessary condition for using the gradient method is that F function should be defined and differentiable in point x. Then, function F has its maximum increase in x if we move from point x in the same direction with the gradient of F and proportional to the gradient value. If this change of variable is written iteratively, then in Eq. (15) we have [25, 26]:

$$x_{n+1} = x_n + \gamma \times \nabla F(x_n) \tag{15}$$

The above recursive relation is convergent to the maximum value of function F if only the coefficient γ is chosen small enough. Hence, $F(a_n) \leq F(a_{n+1})$ and the value of objective functions becomes larger as we move forward at each step. In other words, the value of $\gamma \times \nabla F(x_n)$ is added to x_n so the problem variable could changes aligned with the function gradient and in proportion with it.
Regarding the above iterative equation and the initial guess a_0 for the search variable, the values of a_1, a_2, and ... are obtained where, $F(a_0) \leq F(a_1) \leq F(a_2) \leq \ldots$ and finally, the value of function F will reach it maximum in a desired number of iterations. Fig. 10 shows the steps for the algorithm where the depicted circles represent a level set.

4.4.3 Debugging of the Incremental Gradient Method Using Secant Method

One disadvantage of incremental gradient method is that it needs the first-order derivative term of the function. However, in solving for the MPP, there is no given term available for derivation or for presenting the gradient term. Therefore, to overcome this issue, the slope of the line crossing two points of the function is used instead of using the explicit derivation Of the fitness equal to line slope tangent to the function. Under such conditions for the iterative solution of the problem, two iterative points will be used, where the derivative of the function can be approximated in Eq. (16) according to [27, 28]:

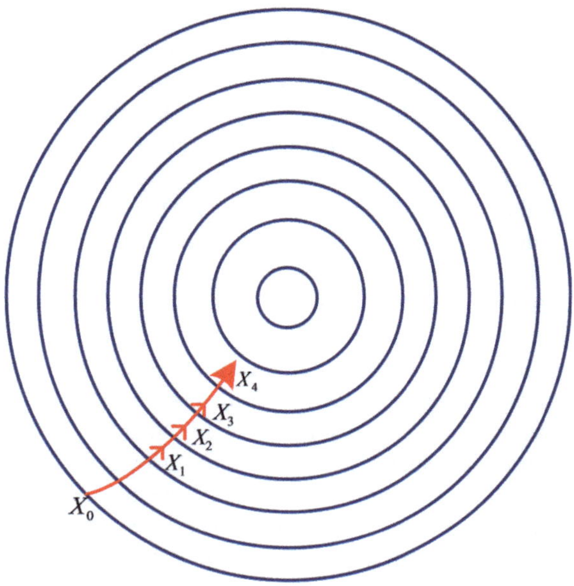

Fig. 10 the change of variable x in the direction of function gradient going toward the extremum point of function $F(x)$

$$\nabla F(mpp) = df(x_n) = \frac{f(x_n) - f(x_{n-1})}{x_n - x_{n-1}} \tag{16}$$

Hence, to start the iteration algorithm, at first, two random points should be specified on the functions curve as the initial guess. Equation (17) shows that the iteration algorithm of incremental gradient method is generalized to the iteration algorithm of secant incremental gradient method [27, 28]:

$$x_{n+1} = x_n + \gamma \times \frac{f(x_n) - f(x_{n-1})}{x_n - x_{n-1}} \tag{17}$$

Now, if one considers the above iteration algorithm, it can be seen that there is no need for explicit derivative of the function and only two values of the function in the given two points are required to guess the new value of x. under these conditions, this iteration method can be employed for finding the MPP.

4.4.4 MPPT Based on Secant Incremental Gradient Method

Application of secant incremental gradient method to reach the maximum produced power for the PV cells demands utilization of some complementary approaches. It should be noted that if the function under study is convex, then any value can be used as the initial guess for start of the iteration process and it is ensured that the algorithm finally converges to the extremum of the objective function in sequential iterations. For instance, when there is only one single PV cell in the circuit, and P-V curve of the PV system is completely continuous and differentiable on all over the points, any value of V_{pv} can be considered as the initial guess and finally reach the MPP. Nonetheless, when multiple PV cells under different solar radiation are joined series in shading condition, it is probable that some initial values for start of the algorithm converge to the local maximum or one of the adjacent peaks of the power curve and do not result in extraction of the maximum pure power. Consequently, under partial shading conditions or in case there are two or more peaks for the P-V curve, the value of the first guess is very effective, and it might lead to undesired solutions.

For this reason, to determine a suitable initial guess before the start of the algorithm, the range of allowed PV voltage is divided into several sections and the amount of power generation by the PV system at each of these sections are separately calculated. The point with the maximum generated power is closer to the MPP and is chosen as the initial value, V_{PV0} , to start the iteration algorithm of incremental gradient method. Thereby, it is ensured that the initial point is suitable and independent before the stathe rt of the algorithm. However, the number of search points should not be less than a given limit, and experimentally it is suggested that the number of search points for V_{PV0} , should be at least equal to the number of series-connected cells (the number of peaks in the P-V curve).

The second point to mention is a determination of derivative or gradient of the power function concerning voltage in different points of the iteration. After setting the initial voltage value, the start of iteration algorithm requires the calculation of gradient in the initial point using secant derivative method. Secant method at each step needs the power values of the current and previous steps. However, since there are no previous points at the first step, a value of $0.99 \times V_{PV0}$ is used as a supporting point for estimation of gradient or derivative and this means that the voltage value in the second step is determined through scheduling and without using the gradient method in order to make it possible to start the iteration method for the third step. In order to the slope of chord under study present a suitable approximation of the function derivative in the initial point, the distance between the two points that create the chord for calculation of the chords slope should be close enough to be as close and similar to the tangent line on point V_{PV0} as possible. For this, the desired voltage value in the second step deviates only 1% from the initial guess point. Flowchart of the proposed Approach is illustrated in Fig. 11. The implementation steps of secant incremental gradient method are defined as: $[0.3 \times V_{OC}\ 0.6 \times V_{OC}]$

Step (1) To determine a suitable initial point for secant incremental gradient algorithm, the range of PV output voltage in $[0.3 \times V_{OC}\ 0.6 \times V_{OC}]$ is divided into Ns, as PV cell numbers with series configuration. These values are named $V_1, V_2, \ldots$, and V_{Ns}. Then, the PV output voltages for each of these points are calculated and a, voltage with the maximum produced power is selected as V_{PV0}.

Step (2) Specify the supporting point as $0.99 \times V_{PV0}$ and calculate the output power corresponding to this point.

Step (3) The initial value of V_1 is called V_{PV0} and the value of supporting point V_2 is called $0.99 \times V_{PV0}$. The corresponding output power of these points are named $P_pV_1 = f(V_1)$ and $P_pV_2 = f(V_2)$.

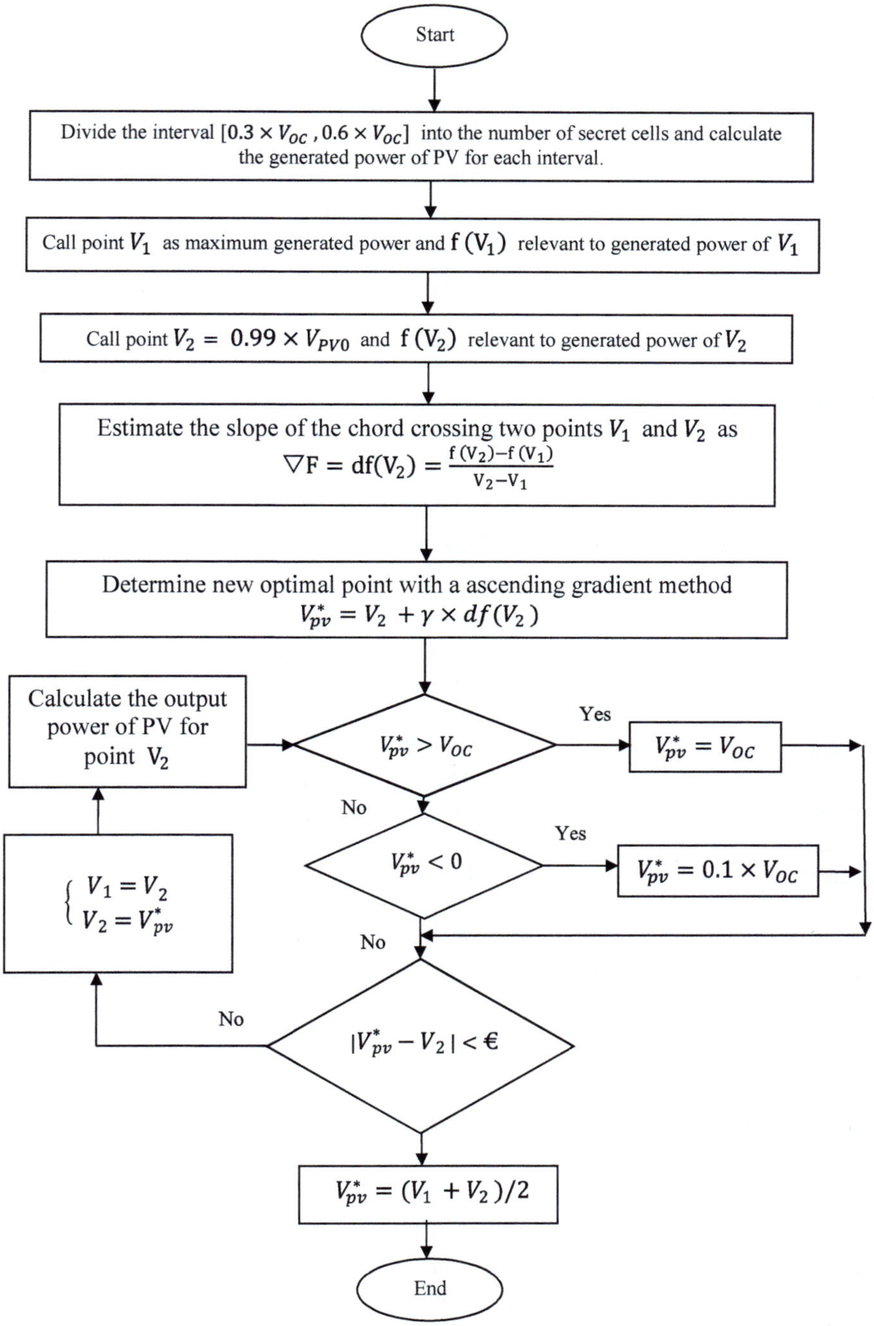

Fig. 11 Flowchart of MPPT algorithm based on secant incremental gradient

Step (4) the slope of the chord crossing two points V_1 and V_2 is calculated as:

$$\nabla F = df(V_2) = \frac{f(V_2) - f(V_1)}{V_2 - V_1}.$$

Step (5) the desired voltage of the next step is determined by $V_{pv}^* = V_2 + \gamma \times df(V_2)$ and if $V_{pv}^* > 0$ then $V_{pv}^* = V_{OC}$, and if $V_{pv}^* < 0$ then $V_{pv}^* = 0.1 \times V_{OC}$, to prevent the produced reference voltage from violating the acceptable PV output voltage range.

Step (6) If $|V_{pv}^* - V_2| < \varepsilon$ then go to Step 7, otherwise update the voltages of the previous Steps and go to Step 2: $\begin{cases} V_1 = V_2 \\ V_2 = V_{pv}^* \end{cases}$.

Step (7) to avoid the power fluctuations around the optimum point, $V_{pv}^* = (V_1 + V_2)/2$ is considered and the algorithm is terminated.

5 Simulation Results and Discussion

In this section, the MPPT simulation results of the PV system under different partial shading conditions using the proposed SIGBNR method are proposed. Also, to verify the proposed method, the MPPT problem was solved using P&O, PSO and GWO methods and the results were compared and analyzed in terms of convergence speed, extracted energy, energy efficiency, and GMPP tracking efficiency under various partial shading patterns. The variable to be optimized is duty cycle, d, of the DC/DC converter, and parameters of P&O, PSO, and GWO methods are according to Refs. [16, 22]. The simulations were implemented in such a way that the transient response of each one of the methods is depicted while the shading pattern was varying from one type to another. The simulation duration time was assumed 40 s, where each 20 s is associated with the implementation of one specific shading pattern. Moreover, the simulations were developed in MATLAB/SIMULINK environment to achieve the GMPP.

5.1 Simulation Results Under Partial Shading Patterns 1 and 2

Figure 12 illustrates the MPPT results of the PV system under partial shading patterns 1 and 2 that are associated with the 6S arrangements based on different methods. The P&O method initiates the MPPT action by estimating the initial duty cycle equal to 0.5. Unfortunately, under all partial shading conditions, the P&O method is not able to extract the maximum power and is trapped in the LMPP.

The solution results for the MPPT problem according to the PSO algorithm show that it has found the MPP and converged around the GMPP. However, there is still a

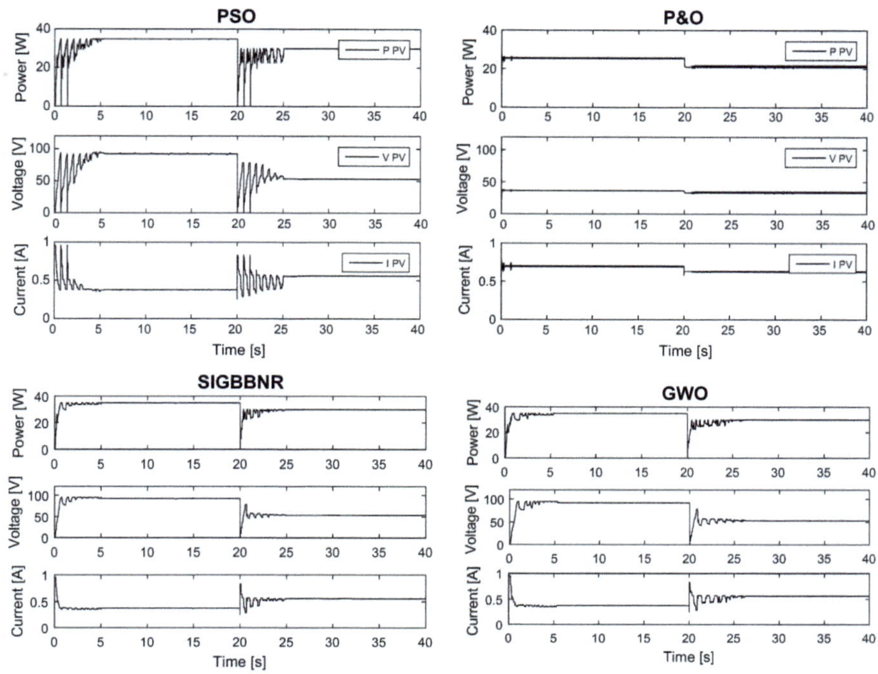

Fig. 12 Responses of different algorithms in solving the MPPT problem under shading patterns 1 and 2

small amount of error. The maximum power for the first pattern is obtained as 34.887 W, and the PSO algorithm has converged to 34.32 W, and in the second pattern it has converged from 29.89 W to 29.79 W. Although the convergence has been performed close to the GMPP, large power, and voltage oscillations during the convergence process and large settling time are considered as its disadvantages.

Additionally, the obtained results based on the GWO algorithm show that it has converged to the GMPP in an acceptable speed and desired accuracy. It has reached the GMPP at $t = 5.2$ s and 6.7 s for the first and second patterns, respectively. Compared to the PSO algorithm, the GWO algorithm has reduced the convergence error form 1.61% to 0.028%. The settling time has also been halved. The solution results of the MPPT problem based on the proposed SIGBNR method in tracking process for pattern 1 show that the algorithm reaches the GMPP at only $t = 2.2$ s. In partial shading changing to pattern number 2 at t = 20 s, then the MPP tracking has started once more, and the new GMPP is tracked. To compare the performance of the proposed method with other methods, the curves for transient responses related to tracking the maximum point are shown. These curves prove that the PSO-based algorithm can achieve the GMPP but the convergence times for patterns 1 and 2 are 10.49 s and 10.02 s, respectively. Also, compared to GWO algorithm, the PSO-based algorithm makes some oscillations in the PV output power for longer time duration. Also, the P&O algorithm is not able to reach the GMPP and settles in the local maximum. On the other hand, the convergence capability of the GWO method to reach the GMPP is less than the proposed method in this chapter. Therefore, it is crystal clear that the proposed SIGBNR method extracts more power in a considerable high convergence speed, i.e., mitigates the power oscillations in a short period and reaches the global optimum.

5.2 *Simulation Results Under Partial Shading Patterns 3 and 4*

The MPPT results for the PV system under shading patterns 3 and 4 related to 3S2P arrangements based on different methods are shown in Fig. 13. The GMPP value for patterns 3 and 4 is 67.08 W, and 27.52 W, respectively, and the GMPP is placed in right side of PV power-voltage curve. The optimization approaches are applied for 20 s for patterns 3 and 4. The results of the proposed method prove that it extracts more power in pattern 3 compared to all other methods. Also, it has less transient oscillations in comparison to other methods. GWO algorithm finds the GMPP in pattern 3 at 5.25 s, and when the shading pattern changes to pattern 4 it takes 5.05 s for GWO algorithm to converge to the GMPP. As one can see, the PSO method takes a longer time to reach the GMPP and the stable oscillations for the PV output power takes longer time. P&O algorithm does not identify the global optimum for pattern 3, but it converges to the GMPP in pattern 4 because the GMPP is located on the P-V curve and the initial value of $d = 0.5$ has guided the P&O algorithm around the GMPP. If the initial value of d is changed and simulation is run again, then the P&O method is no longer able to achieve the GMPP.

5.3 *Comparison of Results Using a Different Algorithm*

The mentioned methods performance and the proposed SIGBNR method in terms of maximum power, convergence rate, energy efficiency, and GMPP tracking efficiency are evaluated and listed in Table 2, 3, 4 and 5. According to these tables, the results show that the proposed algorithm converges to the GMPP very quickly compared to other methods. Hence, the proposed method has extracted the maximum power of the PV system in the least possible time. The performance comparison of MPPT methods in view of method convergence speed that the proposed method has higher convergence speed with regard to other methods. Also, performances of the MPPT methods in terms of tracking the GMPP show that the proposed method has once more a better performance with respect to other methods. Consequently, numerical results verify the desired capability and superiority of the proposed algorithm.

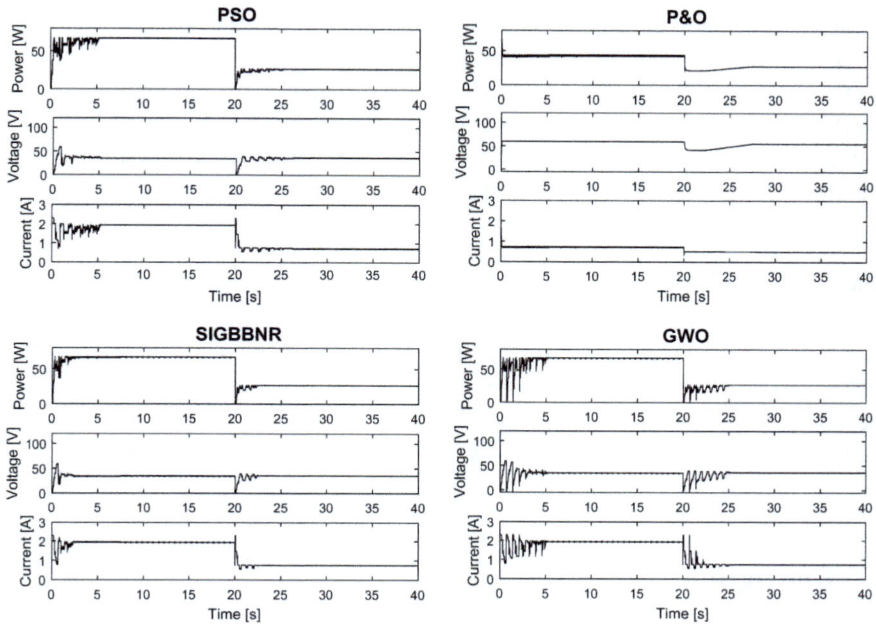

Fig. 13 Responses of different algorithms in solving the MPPT under shading patterns 3 and 4

Table 2 Performance comparison of MPPT methods in terms of maximum power

MPPT algorithm	Pattern1	Pattern2	Pattern3	Pattern4
Maximum power	34.88	29.89	67.07	27.52
PO	25.59	21.77	44.16	26.57
PSO	34.32	29.79	66.69	26.8
GWO	34.87	29.85	66.96	26.78
SIGBNR	**34.85**	**29.86**	**67.01**	**26.86**

Table 3 Performance comparison of MPPT methods in terms of convergence speed

MPPT Algorithm	Pattern1	Pattern2	Pattern3	Pattern4
PO	10.98	10.37	10.06	6.08
PSO	10.49	10.02	10.15	11.18
GWO	5.32	6.22	5.25	5.46
SIGBNR	**5.02**	**4.9**	**2.51**	**2.55**

Table 4 Performance comparison of MPPT methods in terms of energy efficiency

MPPT algorithm	Pattern1	Pattern2	Pattern3	Pattern4
PO	73.245	72.751	65.784	69.493
PSO	97.534	98.833	98.592	96.476
GWO	99.528	99.349	99.399	96.868
SIGBNR	**99.582**	**99.592**	**99.806**	**98.669**

Table 5 Performance comparison of MPPT methods in terms of GMPP tracking efficiency

MPPT algorithm	Pattern1	Pattern2	Pattern3	Pattern4
PO	73.366	72.834	65.832	96.547
PSO	98.394	99.665	99.419	97.384
GWO	99.828	99.86	99.732	97.311
SIGBNR	**99.91**	**99.89**	**99.91**	**97.601**

6 Conclusion

In this chapter a novel method is presented due to solving the MPPT problem in photovoltaic systems under PSC. According to the Newton-Raphson method and based on the chord gradient method, the proposed SIGBNR method has advantages such as high convergence rate and efficiency with fewer computations and high convergence accuracy in low repetitions. Due to verifying SIGBNR method, its results are compared with different patterns of partial shading such as P & O, PSO, and GWO methods. The simulation results show that the proposed algorithm has a higher convergence rate and less error in the computing rather than other methods. In SIGBNR method, in order to use of function derivative and due to having high power and accuracy it can converge to a MPP without any voltage or power fluctuations around the MPP. Thus, energy losses are decreased. The results showed that the SIGBNR converges to GMPP faster than other methods and its tracking efficiency is higher than other methods.

References

1. Subudhi Bidyadhar, Pradhan Raseswari (2013) A comparative study on maximum power point tracking techniques for photovoltaic power systems. IEEE Trans Sustain Energy 4(1):89–98
2. Verma D, Nema S, Shandilya AM, Dash SK (2016) Maximum power point tracking (MPPT) techniques: recapitulation in solar photovoltaic systems. Renew Sustain Energy Rev 54:1018–1034
3. Kheldoun A, Bradai R, Boukenoui R, Mellit A (2016) A new golden section method-based maximum power point tracking algorithm for photovoltaic systems. Energy Convers Manag 111:125–136

4. Seyedmahmoudian M, Rahmani R, Mekhilef S, Oo AMT, Stojcevski A, Soon TK, Ghandhari AS (2015) Simulation and hardware implementation of new maximum power point tracking technique for partially shaded PV system using hybrid DEPSO method. IEEE Trans Sustain Energy 6(3):850–862
5. Linus RM, Damodharan P (2015) Maximum power point tracking method using a modified perturb and observe algorithm for grid connected wind energy conversion systems. IET Renew Power Gener 9(6):682–689
6. Alajmi BN, Ahmed KH, Finney SJ, Williams BW (2011) Fuzzy-logic-control approach of a modified hill-climbing method for maximum power point in microgrid standalone photovoltaic system. IEEE Trans Power Electron 26(4):1022–1030
7. Elgendy MA, Zahawi B, Atkinson DJ (2013) Assessment of the incremental conductance maximum power point tracking algorithm. IEEE Trans Sustain Energy 4(1):108–117
8. Casadei D, Grandi G, Rossi C (2006) Single-phase single-stage photovoltaic generation system based on a ripple correlation control maximum power point tracking. IEEE Trans Energy Convers 21(2):562–568
9. Husain MA, Tariq A, Hameed S, Arif MSB, Jain A (2017) Comparative assessment of maximum power point tracking procedures for photovoltaic systems. Green Energy Environ 2(1):5–17
10. Montecucco A, Knox AR (2015) Maximum power point tracking converter based on the open-circuit voltage method for thermoelectric generators. IEEE Trans Power Electron 30(2):828–839
11. Punitha K, Devaraj D, Sakthivel S (2013) Artificial neural network based modified incremental conductance algorithm for maximum power point tracking in photovoltaic system under partial shading conditions. Energy 62:330–340
12. Algazar MM, El-Halim HA, Salem MEEK (2012) Maximum power point tracking using fuzzy logic control. Int J Electr Power Energy Syst 39(1):21–28
13. Sundareswaran K, Palani S (2015) Application of a combined particle swarm optimization and perturb and observe method for MPPT in PV systems under partial shading conditions. Renew Energy 75:308–317
14. Daraban S, Petreus D, Morel C (2014) A novel MPPT (maximum power point tracking) algorithm based on a modified genetic algorithm specialized on tracking the global maximum power point in photovoltaic systems affected by partial shading. Energy 74:374–388
15. Mohanty S, Subudhi B, Ray PK (2016) A new MPPT design using grey wolf optimization technique for photovoltaic system under partial shading conditions. IEEE Trans Sustain Energy 7(1):181–188
16. Kamarzaman NA, Tan CW (2014) A comprehensive review of maximum power point tracking algorithms for photovoltaic systems. Renew Sustain Energy Rev 37:585–598
17. Abdelsalam AK, Massoud AM, Ahmed S, Enjeti PN (2011) High-performance adaptive perturb and observe MPPT technique for photovoltaic-based microgrids. IEEE Trans Power Electron 4:26
18. Rajani SV, Pandya VJ (2015) Simulation and comparison of perturb and observe and incremental conductance MPPT algorithms for solar energy system connected to grid. Sadhana 40(1):139–153
19. Eberhart R, Kennedy J (1995) A new optimizer using particle swarm theory. In: MHS'95. Proceedings of the sixth international symposium on micro machine and human science, IEEE, pp 39–43
20. Khaehintung N, Kunakorn A, Sirisuk P (2010) A novel fuzzy logic control technique tuned by particle swarm optimization for maximum power point tracking for a photovoltaic system using a current-mode boost converter with bifurcation control. Int J Control Autom Syst 8(2):289–300
21. Mirjalili S, Mirjalili SM, Lewis A (2014) Grey wolf optimizer. Adv Eng Softw 69:46–61
22. Kamboj VK, Bath SK, Dhillon JS (2016) Solution of non-convex economic load dispatch problem using Grey Wolf Optimizer. Neural Comput Appl 27(5):1301–1316
23. Akram S, ul Ann Q (2015) Newton raphson method. Int J Sci Eng Res 6(7):2229–2241

24. Hosseini SH, Farakhor A, Haghighian SK (2013) Novel algorithm of MPPT for PV array based on variable step Newton-Raphson method through model predictive control. In: 2013 13th international conference on control, automation and systems (ICCAS), IEEE, pp 1577–1582
25. Lan G, Zhou Y (2017) An optimal randomized incremental gradient method. Mathematical programming, pp 1–49
26. Blatt D, Hero AO, Gauchman H (2007) A convergent incremental gradient method with a constant step size. SIAM J Optim 18(1):29–51
27. Geum YH, Kim YI (2009) Cubic convergence of parameter-controlled Newton-secant method for multiple zeros. J Comput Appl Math 233(4):931–937
28. Hernández MA, Rubio MJ (2002) The Secant method for nondifferentiable operators. Appl Math Lett 15(4):395–399

Study on Control of Hybrid Photovoltaic-Wind Power System Using Xilinx System Generator

Nadjwa Chettibi and Adel Mellit

Abstract In this chapter, a grid connected hybrid power system consisting of a Photovoltaic (PV) source and a Wind Turbine (WT) generator is investigated. The main goal is the study of the design procedure of a digital control circuit of an energy generation system for future implementation on the Field-Programmable Gate Array (FPGA) platform. Hence, well-known Maximum Power Point Tracking (MPPT) techniques are adopted in order to extract the maximum energy from the renewable energy sources. Further, for the control of Permanent Magnetic Synchronous Generator (PMSG), the Field Oriented Control (FOC) structure is applied. The Virtual Flux Oriented Control (VFOC) scheme is adopted for the control of the grid connected three-phase inverter based on the backstepping approach. The overall control scheme of the PV-WT power system is established using the Xilinx System Generator (XSG) design tool. The simulation results are provided in order to prove the effectiveness of the developed XSG based control circuit.

Abbreviation

AC :	Alternate Current.
ADALINE:	ADAptive LInear NEuron.
DC:	Direct Current.
dq :	direct quadrature.
FOC:	Field Oriented Control.
FPGA:	Field-Programmable Gate Array.
HDL:	Hardware Description Language.
HEGS:	Hybrid Energy Generation System.
LMS:	Least Mean Square.

N. Chettibi · A. Mellit (✉)
Renewable Energy Laboratory, University of Jijel, Jijel, Algeria
e-mail: adelmellit2013@gmail.com

N. Chettibi
e-mail: chettibi.na@gmail.com

A. Mellit
ICTP, Trieste, Italy

R.-E. Precup et al. (eds.), *Solar Photovoltaic Power Plants*, Power Systems,
https://doi.org/10.1007/978-981-13-6151-7_5

MPP:	Maximum Power Point.
MPPT:	Maximum Power Point Tracking.
OTC:	Optimal Torque Control.
P&O:	Perturb and Observe.
PI:	Proportional Integral.
PMSG:	Permanent Magnetic Synchronous Generator.
PV:	PhotoVoltaic.
PVG:	PhotoVoltaic Generator.
PWM :	Pulse Width Modulation.
RES:	Renewable Energy Sources.
SPWM:	Sinusoidal Pulse Width Modulation.
VF :	Virtual Flux.
VFOC:	Virtual Flux Oriented Control.
VOC:	Voltage Oriented Control.
WT:	Wind Turbine.
WTG:	Wind Turbine Generators.
XSG:	Xilinx System Generator.

1 Introduction

Nowadays, renewable energy sources (RESs) such as photovoltaic panels and Wind Turbine Generators (WTG) become very promising electric generators due to their several advantages like cleanness, durability, zero-fuel emissions, etc [1]. However, the RESs (PV and WTG) exhibit nonlinear electric characteristics, which depend on the varying climatic conditions. Thus, the tracking of the optimal operating points is of the main interest in order to enhance the energy generation efficiency of RESs. In the last decades, several MPPT techniques [2, 3] have been proposed in the literature with different degree of complexity.

The PV panels and WTGs are complementary energy sources [1], which can be combined in a single energy system to increase the efficiency and to optimize the power quality supply. In grid-connected mode, a hybrid power system has to transfer with high efficiency, the available renewable energy from the DC-side to the AC grid side. The DC-AC converter driven by a proper controller plays a primordial role in the supervision of the power flow from the RESs to the utility grid [4]. The control strategy of the inverter should be able to improve the operational efficiency and the quality of power injected to the utility grid.

The Voltage Oriented Control (VOC) scheme [1, 4] is widely used for the control of grid connected inverters. This technique performs indirect control of the real and reactive powers through an inner current control loop implemented in the dq synchronous reference frame. However, the effectiveness of the VOC method depend mainly on the quality of the grid current control [5] particularly, in the presence of system perturbations and uncertainties. In this case, the use of more advanced controller instead of PI regulators seems more suitable. Several alternatives [6] have been suggested in order to enhance the control performance of traditional VOC scheme. Recently, the nonlinear control strategies based on the sliding mode control [7, 8] and the backstepping approach [9–11] have received considerable attention.

The backstepping is a nonlinear control technique appeared in the beginning of 1990 for the control of nonlinear systems with uncertainties [12]. It is based on the Lyapunov theory that assures the global asymptotic stability of the controlled system. The backstepping algorithm consists of a design in systematic manner of a nonlinear control law using virtual control variables and appropriate Lyapunov functions [12, 13]. Recently, several research studies have adopted the backstepping approach for the command of power drives and static converters: in [9], a backstepping control scheme is established for the control of an inverter in a WTG tacking into account the parameter variations and disturbances. The control goals have been to stabilize the DC-link voltage and to control the active and reactive currents. The Backstepping is also applied in [13] to a permanent magnetic synchronous machine to design current and speed regulators. To drive an asynchronous generator based WTG, a backstepping power controller is adopted in [14] with a FPGA implementation.

Nowadays, the FPGA platforms receive a considerable attention in several application fields due to theirs attractive features such us: parallelism, reprogramming capability, flexibility, etc. In the power electronic filed, the FPGA presents an important tool for the implementation of control algorithms of power converters and electric machines [14–17].

The main goal of this chapter is to give some details on the design methodology of a digital control circuit for a hybrid PV-WT power system for future implementation on a FPGA device. The MPPT controllers of the RESs are established based on classical control methods. Besides, a Virtual Flux Oriented Control (VFOC) scheme based on a backstepping current controller is adopted for the grid side converter. For the FPGA implementation purpose, the overall control system is developed in the Matlab/Simulink environment using the XSG tool. The obtained results are presented to validate the correctness of the digital control circuit.

2 Modeling of the Grid-Connected Hybrid System

The structure of the grid-connected Hybrid Energy Generation System (HEGS) is shown in Fig. 1. It is composed of a PV array, a boost converter, a WTG that uses a PMSG, a rectifier, a three-phase DC-AC converter and RL filter.

2.1 PV Panel Modeling

The PV panel consists of a number of PV cells connected in series and parallel. The output current of a PV module using the single diode model (see Fig. 2) is calculated as [18]:

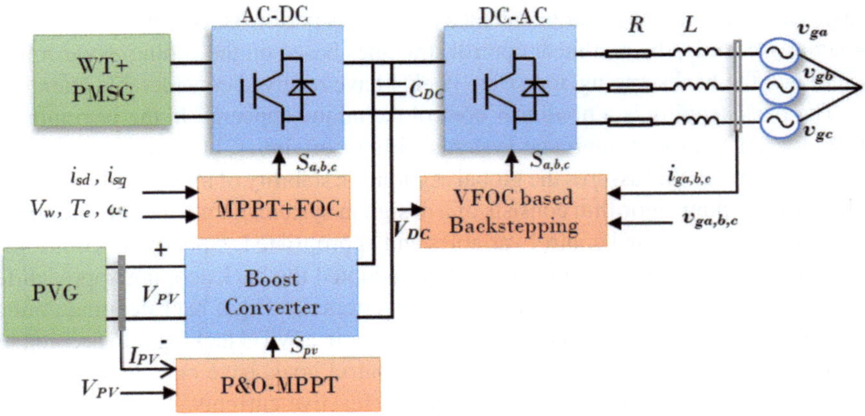

Fig. 1 Overall structure of the grid connected PV-WT system

Fig. 2 Single diode equivalent circuit of PV cell

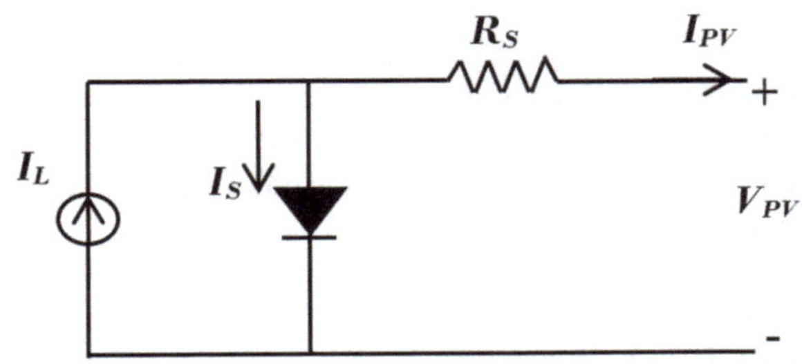

$$I_{PV} = I_L - I_S[exp(\frac{q}{KTA}(V_{PV} + I_{PV}R_S)) - 1] \tag{1}$$

I_{PV} and V_{PV} are the PV output current and voltage respectively, I_S is the saturation current, I_L is the photo-current, R_S is the series resistance, A is the ideality factor, K is the Boltzmann constant ($K = 1.38 \cdot 10^{-23} J/K$), T is the temperature, q is the electron charge ($q = 1.6 \cdot 10^{-19}$).

2.2 *Modeling of the WTG*

In this study, a vertical axis WT of type Darrieus that has a radius R and height H is adopted.

2.2.1 The Wind Turbine

In general, the aerodynamic power (P_t) captured by the WT is expressed in function of the power coefficient C_P as follows [19]:

$$P_t = \frac{1}{2}\rho S C_P(\lambda) V_W^3 \tag{2}$$

ρ, V_W, S are respectively, the air density, the wind speed and the surface belayed by the turbine, such that $S = R.H$. The power coefficient is a nonlinear function of the tip speed ratio (λ) given as [19]:

$$C_P(\lambda) = \sum_{k=0}^{4} C_k \lambda^k \text{ where } \lambda = \frac{R\omega_t}{V_W} \tag{3}$$

Such that: $C_0 = 0.110898$, $C_1 = -0.02493$, $C_2 = 0.057456$, $C_3 = -0.01098$, $C_4 = 0.00054$ [19]. The mechanical torque (T_t) of the WT is calculated as the ratio between P_t and the turbine speed (ω_t) as:

$$T_t = \frac{P_t}{\omega_t} = 0.5\rho S C_P(\lambda)\frac{V_W^3}{\omega_t} \tag{4}$$

2.2.2 The PMSG Modeling

The dynamic model of the PMSG established in the *dq* synchronous reference frame is given by [19]:

$$\begin{cases} v_{sd} = -R_s i_{sd} - L_{sd}\frac{di_{sd}}{dt} + L_{sq}\omega_e i_{sq} \\ v_{sq} = -R_s i_{sq} - L_{sq}\frac{di_{sq}}{dt} - \omega_e L_{sd} i_{sd} + \omega_e \phi_m \end{cases} \tag{5}$$

where v_{sd}, v_{sq}, i_{sd} and i_{sq} are the d-axis and q-axis components of the stator voltage and current, respectively. L_{sd} and L_{sd} are the d-axis and q-axis inductances, ω_e is the rotor angular velocity, p is the number of pole pairs. ϕ_m is the magnet flux, R_s is the stator resistance. For non-salient PMSG, the electromagnetic torque (T_e) can be expressed as follows [2, 19]:

$$T_e = \frac{3}{2} p i_{sq} \phi_m \tag{6}$$

2.3 *Modeling of the Three Phase Inverter*

The $\alpha\beta$ components of the voltage vector at the inverter output (v_α, v_β) can be expressed in function of the switching signals (S_a ,S_b ,S_c) as follows:

$$\begin{cases} v_\alpha = \frac{V_{DC}}{3}(2S_a - S_b - S_c) \\ v_\beta = \frac{V_{DC}}{\sqrt{3}}(S_b - S_c) \end{cases} \tag{7}$$

V_{DC} is the DC-link voltage at the inverter input. The dynamic model of the DC-AC converter in the dq rotating frame is given as [1]:

$$\begin{cases} \frac{di_{gd}}{dt} = \frac{1}{L}(-Ri_{gd} - L\omega i_{gq} - v_{gd} + v_d) \\ \frac{di_{gq}}{dt} = \frac{1}{L}(-Ri_{gq} + L\omega i_{gd} - v_{gq} + v_q) \end{cases} \quad (8)$$

v_d, v_q, i_{gd}, i_{gq} are respectively, the d-axis and q-axis components of the inverter output voltage and current. v_{gd} and v_{gq} are the d-axis and q-axis components of the grid voltage vector. R and L present the filter resistance and inductance, respectively.

3 The XSG Based Control of the HEGS

The Xilinx System Generator is a design tool of control algorithms that have to be implemented on a FPGA board. The XSG assures the automatic generation of HDL (Hardware Description Language) code, which provides a hardware description of the algorithm established in Simulink [14–17]. It offers a library of Xilinx blockset integrated in the Simulink toolbox, which can be used for the controllers design in the Matlab/Simulink environment. The main advantage of the XSG is the facility and the rapidity of the design process without the need for HDL programming [14–17]. In this section, we give some details on the design methodology of the control system based on the XSG tool. We note that the fixed point format is adopted for system data representation in order to reduce the FPGA resources consumption.

3.1 P&O Controller of the PV Generator

The Fig. 3 illustrates the control scheme adopted for the boost converter. In order to reach the MPP of the PV source, the Perturb and Observe (P&O) controller [1, 3] adjusts the voltage reference (V_{PV}^*) based on the measured values of the PV voltage and current. The XSG block sets of P&O algorithm, PI voltage controller and PWM modulator are shown in Figs. 4 and 5:

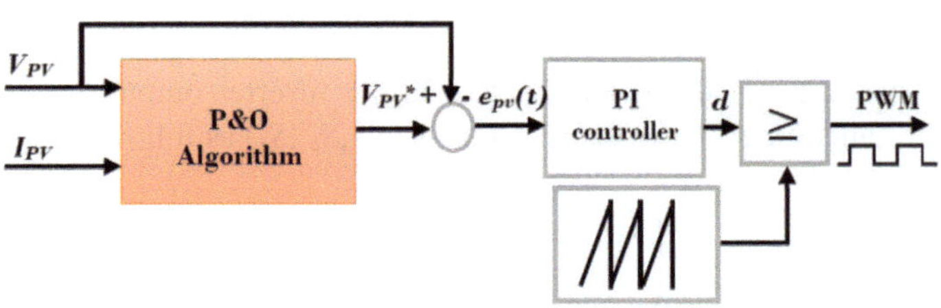

Fig. 3 Control scheme of the boost converter

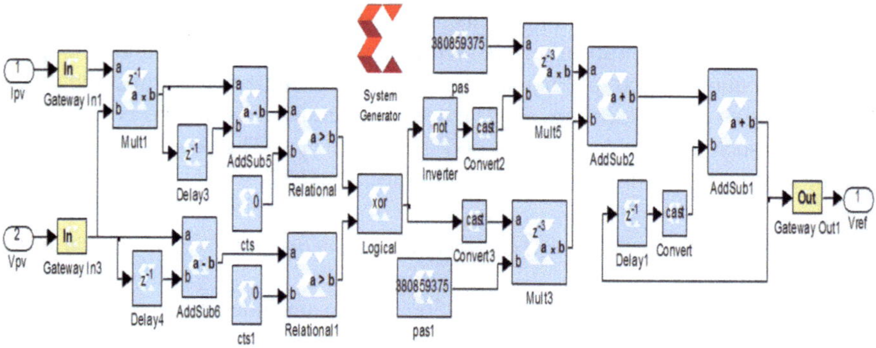

Fig. 4 The XSG implementation of P&O controller

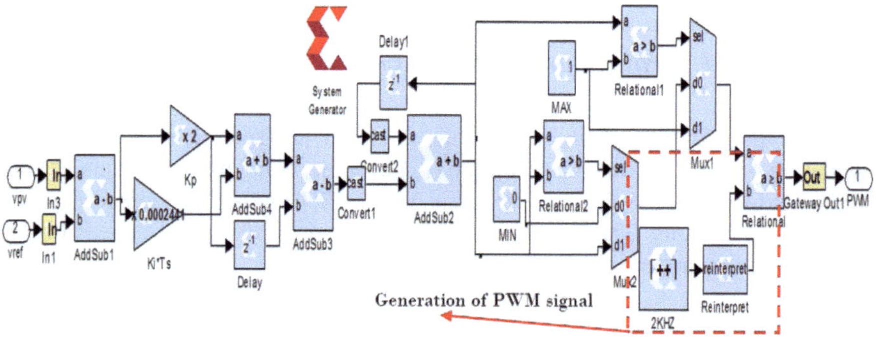

Fig. 5 The XSG implementation of PI regulator with PWM generator for PV voltage control

- As shown in Fig. 4, the digital P&O tracker is established using XSG block sets of addition, multiplication and comparison.
- The XSG block of the PI voltage controller is established by applying the backward Euler approximation method (see Fig. 5). Replacing $\frac{1}{s}$ by $\frac{T_s}{(1-z^{-1})}$, we obtain the following transfer function in the z-domain:

$$T(s) = \frac{D(s)}{E_{pv}(s)} = K_p + \frac{K_i}{s} \Longrightarrow T(z^{-1}) = \frac{D(z^{-1})}{E_{pv}(z^{-1})} = K_p + \frac{K_i T_s}{1 - z^{-1}} \tag{9}$$

D and E_{pv} are, respectively, the duty ratio and the PV voltage error. K_p and K_i present the proportional and integral gains of the PI regulator. By applying the inverse z-transform to Eq. 9, we get:

$$d(k) = d(k-1) + (K_p + K_i T_s) e_{pv}(k) - K_p e_{pv}(k-1) \tag{10}$$

T_s is the sampling time period. $e_{pv}(k) = V_{PV}^*(k) - V_{PV}(k)$. Thus, the Eq. 10 is adopted to build the XSG based digital PI controller.

- The PWM control signal of the boost converter is generated by comparing, at each iteration k, the duty ratio value $d(k)$ with the value of a triangular signal. So, a 8-bits up-ccunter is used to generate an asymmetric triangle waveform at the carrier frequency of 2 kHz (see Fig. 5). The counting sequence is limited between zero and 255 such that the sampling period is fixed to 2 μs. The count maximal value is determined according to the following formula [15]:

$$\textit{Counting limit} = \textit{required triangle period/sampling period} = 5.e^{-4}/2.e^{-6} = 250 \quad (11)$$

The nearest integer to the value 250 is $256 = 2^8$; so, unsigned 8-bits counter should be used. Also, the Reinterpret block is used to obtain at the output a normalized triangular signal.

3.2 *Optimal Torque Control of WTG*

The principal of the Optimal Torque Control (OTC) [2] (see Fig. 6) consists of the regulation of the electromagnetic torque (T_e) to achieve the optimal value (T_e^*) that correspond to the MPP (i.e. $P_t = P_{t-max}$). This MPPT method is adopted here due to its quite simplicity. The expression of the mechanical power of Darrieus turbine (of Eq. 2) can rewritten as follows:

$$P_t = \frac{1}{2}\rho H R^4 C_P(\lambda)\frac{\omega_t^3}{\lambda^3} \quad (12)$$

For a given wind speed, the tip speed ratio is optimal at the MPP (i.e $\lambda = \lambda_{opt}$) and the corresponding value of the power coefficient is maximal ($C_P = C_{P-max}$). In these conditions, an optimal aerodynamic torque (T_{t-opt}) will be obtained, such that:

$$\begin{cases} T_{t-opt} = \frac{P_{t-max}}{\omega_t} = K_{opt}.\omega_t^2 \\ \text{with } K_{opt} = \frac{1}{2}\rho H R^4 \frac{C_{P-max}(\lambda)}{\lambda_{opt}^3} \end{cases} \quad (13)$$

So, this value of optimal torque (T_{t-opt}) is defined as the reference of the generator torque i.e: $T_e^* = T_{t-opt} = K_{opt}\omega_t^2$. (see Fig. 6). In the steady state, where $d\omega_t/dt \approx 0$ under constant wind speed, the OTC ensures that the MPP is reached such that: $T_e^* = T_e = T_t$ [2].

3.3 *The Field Oriented Control Scheme*

In this study, the conventional FOC method [1, 19] is applied for the control of the rectifier that interfaces the PMSG. As shown in Fig. 7, the decoupled control

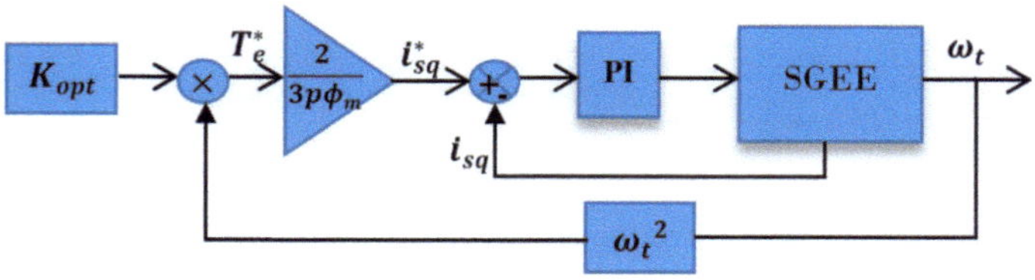

Fig. 6 The principal of the OTC technique [2]

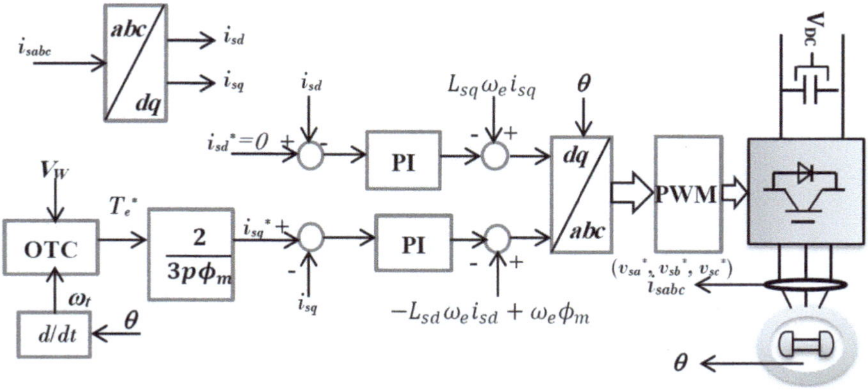

Fig. 7 The FOC scheme applied to the rectifier [19]

of the stator flux and the electromagnetic torque is assured by acting separately on the d and q components of the stator current. The flux is oriented by maintaining the d-axis stator current close to zero ($i_{sd}^* = 0$). Besides, the electromagnetic torque is controlled by regulating the q-axis stator current to track the reference. Hence, the q-axis current setpoint is calculated in function of the torque reference (T_e^*) as follows:

$$i_{sq}^* = \frac{2T_e^*}{3p\phi_m} \tag{14}$$

As depicted in Fig. 7, the signals at the outputs of the d-axis and q-axis current PI controllers (v'_{sd} and v'_{sq}) are calculated as follows:

$$\begin{cases} v'_{sd} = k_p(i_{sd}^* - i_{sd}) + k_i \int (i_{sd}^* - i_{sd})dt \\ v'_{sq} = k_p(i_{sq}^* - i_{sq}) + k_i \int (i_{sq}^* - i_{sq})dt \end{cases} \tag{15}$$

It can be seen from Eq. 5 that there is a coupling between the d-axis and q-axis equations, which affects considerably the control performance. In order to ensure a decoupled control of d and q components of the stator current, the rectifier voltage references (v_{sd}^* and v_{sq}^*) are calculated as follows:

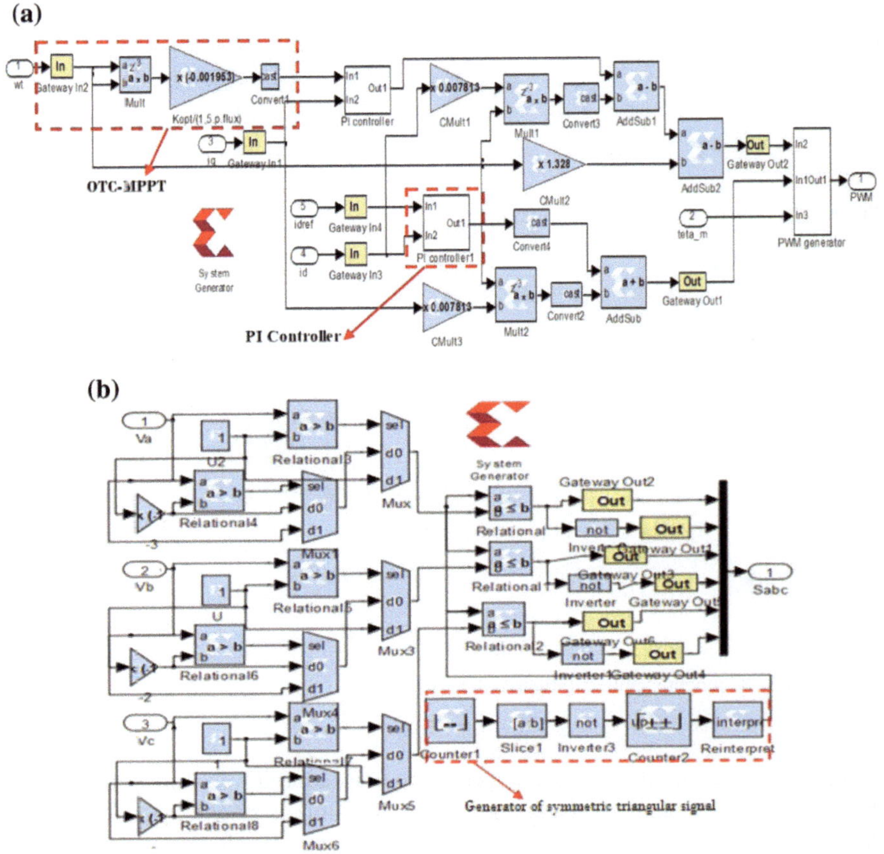

Fig. 8 XSG blocks of **a** the FOC scheme, **b** the PWM modulator

$$\begin{cases} v_{sd}^{*} = -v_{sd}' + L_s\omega_e i_{sq} \\ v_{sq}^{*} = -v_{sq}' - L_s\omega_e i_{sd} + \omega_e \phi_m \end{cases} \quad (16)$$

The Fig. 8a shows the XSG module of the FOC scheme. The XSG based digitized PI controller for q-axis current is designed according to the following expression:

$$v_{sq}'(k) = v_{sq}'(k-1) + (K_{pq} + K_{iq}T_s)e_{iq}(k) - K_{pq}e_{iq}(k-1) \quad (17)$$

K_{pq} and K_{iq} are, respectively, the proportional and integral gains. $e_{iq}(k) = i_{sq}^{*}(k) - i_{sq}(k)$ such that, i_{sq}^{*} is the q-axis current set-point. We note that PI controller for d-axis current component is similar to this of the d-axis.

The Fig. 8b displays the details on the developed XSG based SPWM modulator, which generates the PWM signals for the rectifier switching devices. This block performs a comparison between a triangle carrier signal with the converter voltage references (v_{sa}^*, v_{sb}^*, v_{sc}^*). As presented in [15], five XSG blocks are used to generate the triangular carrier signal at the frequency of $f_{tri} = 5$ kHz. At each period $T_{tri} = (1/f_{tri}) = 0.2$ ms, the 7-bits down counter repeats a counting sequence to generate an asymmetric triangle waveform, which varies between 127 and 0. The 9-bits up/down counter generates a symmetric triangular signal, where the count value varies between −32 and 32.

3.4 *VFOC Scheme Based on the Backstepping Approach*

In this study, the VFOC strategy [20, 21] is adopted for the control of the inverter. As depicted in Fig. 9, the inner current control loop is established based on the backstepping approach [9, 13]. Further, the control of the DC-link voltage is accomplished by means of an ADAptive LInear NEuron (ADALINE) regulator. The VF estimator ensures the synchronization with the utility grid.

3.4.1 Design of Backstepping Current Controller

According to the VFOC strategy, the d-axis and q-axis grid current components (i_{gd} and i_{gq}) should be controlled to track their respective references (i_{gd}^* and i_{gq}^*), in order to ensure a decoupled control of the active and reactive powers injected to the grid [21]. The ADALINE controller determines the set-point of the q-axis grid current (see Fig. 9). Hence, the control goal of the backstepping is to eliminate the static errors of the d-axis and q-axis grid currents, and the Eq. 8 can be rewritten in function of the controlled variables ($i_{gq} = y_1$) and ($i_{gd} = y_2$) as follows:

$$\begin{cases} \dot{y_1} = \frac{1}{L}(-Ry_1 + L\omega y_2 - v_{gq} + v_q) \\ \dot{y_2} = \frac{1}{L}(-Ry_2 - L\omega y_1 - v_{gd} + v_d) \end{cases} \tag{18}$$

The backstepping control law can be now designed step by step using Lyapunov candidate functions, which assure that the control targets will be reached:
Step1: The control error of q-axis grid current is defined as:

$$e_{gq} = i_{gq}^* - i_{gq} = y_1^* - y_1 \tag{19}$$

Taking into account the Eq. 18, the error derivative is calculated as:

$$\dot{e_{gq}} = \dot{y_1^*} - \dot{y_1} = -\dot{y_1} = -\frac{v_q}{L} + \frac{R}{L}y_1 - \omega y_2 + \frac{v_{gq}}{L} \tag{20}$$

We define the following Lyapunov function as:

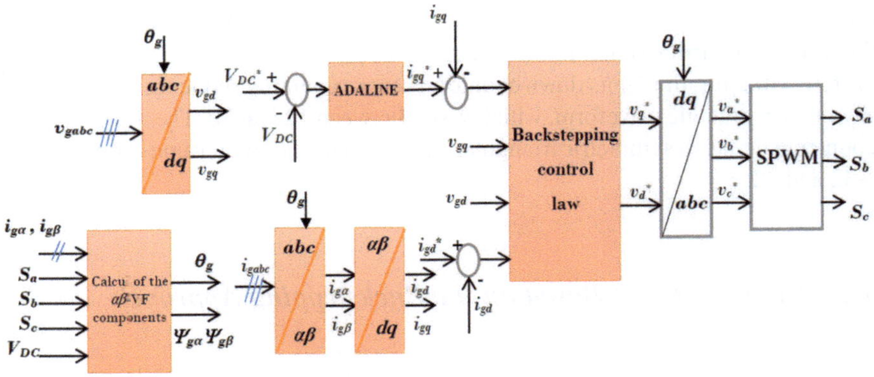

Fig. 9 The backstepping based VFOC scheme

$$V_1 = \frac{1}{2}e_{gq}^2 \tag{21}$$

Its derivative is so computed as:

$$\dot{V}_1 = e_{gq}.\dot{e_{gq}} = e_{gq}\left(-\frac{v_q}{L} + \frac{R}{L}y_1 - \omega y_2 + \frac{v_{gq}}{L}\right) \tag{22}$$

According to the Lyapunov stability condition [13], we should have:

$$-\frac{v_q}{L} + \frac{R}{L}y_1 - \omega y_2 + \frac{v_{gq}}{L} = -g_1.e_{gq} \text{ with } g_1 > 0 \tag{23}$$

In order to have $\dot{V}_1 = -g_1 e_{gq}^2 < 0$ that guarantees the elimination of the q-axis current error. According to the backstepping method [13], we define v_q as the virtual control input:

$$v_q = Ry_1 - \omega L y_2 + v_{gq} + g_1 L e_{gq} \tag{24}$$

Step 2: The d-axis current error is defined as follows:

$$e_{gd} = i_{gd}^* - i_{gd} = y_2^* - y_2 \tag{25}$$

Its derivative is given by:

$$\dot{e_{gd}} = \dot{y_2^*} - \dot{y}_2 = -\dot{y}_2 = -\frac{v_d}{L} + \frac{R}{L}y_2 + \omega y_1 + \frac{v_{gd}}{L} \tag{26}$$

For this sub-system, we define a second Lyapunov function as:

$$V_2 = V_1 + \frac{1}{2}e_{gd}^2 \tag{27}$$

The derivative with respect to the time is given as:

$$\dot{V}_2 = -g_1 e_{gq}^2 + e_{gd}.\dot{e_{gd}} = -g_1 e_{gq}^2 + e_{gd}\left(-\frac{v_d}{L} + \frac{R}{L}y_2 + \omega y_1 + \frac{v_{gd}}{L}\right) \tag{28}$$

By defining:

$$-g_2 e_{gd} = -\frac{v_d}{L} + \frac{R}{L}y_2 + \omega y_1 + \frac{v_{gd}}{L} \text{ with } g_2 > 0. \tag{29}$$

The derivative $\dot{V}_2$ will be negative:

$$\dot{V}_2 = -g_1 e_{gq}^2 - g_2 e_{gd}^2 < 0 \tag{30}$$

So, the backstepping control law v_d is given as:

$$v_d = Ry_2 + L\omega y_1 + v_{gd} + g_2 L e_{gd} \tag{31}$$

With this backstepping control law, we will have $V_2 > 0$ and $\dot{V}_2 < 0$ that assures the asymptotic convergence of the current error to zero. Hence, the Eqs. (24) and (31) are used for the calcul of the inverter voltage references (v_d^* and v_q^*).

On the other hand , the grid VF vector ($\bar{\Psi_g}$) is defined as the integral of the grid voltage vector ($\bar{v_g}$) such that [22]:

$$\bar{\Psi_g} = \int \bar{v_g} dt = \int \left(\bar{v} - L\frac{d\bar{i_g}}{dt}\right) dt \tag{32}$$

$\bar{v}$ is the inverter voltage vector. The estimation of $\alpha\beta$ components of the grid VF vector can be done in function of the $\alpha\beta$ components of the grid current ($i_{g\alpha}, i_{g\beta}$) and the DC-link voltage (V_{DC}) as follows:

$$\begin{cases} \Psi_{g\alpha} = \int \frac{V_{DC}}{3}(2S_a - S_b - S_c)dt - Li_{g\alpha} \\ \Psi_{g\beta} = \int \frac{V_{DC}}{\sqrt{3}}(S_b - S_c)dt - Li_{g\beta} \end{cases} \tag{33}$$

3.4.2 ADALINE Based Control of DC-Link Voltage

As has been mentioned, an adaptive ADALINE regulator is adopted in this study for the stabilization of the DC-link voltage at the desired setpoint. Hence, the inputs of the ADALINE controller are defined as the DC voltage reference ($V_{DC}^*(k)$) and the previous voltage error $e_{DC}(k-1)$(such that $e_{DC}(k) = V_{DC}(k) - V_{DC}^*(k)$). The LMS adaptation law [22] of the ADALINE weights is given by:

$$W(k+1) = W(k) + 2\alpha E_{DC}(k)X(k) \tag{34}$$

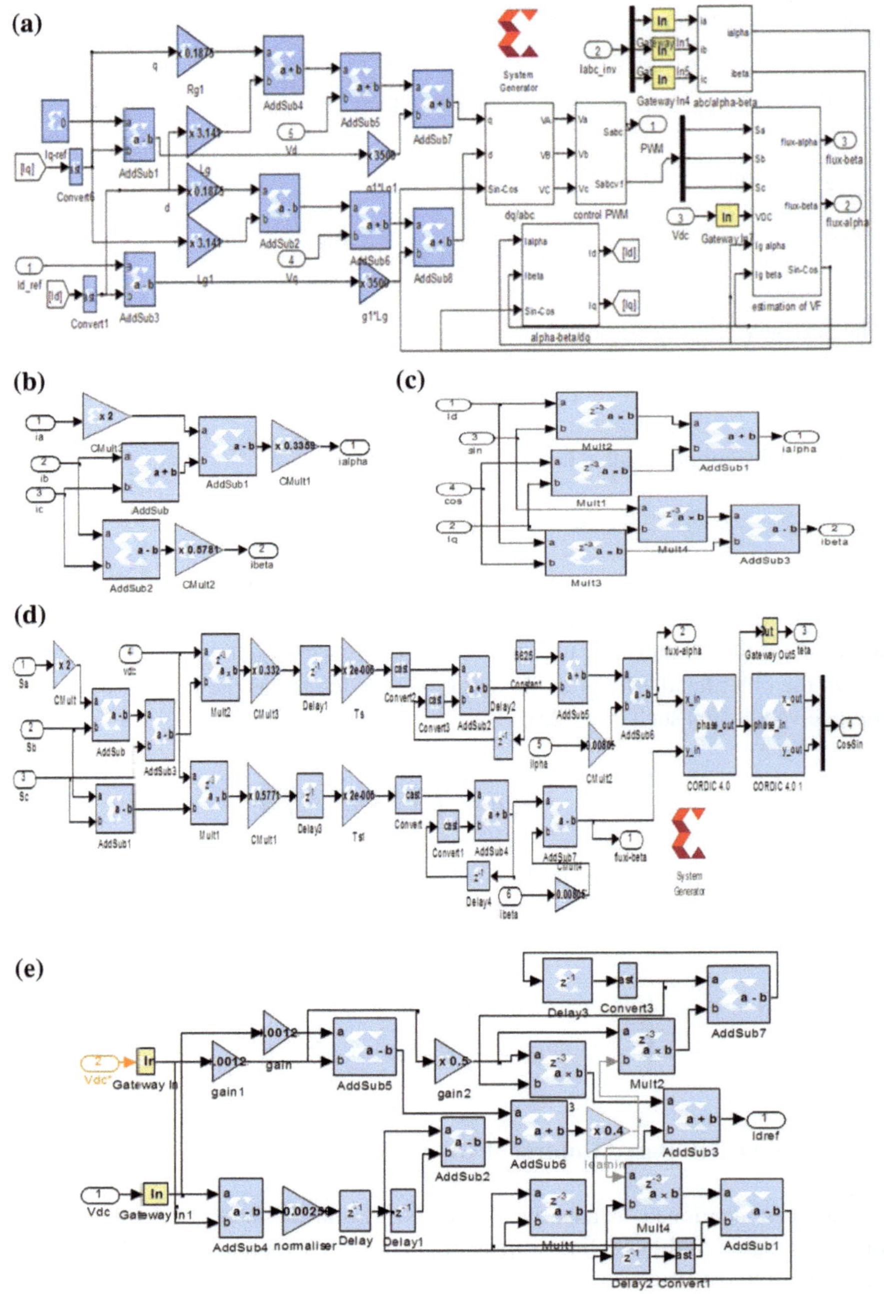

Fig. 10 XSG blocks of: **a** VFOC scheme, **b** $abc/\alpha\beta$ transformation, **c** $dq/\alpha\beta$ transformation, **d** Grid VF estimator, **e** ADALINE controller

Such that $W(k) = [w_1(k), w_2(k)]^T$ and $X(k) = [V_{DC}^*(k), e_{DC}(k-1)]^T$ are respectively the weights vector and the inputs vector. α is the learning rate. $E_{DC}(k)$ is the error function defined as:

$$E_{DC}(k) = 0 - S(k) = 0 - (\lambda_1 e_{DC}(k) + de_{DC}(k)) \tag{35}$$

Such that: λ_1 is a positive constant, de_{DC} is the change of error where, $de_{DC}(k) = e_{DC}(k) - e_{DC}(k-1)$ and $S(k)$ is the sliding surface of the controlled variable. In the steady state, the system state (V_{DC}) tracks accurately the setpoint in such way that $S(k) = 0$.

The Fig. 10a shows the different components of the XSG based inverter control unit. The XSG modules for the Clarke and Park transformations are illustrated in Fig. 10b, c.The Xilinx block of the VF estimator (see Fig. 10d) is established by applying the Euler approximation method as:

$$\begin{cases} \Psi_{g\alpha}(k) = \Psi_{g\alpha}(k-1) + T_s \frac{V_{DC}(k)}{3}(2S_a - S_b - S_c) - Li_{g\alpha}(k) \\ \Psi_{g\beta}(k) = \Psi_{g\beta}(k-1) + T_s \frac{V_{DC}(k)}{\sqrt{3}}(S_b - S_c) - Li_{g\beta}(k) \end{cases} \tag{36}$$

It can be seen from Fig. 10d that we have used two CORDIC blocks of XSG in order to estimate the grid phase angle and to generate both sine and cosine waveforms. For each block, the phase format is selected to be in Radians and the output width is fixed to 16. The Fig. 10e illustrates the XSG block of the discretized ADALINE controller.

4 Simulation Results

The overall PV-WT system is implemented and simulated in the Matlab/Simulink environment using XSG. The simulation parameters are listed in Table 1. In order to validate the accuracy of the adopted control scheme, the following simulation tests are performed:

The first test is accomplished for stabilized climatic conditions (irradiance $G = 1000\ \mathrm{W/m^2}$, temperature $T = 25\,^\circ\mathrm{C}$ and wind speed of $V_W = 12$ m/s). It can be seen that the XSG based control blocks work correctly and ensure a correct operation of the HEGS. As exhibited in Figs. 11 and 12, the MPPT controllers (P&O and OTC) determine correctly the MPPs of both the PVG and WTG. Also, it can be seen from Fig. 11, that there is a strong similarity between the results obtained with the Simulink blocks and the XSG block set of PVG controller (P&O tracker+PI controller). The PV output voltage is well regulated and tracks accurately the setpoint signal. However, the Fig. 12 shows that the static oscillations of the electromagnetic torque with the Xilinx module are larger due to the quantization error. The Fig. 12 demonstrates that the maximum power is extracted from the WTG for nominal wind speed. On the other hand, the decoupled control of the active and reactive powers is assured with

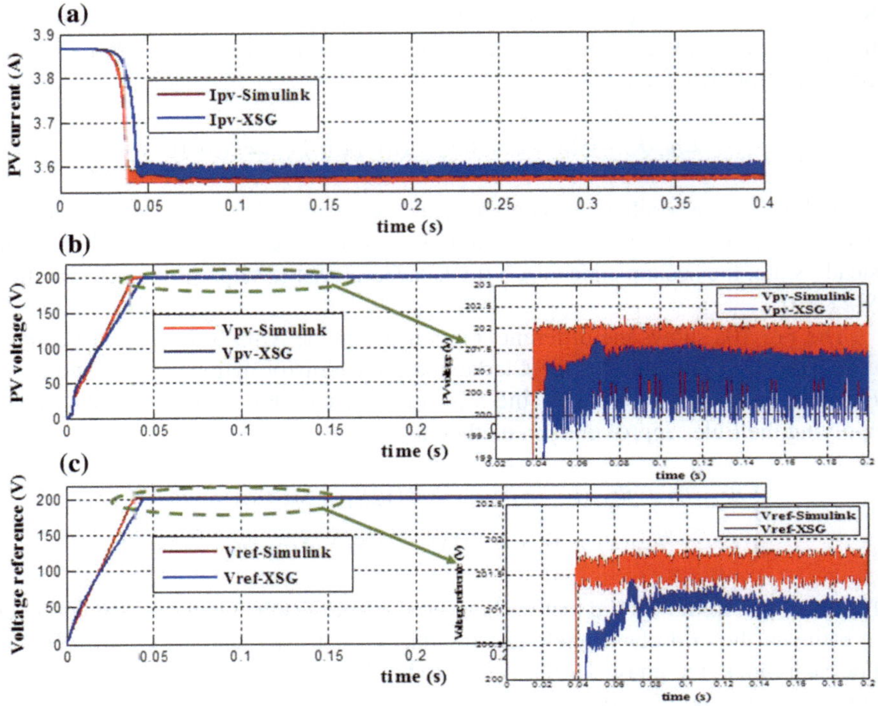

Fig. 11 The PVG performance in the first simulation case

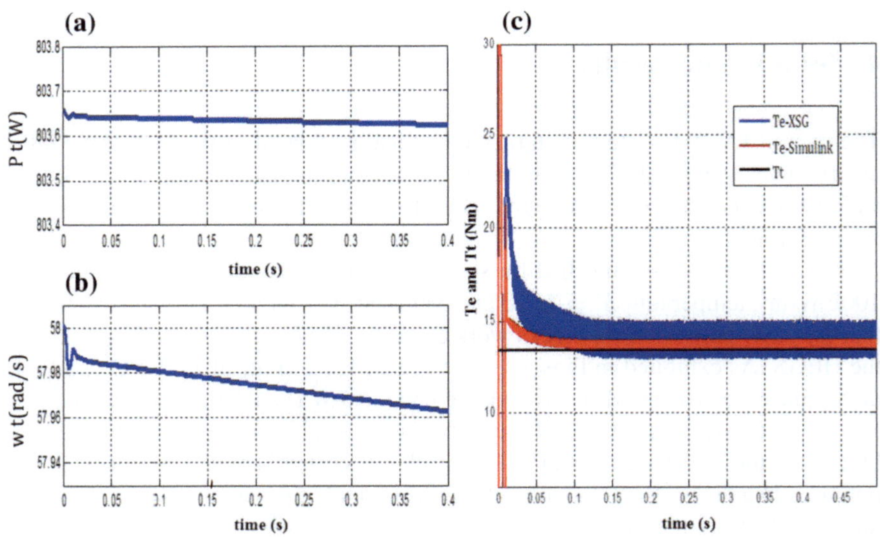

Fig. 12 The MPPT performance for the WTG in the first simulation case

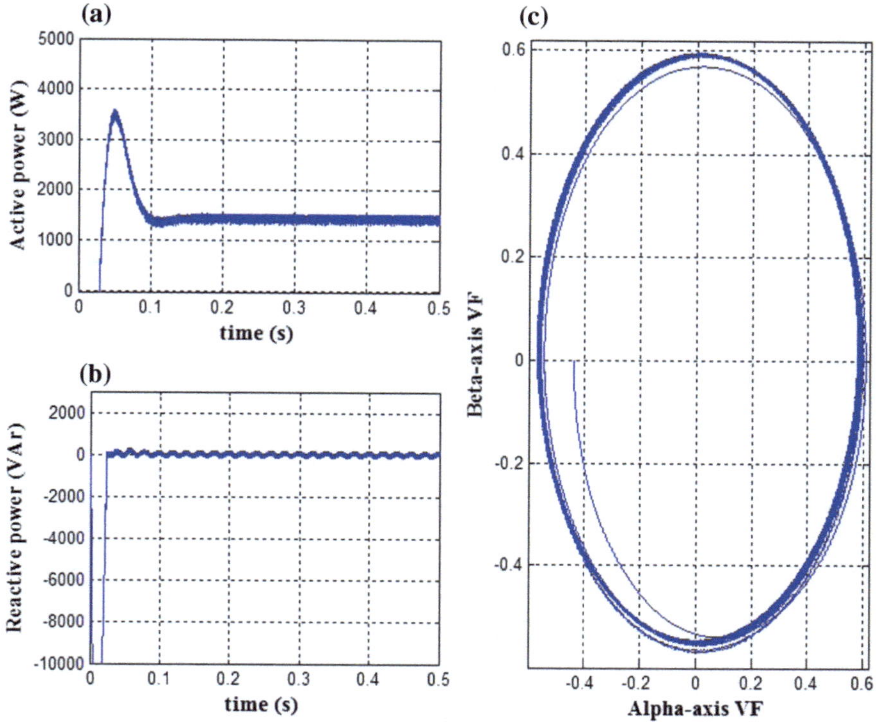

Fig. 13 Variation in the first test of: **a** Inverter active power **b** Inverter reactive power, **c** $\alpha\beta$ components of the grid VF vector

Table 1 Simulation parameters

System parameters	Value
PV voltage at the MPP	33.7 V
PV current at the MPP	3.56 A
Number of series cells in the BPMSX120 module	72
Darrieus Rotor radius	1 m
Pole pairs number of the PMSG	8
Stator inductance of the PMSG	8 mH
PMSG Inertia	5 kg.m^2

a unity power factor operation as depicted in Fig. 13. The total harmonic distortion of the grid current is acceptable, which is about 2.71%. Besides, it can be deduced from Fig. 13c that the grid VF estimation is correctly performed.

In the second test, the irradiation level varies gradually from 900 W/m^2 to 600 W/m^2 and then, changes to 200 W/m^2 at the instant $t = 0.9$ s. The wind speed

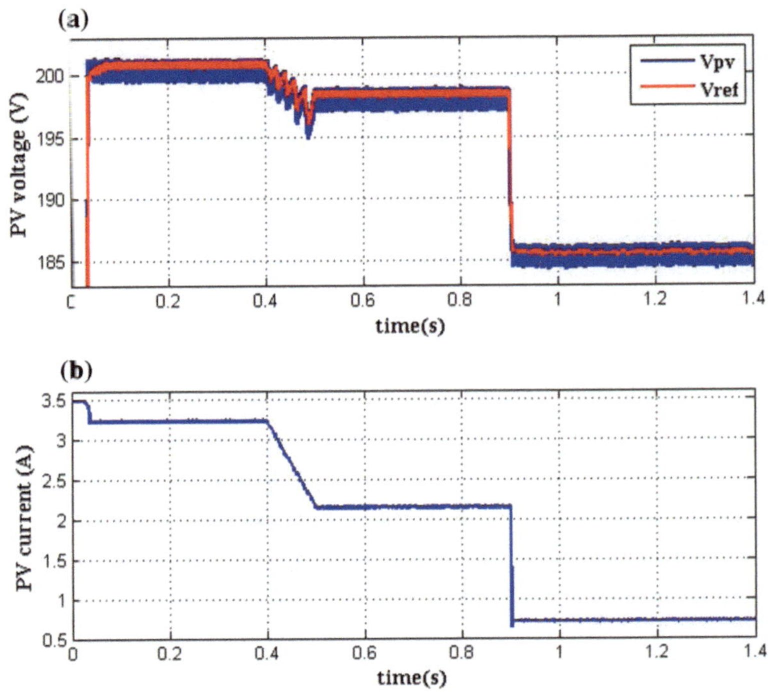

Fig. 14 The MPPT performance of the PVG in the second simulation test

Table 2 The Virtex-5 FPGA resources used by the PV local controller (P&O tracker with PI controller)

Device resources	Used	Utilization (%)
Number of slice registers	196	0
Number of bonded IOBs	54	5
Number of slice LUTs	71	0
Number of BUFG/ BUFGCTRLs	1	3
Number of fully used LUT-FF pairs	12	4

is maintained constant at the value $V_W = 10$ m/s. As illustrated in Fig. 14, the P&O controller converges rapidly to the correct MPP that correspond to each irradiation level. This result validates the fixed point representation and the precision value adopted for PVG data. Moreover, the Fig. 15 confirms that the static errors of the d-axis and q-axis grid currents are eliminated thanks to the Backstepping control action.

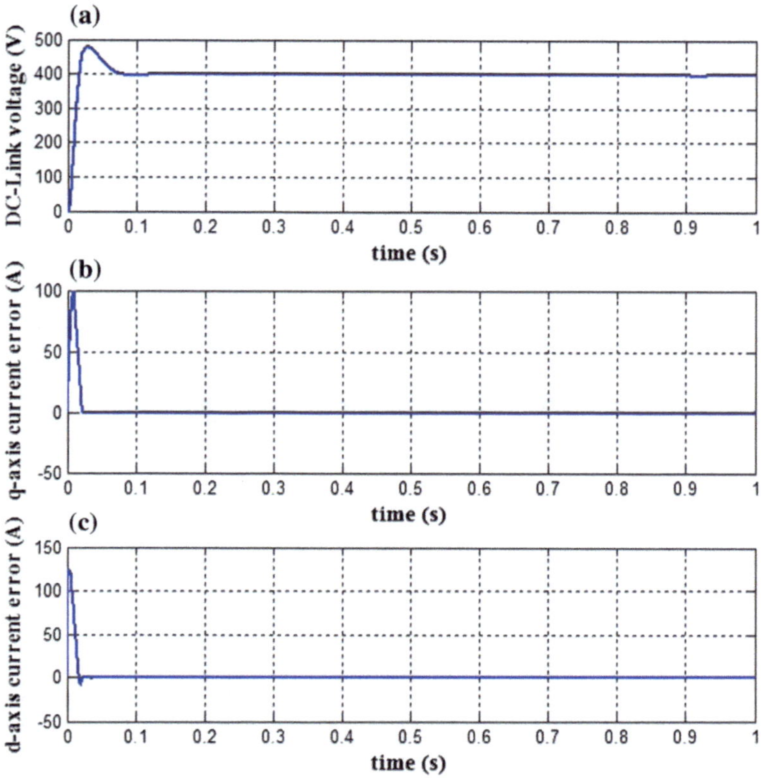

Fig. 15 Performance of the VFOC control scheme in the 2nd test

Table 3 The Virtex-5 FPGA resources used by the PI controller of the q-axis stator current

Device resources	Used	Utilization (%)
Number of slice Registers	98	0
Number of bonded IOBs	79	8
Number of slice LUTs	72	0
Number of BUFG/ BUFGCTRLs	1	3
Number of fully used LUT-FF pairs	23	15

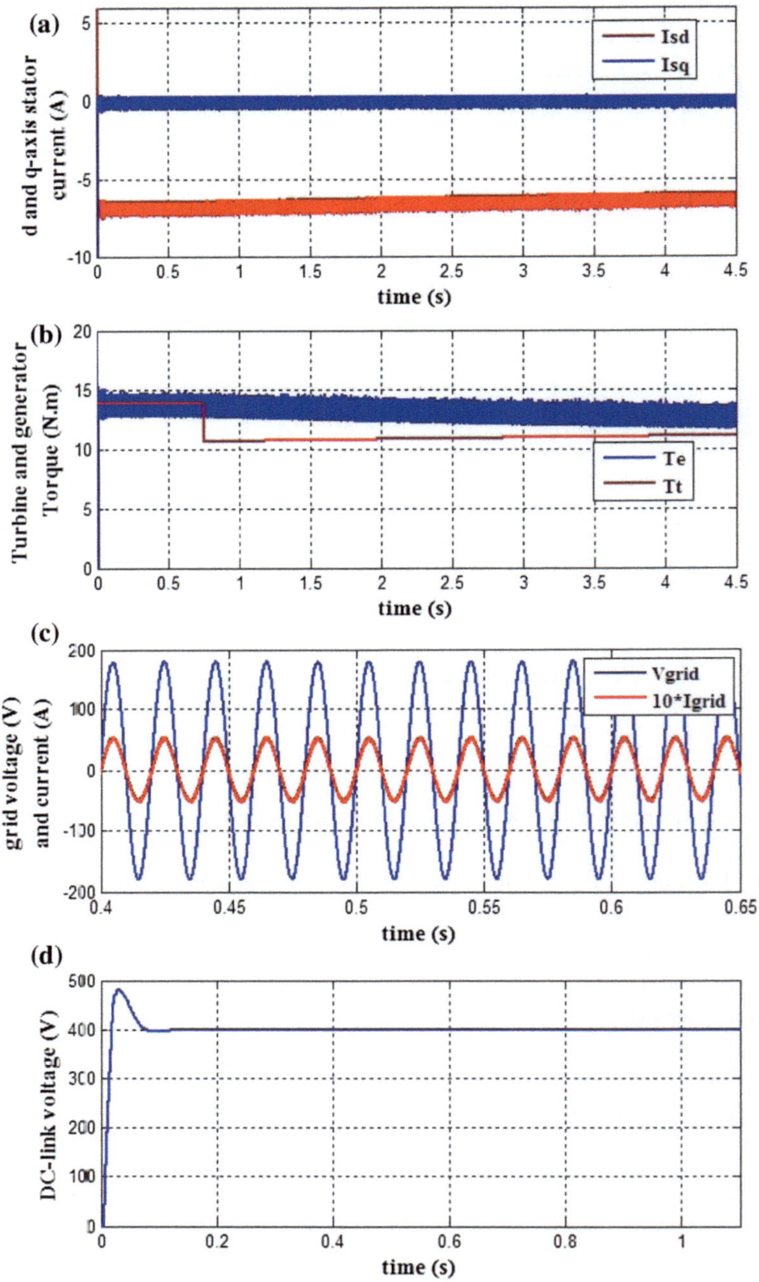

Fig. 16 Performance of the FOC and the VFOC schemes in the 3rd test

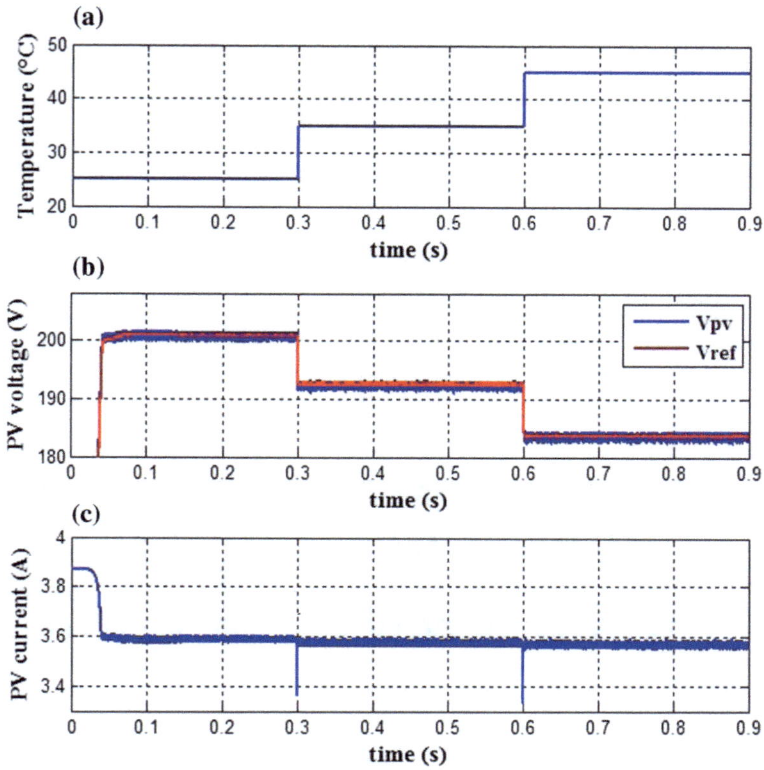

Fig. 17 The MPPT performance of the PVG in the fourth simulation test

Table 4 The Virtex-5 FPGA resources used by the inverter control unit

Device resources	Used	Utilization (%)
Number of slice registers	1535	0
Number of bonded IOBs	207	21
Number of slice LUTs	2374	1
Number of BUFG/ BUFGCTRLs	1	3
Number of DSP48Es	54	28
Number of fully used LUT-FF pairs	1050	36

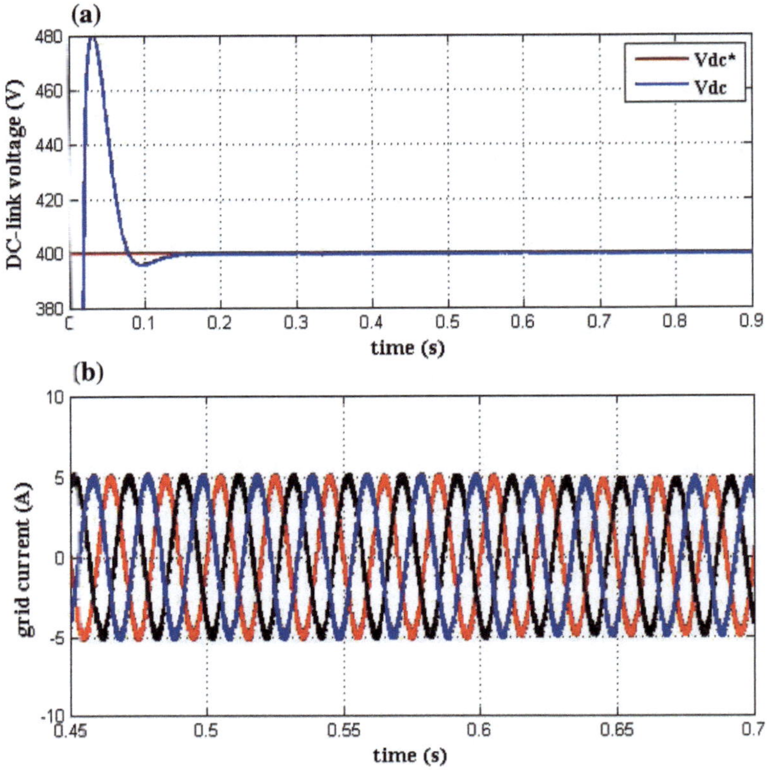

Fig. 18 The performance of the VFOC scheme in the fourth simulation test

In the third test, the wind speed varies from 12 m/s to 11 m/s (the irradiance is maintained at the level of $G = 1000$ W/m^2). The Fig. 16a proves the accuracy of the Xilinx block sets of the FOC scheme, which gives good static and dynamic performance. The decoupled control of the d-axis and q-axis stator current is achieved as shown in Fig. 16a. Besides, the Fig. 16b illustrates that the PMSG torque converges slowly to the optimal value that correspond to the MPP of the WTG. The VF estimator ensures the synchronization of the inverter current with the grid voltage as depicted in Fig. 16c.Besices, the Fig. 16d shows that the ADALINE controller has successfully stabilized the DC-link voltage to the desired set-point (400 V).

In the fourth test case, the PVG is simulated under varying temperature (see Fig. 17a) and nominal insolation ($G = 1000$ W/m^2). The wind speed stills constant the value of 9 m/s. It can be seen from Fig. 17 that the adopted MPPT tracker based on XSG perfcrms very well. For each temperature, the optimal PV voltage value is rapidly reached with acceptable steady-state oscillations. The results shown in Fig. 18 validate the adopted VFOC scheme and the ADALINE regulator.

The Tables 2, 3 and 4 recapitulate the resources of Virtex-5(XC5VLX330T)FPGA device used by some components of the studied control scheme.

5 Conclusion

The goal of this chapter is to study the XSG based design method of a digital control circuit of a HEGS for future FPGA implementation. The local controllers of the RESs based on classical MPPT algorithms are developed. Furthermore, the inverter control tasks are assured using the VFOC scheme based on the Backstepping technique and adaptive ADALINE regulator. The simplicity and the low consumption of FPGA resources justify the choice of the adopted control techniques.

The XSG/Simulink simulation is accomplished in order to verify the correctness and the validity of the developed control circuit. The simulation results show satisfactory static and dynamic performance for different operating conditions. After this study, we can pass directly to the experimental validation of the studied system based on a FPGA platform.

References

1. Abu-Rub H, Malinowski M, Al-Haddad K (2014) Power electronics for renewable energy systems, transportation and industrial applications. Wiley, IEEE, New York
2. Kumar D, Chatterjee K (2016) A review of conventional and advanced MPPT algorithms for wind energy systems. Renew Sustain Energy Rev 55:957–970
3. Coelho RF, Concer FM, Martins DC (2010) A MPPT approach based on temperature measurements applied in PV systems. In: 2010 9th IEEE/IAS international conference on industry applications (INDUSCON), pp 1–6
4. Blaabjerg F, Teodorescu R, Liserre M, Timbus AV (2006) Overview of control and grid synchronization for distributed power generation systems. IEEE Trans Ind Electron 53:1398–1409
5. Malinowski M, Jasinski M, Kazmierkowski MP (2004) Simple direct power control of three-phase PWM rectifier using space-vector modulation (DPC-SVM). IEEE Trans Ind Electron 51:447–454
6. Parvez M, Elias MFM, Rahim NA, Osman N (2016) Current control techniques for three-phase grid interconnection of renewable power generation systems: a review. Sol Energy 135:29–42
7. Kim Il-Song (2007) Robust maximum power point tracker using sliding mode controller for the three-phase grid-connected photovoltaic system. Sol Energy 81:405–414
8. Naouar MW Ben, Hania B, Slama-Belkhodja I, Monmasson E, Naassani AA (2013) FPGA-based sliding mode direct control of single phase PWM boost rectifier. Math Comput Simul 91:249–261
9. Wang GD, Wai RJ, Liao Y (2013) Design of backstepping power control for grid-side converter of voltage source converter-based high-voltage dc wind power generation system. IET Renew Power Gener 7:118–133
10. Aouadi C, Abouloifa A (2014) Backstepping based control of PV system connected to the grid. Int J Comput Inf Technol 3
11. Naoufel K, Zazi M, Mahmoudi H (2015) Grid-connected photovoltaic system using an advanced backstepping approach. Theor Appl Inf Technol 79:330–337
12. Zhou J, Wen C (2008) Adaptive backstepping control of uncertain systems: nonsmooth nonlinearities, interactions or time-variations. Springer, Berlin (2008)
13. Chen CX, Xie YX, Lan YH (2015) Backstepping control of speed sensorless permanent magnet synchronous motor based on slide model observer. Autom Comput 12:149–155
14. Bossoufi B, Karim M, Lagrioui A, Taoussi M, Derouich A (2015) Observer backstepping control of DFIG-Generators for wind turbines variable-speed: FPGA-based implementation. Renew Energy 81:903–917

15. Selvamuthukumaran R, Gupta R (2014) Rapid prototyping of power electronics converters for photovoltaic system application using Xilinx System Generator. IET Power Electron 7:2269–2278
16. Krim S, Gdaim S, Mtibaa A, Mimouni MF (2015) Design and implementation of direct torque control based on an intelligent technique of induction motor on FPGA. Electr Eng Technol 10:1527–1539
17. Murugesan K, Muthu R, Vijayenthiran S, Mervin JB (2015) Prototype hardware realization of the DSTATCOM for reactive power compensation. Electr Power Energy Syst 65:169–178
18. Walker G (2001) Evaluating MPPT converter topologies using a Matlab PV model. Electr Electron Eng 21
19. Aubree R, Auger F, Mace M, Loron L (2016) Design of an efficient small wind-energy conversion system with an adaptive sensorless MPPT strategy. Renew Energy 86:280–291
20. Teodorescu R, Liserre M, Rodriguez P (2011) Grid converters for photovoltaic and wind power systems. Wiley, IEEE, New York (2011)
21. Malinowski M, Kazmierkowski MP, Trzynadlowski AM (2003) A comparative study of control techniques for PWM rectifiers in AC adjustable speed drives. IEEE Trans Power Electr 18:1390–1396
22. Chettibi N, Mellit A, Sulligoi G Massi, Pavan A (2018) Adaptive neural network-based control of a hybrid AC/DC microgrid. IEEE Trans Smart Grid 9:1667–1679

Artificial Intelligence for Photovoltaic Systems

Rami Ghannam, Paulo Valente Klaine and Muhammad Imran

Abstract Photovoltaic systems have gained an extraordinary popularity in the energy generation industry. Despite the benefits, photovoltaic systems still suffer from four main drawbacks, which include low conversion efficiency, intermittent power supply, high fabrication costs and the nonlinearity of the PV system output power. To overcome these issues, various optimization and control techniques have been proposed. However, many authors relied on classical techniques, which were based on intuitive, numerical or analytical methods. More efficient optimization strategies would enhance the performance of the PV systems and decrease the cost of the energy generated. In this chapter, we provide an overview of how Artificial Intelligence (AI) techniques can provide value to photovoltaic systems. Particular attention is devoted to three main areas: (1) Forecasting and modelling of meteorological data, (2) Basic modelling of solar cells and (3) Sizing of photovoltaic systems. This chapter will aim to provide a comparison between conventional techniques and the added benefits of using machine learning methods.

1 Introduction

According to the late Nobel laureate, Prof. Richard Smalley, *energy is the most important challenge facing humanity today* [1]. Not only can solar energy help in the democratization of energy, but it also has the potential to profoundly improve the lives of communities worldwide. The Sun provides a tremendous source of energy and has an important role to play in the energy generation mix of many nations. In particular, photovoltaic (PV) technology is a mature, proven and reliable method for converting the Sun's vast energy into electricity.

R. Ghannam (✉) · P. V. Klaine · M. Imran
School of Engineering, University of Glasgow, University Avenue, Glasgow, UK
e-mail: rami.ghannam@glasgow.ac.uk

P. V. Klaine
e-mail: p.valente-klaine.1@research.gla.ac.uk

M. Imran
e-mail: muhammad.imran@glasgow.ac.uk

R.-E. Precup et al. (eds.), *Solar Photovoltaic Power Plants*, Power Systems,
https://doi.org/10.1007/978-981-13-6151-7_6

The Sun therefore provides a free source of energy, which can be harnessed and converted into electricity using photovoltaic (PV) technology. PV technology has the benefit of being modular and scalable. It can therefore be very quickly installed in a wide range of locations. These locations can vary from conventional ground installations to domestic and commercial buildings. However, among the challenges in achieving widespread use of this technology is the price of solar electricity in comparison to conventional sources of energy. Innovation into the development of new materials and solar cell architectures is therefore important in order to help drive the cost of solar electricity down and to develop new solar cells that can generate more electricity per unit area. However, thanks to advancements in computation capacity and speed, artificial intelligence is now emerging as another effective technique to help achieve these targets. We will focus on how AI can be applied to the field of PV in three main areas, which are (1) Forecasting and modelling of meteorological data, (2) Basic modelling of solar cells and (3) Sizing of photovoltaic systems.

Machine learning (ML) is an artificial intelligence technique that involves feeding data to algorithms, which aim to figure out patterns in the data. Examples of AI algorithms include Neural Networks (NN), Fuzzy Logic (FL), Simulated Annealing (SA), Genetic Algorithm (GA), Ant Colony (ACO), Particle Swarm Optimization (PSO) and Hybrid Techniques (HT). Consequently, the aim of this chapter is to provide an overview of these AI techniques and to demonstrate how some of them can be used to improve PV system performance in three areas.

2 Brief Introduction to Artificial Intelligence Techniques

Artificial Intelligence (AI) is a computational technique that is concerned with designing systems, which are able to understand reason and solve problems in a similar way to humans [2]. Nowadays, intelligent computing technologies are either replacing conventional techniques or are being integrated into existing systems.

AI is a vast subject containing many topics and subdivisions. One particular topic that has attracted increased attention is the field of Machine Learning (ML), whereby algorithms are designed with the ability to learn without being explicitly programmed to [3]. By using statistical techniques, these algorithms are capable of analysing an input dataset in order to make useful predictions about missing or future data.

In general, ML solutions can be divided into the ways in which learning is performed by the algorithms. There are three major branches, which are supervised learning, unsupervised learning and reinforcement learning. The following sections will provide a brief introduction to each ML branch, as well as an explanation into how other AI algorithms have been used for various PV applications.

2.1 *Supervised Learning*

With this type of learning, a supervisor or teacher is required to assist the algorithm in learning its parameters. These algorithms require a dataset that has information about both the input data as well as the output. During its learning phase, as the algorithms try to make predictions about the dataset, the teacher corrects and guides the algorithms in the right direction, making them improve over time.

In addition, supervised learning methods can be divided into two main categories, depending on the output variable that they are trying to predict. If the output data is a discrete variable, such as trying to determine if the next day will be sunny, cloudy or rainy (Class 1, 2 or 3), then these cases are said to be a classification problem. On the other hand, if the output required is a continuous or real value, such as trying to predict the irradiation levels of a city during a specific time, or trying to determine the best size of a PV panel, then the case becomes a regression problem [4].

Some examples of supervised learning algorithms include linear and logistic regression, k-Nearest Neighbours, Neural Networks, as well as more robust algorithms such as deep neural networks and their variations. Figure 1 summarises the concept of an Artificial Neural Network (ANN), which is inspired by biological networks in the brain [5–9]. An ANN therefore contains three layers (input, hidden and output), connections, biases, weights, an activation function and a summation node. These weights and biases are important parameters that influence the output function.

2.2 *Unsupervised Learning*

Unsupervised learning algorithms, on the other hand, do not re-quire a supervisor in order to learn or make predictions about the input data. In this case, these types of algorithms require only a dataset with input data and their goal is to correctly learn a model that best represents the given data [4]. Consequently, since these algorithms rely on finding patterns in the input data, unsupervised learning methods mainly consist of clustering algorithms, such as K-means and self-organizing maps.

2.3 *Reinforcement Learning*

Lastly, the third major branch of ML is the field of reinforcement learning. In contrast to the previous two fields, reinforcement learning algorithms rely on a goal seeking approach, whereby the learner tries different actions in order to discover which ones are best in achieving a certain goal [10]. Some examples of reinforcement learning algorithms include Q-Learning and Monte Carlo methods.

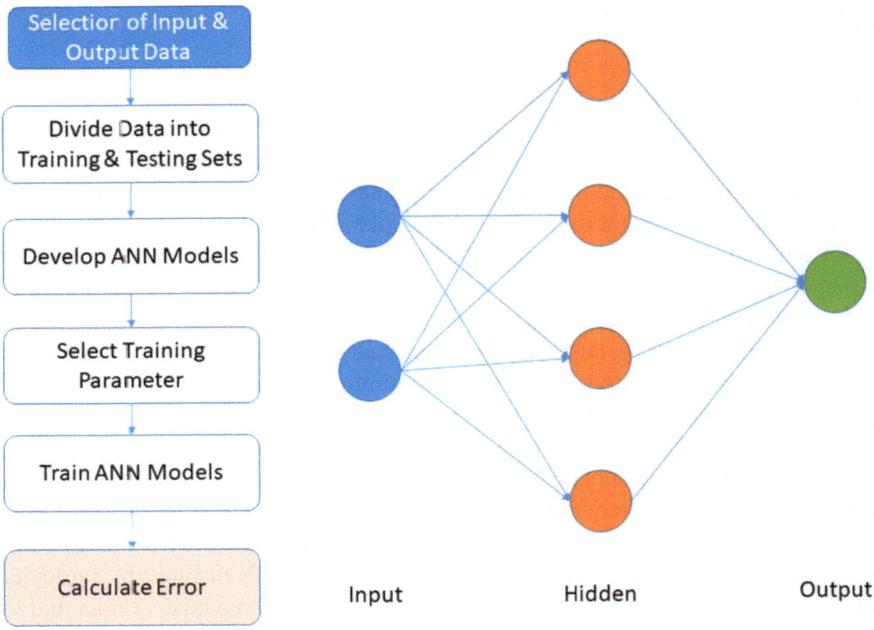

(a) ANN Prediction Methodology

(b) Generalised ANN Model

Fig. 1 General ANN concept

2.4 Other Techniques

Other intelligent ML approaches include Expert Systems (ES), Fuzzy Logic (FL), Simulated Annealing (SA), Genetic Algorithms (GA), Ant Colony (ACO) and Particle Swarm Theory (PSO). In contrast to conventional software programs that solve specific tasks within a range of boundary conditions, Expert Systems (ES) are designed to solve problems using the same approach as humans. An ES therefore consists of two main parts: an inference engine and a knowledge base. The knowledge base contains facts and rules, whereas the inference engine aims to apply these rules and facts to infer new facts [11, 12].

Similarly, a number of optimisation techniques have been developed that were inspired by nature. These include Genetic Algorithms (GA), which were first developed by Holland in 1975 and are based on the principles of genetics and evolution [13, 14]. Furthermore, Ant Colony (ACO) is another computational optimisation problem that was inspired by the behaviour of ants in finding the shortest path from their nests to their food. It was first formulated by Marco Dorigo in 1992 [15]. Here, ants lay down there pheromones as they randomly move towards their food. The stronger the pheromones, the more likely ants will follow that particular path. This technique is nowadays used to optimise machine scheduling and telecommunications networks

[16]. Other nature-inspired techniques include Particle Swarm Optimisation (PSO), which is influenced by the swarm and flocking of birds [17–20].

Another optimisation technique that was introduced by Zadeh in 1965 includes Fuzzy Logic (FL), which is a branch of computer logic that is different from Boolean or classic logic [21]. In contrast to classic logic, which has binary values of 1 (true) or 0 (false), Fuzzy Logic permits multiple intermediate possibilities within that set. Fuzzy Logic is often used in combination with Expert Systems and Artificial Neural Networks [22]. Moreover, Simulated Annealing (SA) is an effective optimisation technique that was inspired by the process of heating and slow cooling of solids and can be used for maximizing or minimizing a function [23, 24].

3 AI in PV Systems

We will demonstrate how AI has been successfully applied in three different applications of photovoltaics. With each of these applications, a comparison between conventional and AI techniques is presented. The first step in PV system sizing and modelling is weather forecasting. Consequently, it is only appropriate that this section starts with the application of AI in weather forecasting.

3.1 Forecasting of Meteorological Data

Predicting the weather is of critical importance for determining the power output of a PV system. Meteorological data such as solar radiation, ambient temperature, humidity, wind speed and sunshine duration are among the vital input parameters. These parameters play an important role in PV system performance. Instruments such as a Pyranometer, Pyrheliometer and two-axis tracker must be used to measure global and direct solar radiation. However, in some cases these parameters are impractical to obtain due to two main reasons:

1. The PV system is located in a remote or isolated area, where the required input data is not available.
2. The high cost and complexity of the equipment needed to measure these parameters.

A review of the role of AI in weather forecasting will be provided in this section. The main objective is to review some of the successfully implemented techniques in the literature and to present some of our own techniques for predicting solar radiation.

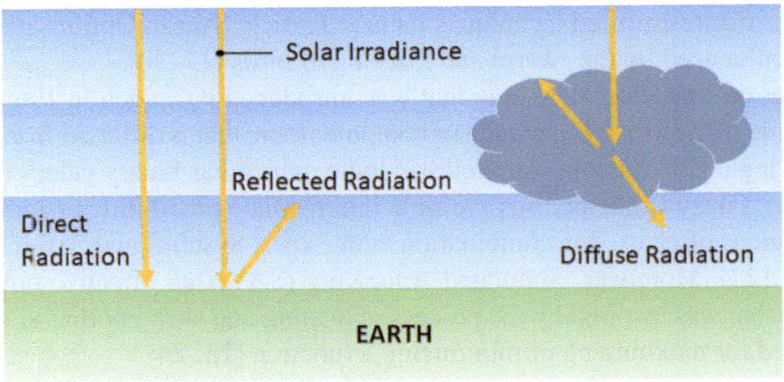

Fig. 2 Different components of solar radiation

3.1.1 Important Concepts

First, it is perhaps important to introduce a few key concepts related to the Sun, which is often regarded as a giant thermonuclear reactor that runs on hydrogen fuel. The Sun radiates energy in all directions in the form of electromagnetic radiation [25]. When describing the Sun's energy, there are four commonly used parameters in the PV community:

- Solar Irradiance - This is a term that describes the intensity of solar power per unit area. Its units are therefore in W/m^2.
- Solar Irradiation - This is the total amount of solar energy collected per unit area over time (Wh/m^2).
- Insolation - This describes the amount of solar irradiation collected during one day ($kWh/m^2/day$).
- Solar Constant - This is the average amount of solar irradiance that arrives above the Earth's atmosphere, which is approximately (1353 W/m^2) [25].

Consequently, due to atmospheric effects, there are four main types of solar radiation, which are Direct, Diffuse, Reflected and Global radiation. Figure 2 illustrates the various types of solar radiation. The global solar irradiation (G_T) on the Earth's surface is the sum of three main components:

$$G_T = G_B + G_D + G_R \tag{1}$$

where G_R, G_D and G_B are the components of reflected, diffused and direct solar radiation. At normal incidence to the Earth's surface, G_R can be neglected, which means that $G_T = G_B + G_D$.

On a clear day and when the Sun is directly overhead, almost 70% of the incident solar radiation reaches the Earth's surface. The magnitude of solar radiation that is scattered or absorbed depends on the amount of atmosphere it must travel before reaching the Earth's surface [25]. Consequently, Air Mass (AM) depicts the relative distance that solar radiation must travel to reach the Earth's surface. Thus, $AM =$

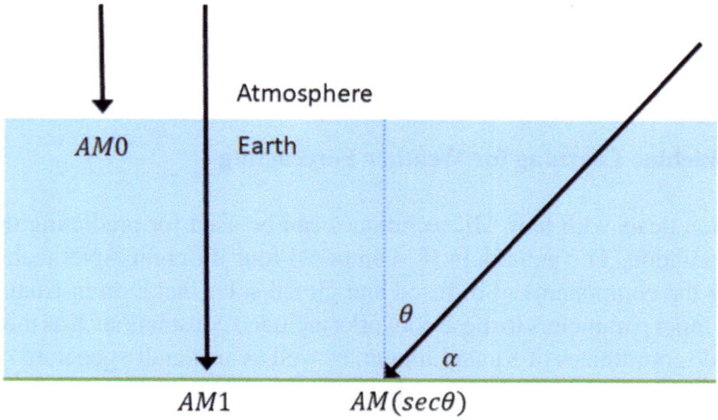

Fig. 3 Explanation of AM0, AM1 and AM($sec\theta$)

$1/cos\theta = sec\theta$, where θ is the zenith angle, as shown in Fig. 3. Similarly, Air Mass One (AM1) refers to the thickness of the atmosphere a sunbeam passes through at normal incidence to the Earth's surface. A list of accepted values of solar flux at AM1 can be found in [25], where the direct solar radiation at standard sea level is considered 0.930 kW/m^2.

Mathematically, we are able to predict the intensity of the Sun's energy arriving at a point on Earth using the Meinel and Meinel approximation [26]:

$$I = 1.353 \times 0.7^{(cosec\alpha)^{0.678}} \tag{2}$$

where $cosec\alpha = 1/sin\alpha$. The solar latitude, α, is the angle between the Sun's rays and the horizon. Thus, the light intensity is maximum when $\alpha = 90°$. Therefore, from Eq. 2, $I_{max} = 0.95\,\text{kWm}^{-2}$ [27].

Experimentally, the measurement of solar irradiance requires specialist equipment. For example, to measure global solar radiation, a Pyranometer is needed, which is usually mounted horizontally away from tall objects that may obstruct its field of view. Since the Sun contains a spectrum of different wavelengths, the best instruments are designed to respond equally to all wavelengths. Due to its ease of use, Pyranometers are widely used for collecting the vast majority of solar insolation data [25].

Furthermore, a ring-shaped hoop may be added to the Pyranometer in order to exclude direct sunlight. This process permits the measurement of diffuse sunlight. When this reading is subtracted from the data collected by a standard Pyranometer, the result is the direct solar radiation. This instrument is called the Shaded-ring Pyranometer [25].

Similarly, to measure direct solar radiation, a Pyrheliometer is used. The Pyrheliometer has a small field of view, which is approximately 6°. It is usually mounted onto a dual axis Sun tracker in order to continuously measure the Sun's direct normal radiation. Pyrheliometers are often needed to predict the performance of concentrator photovoltaic systems, which also require the use of dual axis trackers [25].

3.1.2 Machine Learning for Weather Forecasting

This section deals with how ML techniques can be used for predicting the amount of solar radiation. The authors in [8] compared four different ANN techniques for predicting the components of diffused and global solar radiation in Kuala Lumpur. The main input parameters to these networks included constants such as the longitude and latitude coordinates of Kuala Lumpur, as well as artificially generated variations in humidity, temperature and daily sunshine ratio. 80% of this data was used for training, while the remaining 20% was used for testing. According to their study, the GRNN technique provided the most accurate prediction results in comparison to the Feed-Forward back propagation Neural Network (FFNN), the Elman Back propagation Neural Network (EBNN) and the cascade-forward back propagation neural network (CFNN).

The literature provides more examples of using neural networks for prediction purposes. In each case, four parameters were chosen as inputs to a neural network for predicting a particular aspect of the weather. For example, the authors in [28, 29] used ANNs for predicting global solar radiation in Saudi Arabia using data collected from 41 weather stations. The input parameters to their network included the latitude, longitude and altitude, as well as the sunshine duration. Their network also consisted of 10 hidden layers and was tested using data from 31 cities. The prediction results for the remaining 10 cities showed a Mean Absolute Percentage Error (MAPE) variation between 6.5 and 19.1%. Similarly, the authors in [30] trained a neural network with 77% of data from a weather station in La Sirena. Wind speed, air temperature, soil temperature and humidity were used as inputs and the Average Absolute Relative Deviation (AARD) was <9%. Similarly, the feed forward multilayer perception model was used to predict global and diffuse radiation in Malaysia using longitude, latitude, day number and sunshine ratio [31]. 82% of data from 28 different weather stations was used for training, while the remainder was used for testing. Their network consisted of one hidden layer and the calculated Root Mean Square Error (RMSE) was 7.96%. More accurate prediction can be obtained by increasing the number of inputs and the number of hidden layers in a neural network. This was achieved by Mellit and Pavan in [32], whereby the correlation coefficient between experimental and predicted values varied between 95 and 98%. A thorough review of these ML techniques can be found in in [7, 33].

Similarly, we have applied a supervised learning technique for predicting the solar irradiation levels for a location in the city of Beijing, China. Beijing Sunda Solar Energy Technology Company collected the data over the period of two years [34]. Consequently, we were able to design a Support Vector Machine (SVM) regression model that is able to predict, to a certain extent, the irradiation levels for the next couple of days. The data collected consisted of solar radiation levels for the months of January and July of 2007, with measurements taken for the whole 31 days of each

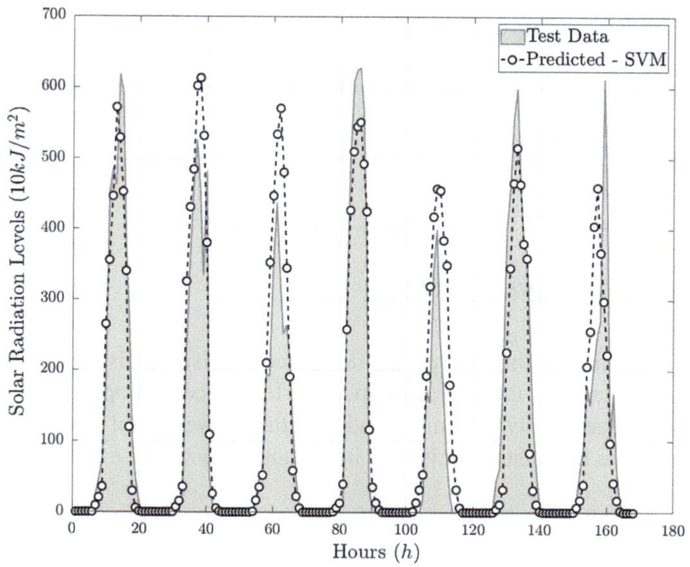

Fig. 4 Predictions of the SVM model using data from 3 previous days to predict the solar radiation levels of the next day

month in intervals of one hour. Based on that, we designed a SVM model that took as inputs the radiation level of previous days in order to predict the radiation levels of the next day.

Given this model, different time horizons were tested, in which the algorithm could consider the solar radiation levels of either one, three, or five previous days in order to predict the levels of the next day. The approach that yielded the best results was considering the previous three days. In addition, the dataset was split into training and test data, in which 51 days were utilized to train the network and the remaining 7 days were used to test it (some days were not accounted because of the time horizon).

Figure 4 shows our results of the proposed solution using the best approach for one week of test data. Both the real and predicted values of solar radiation are shown. A comparison with other regression techniques was performed, including the Regression Tree and the Tree Ensemble. It was found that the SVM technique provided best fit with the actual data, whereby the Root Mean Square Error (RMSE) was 12.41% and the Mean Absolute Error (MAE) was 6.95%. A summary of the results is presented in Table 1.

Table 1 Comparison between different regression methods

Model	RMSE (%)	MAE (%)
SVM	12.41	6.95
Regression tree	21.36	11.72
Tree ensemble	22.46	11.71

3.2 Modelling of Solar Cells

A mathematical model that accurately describes a solar cell is an invaluable tool for better understanding the characteristics, performance and optimization of a PV cell system. Consequently, having introduced the conventional and AI techniques for solar radiation forecasting, it is now important to familiarise the reader with the concept of a *solar cell.*

3.2.1 Solar Cell Theory

A solar cell (SC) is a basic device that is used to convert the Sun's energy into electricity. Ultimately, semiconductor materials are commonly used for the purpose of producing currents and voltages as a result of the absorption of sunlight, which is a phenomenon known as the photovoltaic effect. Most solar cells are fabricated from either monocrystalline or polycrystalline silicon (Si) materials. In its most basic form, a solar cell consists of a *pn* junction diode. Typical solar cell efficiencies range from 18% for polycrystalline to 24% from highly efficient monocrystalline technologies. These high end devices typically include special light trapping structures that absorb as many of the incident photons as possible.

Figure 5 shows the basic operation principles of a solar cell, which was adopted from the literature in [27]. Both drift and diffusion of carriers takes place across the depletion region of width, W. The built-in electric field E_0 in this depletion layer prevents further diffusion of minority carriers. The finger electrodes on the surface of the n-type semiconductor material allows light to penetrate into the device. Furthermore, these electrodes result in a small series resistance. The photogenerated electron hole pairs in the depletion region become separated by the built-in electric field, E_0. Through the process of *drift*, electrons reach the neutral n-region and make it negative by an amount of charge $-q$. Similarly, holes drift to the p-region, which effectively turns that region more positive. Consequently, an open circuit voltage, V_{oc}, develops between the terminals of the device, whereby the p-region is positive with respect to the n-region. The total current flowing through a solar cell can determined using Kirchhoff's law, whereby:

$$I = I_{ph} - I_d \tag{3}$$

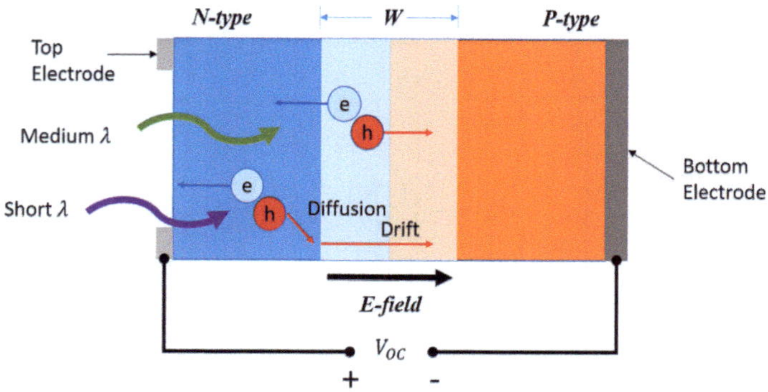

Fig. 5 The basic principle of operation of the solar cell, as depicted by Kasap in [27]

Here, I_{ph} is the photocurrent, I_d is the diode current, which is proportional to the saturation current by the following relationship:

$$I_d = I_o \times (e^{V/\eta V_t} - 1) \tag{4}$$

where I_o is the reverse saturation current, V is the voltage imposed on the diode and V_t is the thermal voltage, $V_t = kT/q$, where k is the Boltzmann constant, q is the charge and T is the temperature. Similarly, the diode ideality factor, η, typically depends on the type of solar cell technology used. For the case of monocrystalline silicon, this is usually $\eta = 1.2$ [35].

In reality, photogenerated electrons need to travel across a semiconductor region in order to be collected by the nearest electrode. Consequently, an effective series resistance, R_s, is introduced in the photovoltaic circuit. Similarly, photogenerated carriers flow through the crystal surfaces or through grain boundaries in polycrystalline devices. These effects can be described in terms of a shunt resistance, R_p, which drives photocurrent away from the load, R_L. Consequently, the equivalent electrical circuit representation of a typical solar cell can be modelled as shown in Fig. 6 [27, 36].

Thus, from Fig. 6, an expression for the total output current of the cell, I, can be deduced:

$$I = I_{ph} - I_0(e^{q(V+IR_s)/\eta kT} - 1) - ((V + IR_s)/R_p) \tag{5}$$

where q is the electric charge, V is the voltage, k is the Boltzmann constant and T is the cell temperature in Kelvins, K. Consequently, we can determine the $I - V$ characteristics of a SC as a function of input solar radiation, series resistance and shunt resistance.

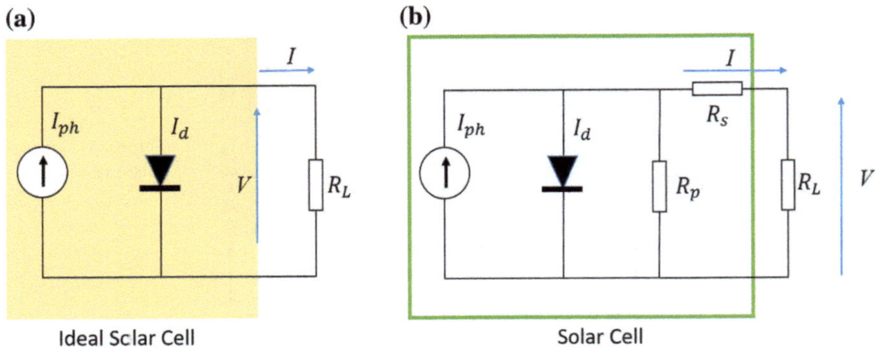

Fig. 6 Electrical circuit representation of **a** an ideal solar cell and **b** practical solar cell with series and shunt resistances

3.2.2 Machine Learning for Solar Cells

The accurate modelling of a solar cell involves the prediction of five important parameters, which are η, I_0, R_s, R_p and I_{ph}. Numerical and analytical techniques have been proposed to extract these parameters. However, since the $I - V$ relationship is highly non-linear, many algorithms suffered from drawbacks. For example, the non-linear least error squares approximation approach by the authors in [37] was strongly dependent on the choice of initial values. Other analytical methods suffer from heavy computation complexity, algebraic manipulation and curve fitting [38]. Consequently, we will examine the machine leaning techniques that can be used for parameter extraction. We will focus on three techniques, which are the Simulated Annealing approach, the Harmony Search approach and the use of ANNs.

Among the successful techniques that have been proposed is a neural network method by the authors in [39]. Their research relied on two different approaches. The first involves generating $I - V$ curves using the Sandia National Laboratory (SNL) PV performance model for different operating temperatures and solar irradiation [40]. Later, five points are chosen from these $I - V$ curves, which are located at $V = 0$, $V_{oc}/2$, V_{mp}, $((V_{oc} + V_{max})/2)$, V_{oc}. Here, V_{oc} is the open-circuit voltage and V_{max} is the maximum voltage. According to Sandia, these five points represent an accurate representation of the $I - V$ curve.

The authors in [39] then trained the neural network with 191 operating conditions of temperature and radiation. The five solar cell parameter were determined using Eq. 5 and the solar cell equations in SNL. Subsequently, the five extracted parameters were then fed into the one diode solar cell model to obtain the $I - V$ characteristics of a PV module, as depicted in Fig. 7.

The second approach involved generating the $I - V$ curves experimentally and determining the five operating points using the SNL model. This was done using a solar panel, an electronic load, a computer controlled $I - V$ tracer and a weather

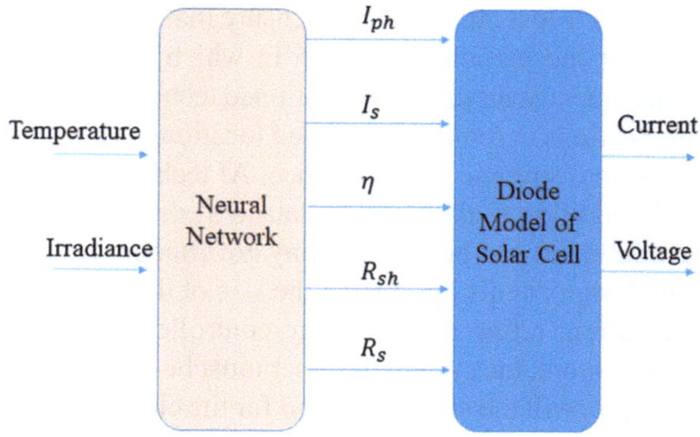

Fig. 7 PV model approach that was adopted by [38]

station. Different values of irradiance and temperature were obtained by triggering the $I - V$ tracing at different times throughout the day.

The neural network was trained to identify the five equivalent circuit parameters from 41 $I - V$ curves. The network was tested on 8 curves. Despite the ease of setting up this experimental facility, a drawback of this technique is that it was not possible to decouple the effects of temperature and radiation. For example, it was not possible to determine the percentage error at a module temperature of 25 °C and an irradiation level of 850 W/m^2, since irradiation levels caused an increase in module temperature. Nevertheless, a percentage error of 1.2% in V_{mp} at an irradiation level of 851 W/m^2 and a temperature of 46.7 °C was achieved, which is approximately three times more precise than the conventional model described by the Townsend equations in [41, 42].

Other machine learning techniques include the simulated annealing (SA) approach by the authors in [43], which is compared with the GA and PSO prediction techniques. An objective function was defined and minimised. Their results were compared with experiments and a Root Mean Square Error (RMSE) of 0.0017 was achieved for the single diode solar cell model.

In comparison, the authors in [44] showed that the Harmony Search (HS) optimisation process provides better results. Here, HS is an optimization technique that aims to imitate the improvisation process of musicians. According to their research, an objective function based on the single diode model was minimised with respect to a particular range. Consequently, the HS technique was able to extract the five solar cell parameters with a lower RMSE than the SA technique by a factor of 0.075.

3.3 Sizing of PV Systems

Determining the size of a PV system for a particular application in order to yield the best return on investment (ROI) is of paramount economic importance. Generally, the main techniques for sizing a PV system rely on empirical, analytical and numerical methods. Most importantly, to ensure that the sizing method is accurate, the weather data for the particular location in which the system will be installed is essential. Consequently, all the aforementioned techniques are ineffective for PV system sizing applications in remote or isolated locations, where the required data is not available. Thus, to overcome this hindrance, AI techniques may be used.

In addition to the weather data, a system designer must understand the end user requirements and be aware of any government incentives or policies. Generally, system design entails optimising the size of the solar array, the size of the battery bank, the current rating of the charge controller and the size of the DC to AC inverter. Furthermore, the system designer must be aware of how PV system will be used. For example, will the system be used for the electrification of a village, for a telecommunications application, or for water pumping? To avoid economic waste, PV system sizing is an application oriented optimisation process that aims to meet a certain load requirement for the expected lifetime of the system [45]. In this section, particular attention is attributed to determining the Loss of Load Probability (LLP) of a stand-alone PV system. Comparisons between AI predicted results with other sizing methods will be made.

3.3.1 Introduction to PV Systems

A photovoltaic array (or string) is a combination of photovoltaic modules that are connected in series in order to obtain the desired system voltage. Subsequently, these strings are connected in parallel to increase the system's output power [46]. A photovoltaic installation mainly consists of an array of photovoltaic modules or panels, an inverter, batteries (for off grid) and interconnection wires. However, the balance of system (BOS) components in a PV system include mounting materials for the modules, wires, distribution panel, junction box, lighting protectors, grounding connections, battery fuses, battery cables and battery containers. In general, PV systems can be grouped into grid connected or autonomous (or stand-alone) systems.

A standalone or autonomous PV system is not connected to the national electricity grid. Such systems were the first application of PV, where there was no electricity supply from a national grid. Applications of these systems range from pocket chargers to large water pumping systems. Figure 8 illustrates the main components of this system, which are:

- PV Module - An interconnected array of solar cells.
- Charge Controller - Also known as a charger regulator, which aims to control the rate of current flow into and out from storage batteries. This is done to prevent overcharging and deep discharging of the battery, which can severely reduce battery performance and lifetime.
- Battery Bank - Batteries are the heart of an autonomous solar electric system. They are the reservoir for storing electrical energy. The size of a battery is measured in terms of its storage capacity in Ampere-Hours (Ahr). There are different types of battery technologies, including Lithium Ion, Lead Acid, Nickel Cadmium

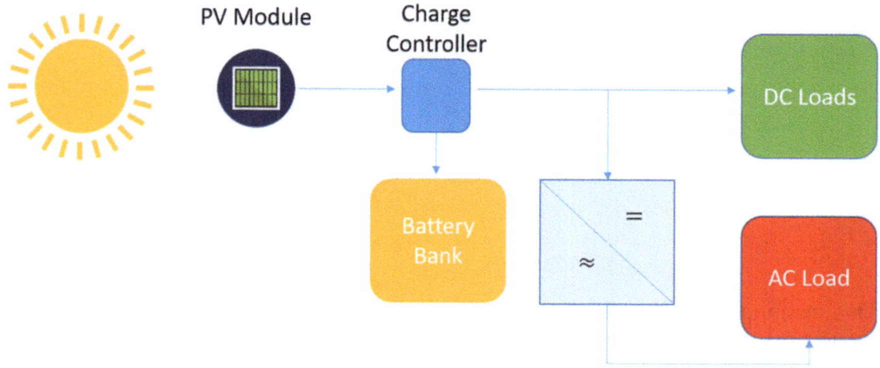

Fig. 8 Stand-alone or autonomous PV System. The illustration shows two possible configurations. In the DC coupled configuration, DC loads are directly connected to the charge controller. In contrast, AC loads require an inverter

and many others. The depth of discharge (DOD) is the amount a solar battery is discharged.

- Inverter - This device is responsible for converting the PV panel's direct current (DC) into an alternating current (AC) that can be fed into a commercial electrical grid, or to be used in a stand-alone system.
- Loads - These can be a combination of either DC or AC appliances that are connected to the system. The vast majority of present day appliances require AC power. DC coupled systems do not require AC converters, which reduces overall system cost and increases system efficiency.

3.3.2 Sizing Autonomous PV Systems

In this chapter, particular attention will be devoted to sizing autonomous or off-grid PV systems. This can be achieved in a number of different ways, as depicted in Fig. 9. The first is called the Intuitive Method, which relies on the PV designer's experience in sizing the PV array and the battery bank. Another is called the Numerical Method, which is more accurate than the Intuitive Method, but requires a large dataset of solar radiation values, which makes them complex to use [47]. Similarly, the Analytical Method involves describing the sizing problem as a function of the Loss of Load Reliability (LLR). In this section, we shall illustrate the most commonly used techniques in the literature for PV system sizing.

In brief, the Intuitive Method involves estimating the daily load demand, optimizing the tilt angle, calculating the size of the PV array and determining the size of the battery bank. By using simple mathematical equations, the size of the PV generator and battery bank can be determined [45]. Similarly, the size of the inverter can be determined by considering its efficiency and the maximum AC power it can deliver.

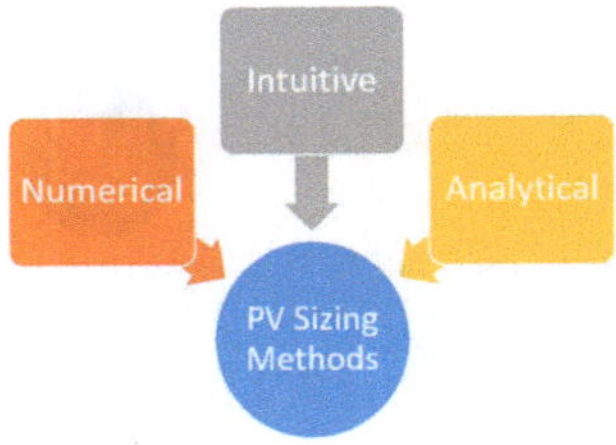

Fig. 9 Standalone PV System sizing techniques

This technique has been used for the optimum sizing of residential PV systems in Egypt [48] and in Bangladesh [49].

The Intuitive Method first relies on estimating the daily load demand and optimizing the tilt angle before calculating the PV array size and battery capacity. The equations for calculating these two parameters are similar to those in [45]. Figure 10 provides a summary of the main steps involved in sizing a stand-alone PV system intuitively. From Fig. 10 and assuming that all the modules in the system are identical, N is the number of modules, η_{pv} is the efficiency of the PV module, APV is the effective area of a single PV module and G_t is the global incident irradiation. Moreover, N_{aut} is the number of days of autonomy and DOD_{max} is the maximum depth-of-discharge of the batteries. The number of batteries (parallel) is N_{batt} and S_{pk-hr} is the peak Sun hours for the worst calendar month of the year.

Having determined the size the PV generator and the capacity of the battery bank, the remaining PV system components need to be configured. For example, the simplest method to calculate the size of an inverter is to take the PV module peak power and divide this by the nominal AC efficiency of the inverter, $P_{inv,AC} = P_{pv}/\eta_{inv}$ [26]. Similarly, charge controllers and cable cross-sections need to be sized accordingly. In this case, cable cross-sections need to be designed to deliver the least amount of voltage drop. Article 690 in the National Electric Code (NEC) describes the recommendations for PV wire gauge sizing the current protection mechanisms necessary in a PV system [50].

Nevertheless, the Intuitive Method is best used as a quick technique for approximating or validating the size of a PV system. It needs to be combined with other techniques to avoid over (or under) sizing of an autonomous system. Consequently, it can be combined with analytical and experimental methods, as demonstrated by the authors in [51] for a system in India. Their work relied on developing an iterative algorithm that takes into consideration the number of days of autonomy, system efficiencies, DOD_{max}, battery cut-off voltages, cable losses, PV mismatch losses and the dust factor. Depending on the load profile, it was possible to reduce system investment requirements by 14%.

Advanced simulation software can also be used for PV system sizing. Currently, there are software tools that help engineers design PV systems. Examples of these tools include RETScreen, Homer, PVSyst, PV*Sol and Polysun. The majority of these tools rely on a combination of mathematical and analytical methods for PV

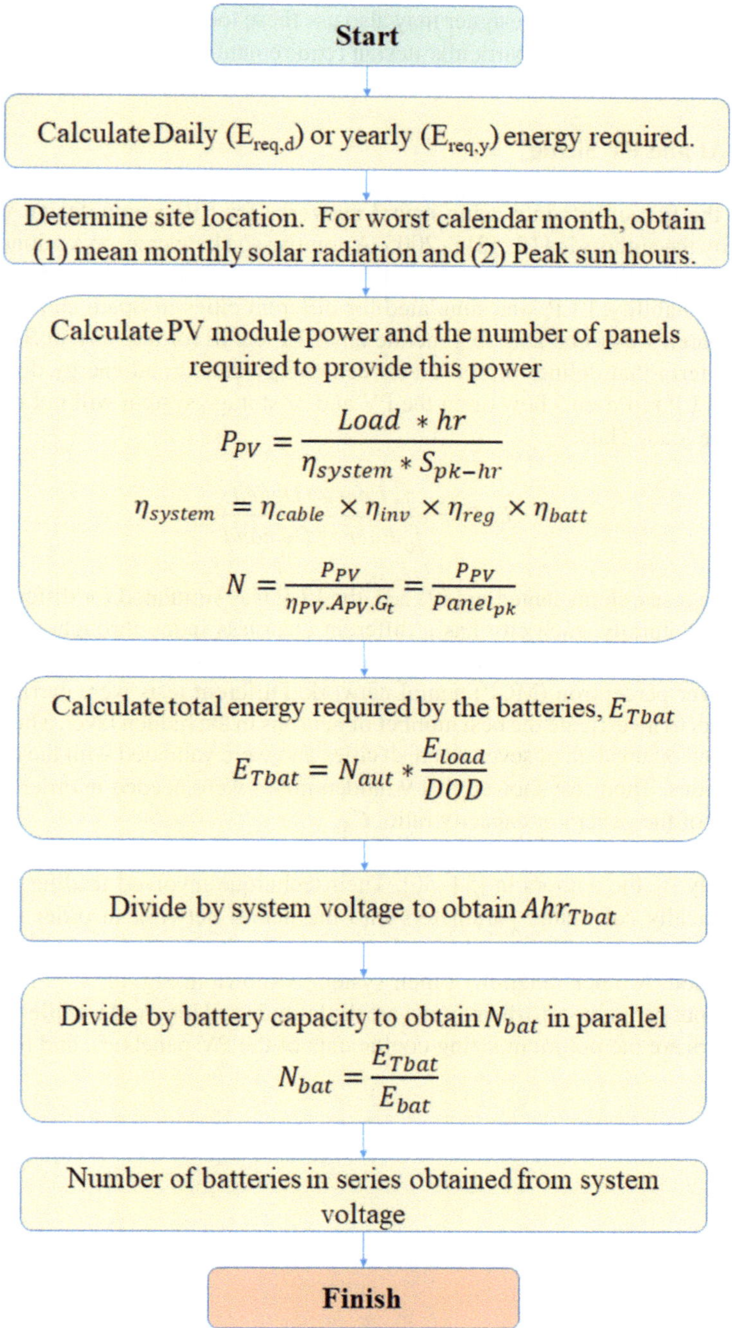

Fig. 10 Flow chart for the intuitive method of PV sizing

Systems sizing. A system designer may also use these tools in combination with one another in order to meet a particular design requirement.

3.3.3 AI and PV Sizing

Among the first efforts into using neural networks for PV system design was performed by the authors in [19, 52] in 2005. A comparison between conventional sizing methods and ANN is provided in [53]. Furthermore, a parameter known as the loss of load probability, LLP, was simulated for different cities in Spain and the aim of their research was to be able to generate any LLP curve for any city in Spain. The LLP is a term that defines the ratio between energy deficit and energy demand. In fact, the LLP represents how often the PV and/or storage system will not be able to satisfy the load. Thus,

$$LLP = \frac{\int_0^t Energy\,Deficit}{\int_0^t Energy\,Demand} \tag{6}$$

For the research presented in [52, 53], the LLP was simulated for different cities in Spain. Naturally, each city has a different clearness index throughout the year. Consequently, these two parameters in addition to the battery capacity were fed into a multilayer perceptron (MCP) neural network. Different tests were carried out by the authors to investigate the best number of neurons in the hidden layer. The network was trained on the data of seven Spanish cities and were validated with the data from 3 other cities. Their test showed that 9 hidden nodes were needed in order to obtain the value of the generator capacity ratio, C_A.

The use of neural networks for SAPV system sizing has also been investigated extensively by the authors in [54–56]. Their technique involved feeding a number of numerically calculated parameters into the neural network in order to predict the number of solar PV modules and storage capacity needed to satisfy a given consumption. A block diagram of their system is shown in Fig. 11.

The input parameters to their model are the latitude and longitude, while the output parameters are the optimum sizing coefficients of the PV panels, f_p, and the battery

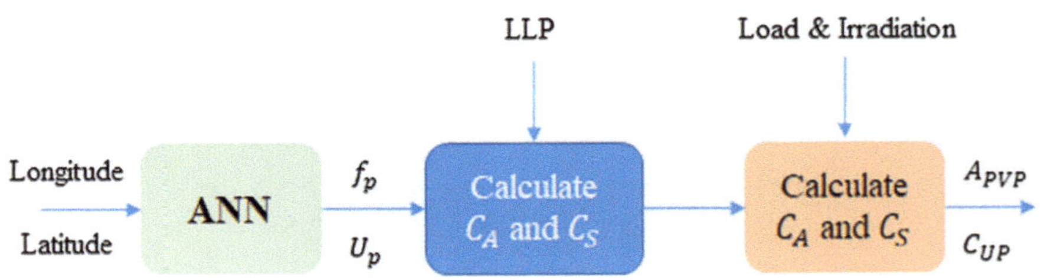

Fig. 11 Method of obtaining the sizing parameters of an autonomous PV system, as defined by Mellit in [54]

capacity, u_p. Furthermore, The LLP is the Loss of Load Parameter, whereas C_A is the ratio of average PV generator production during the worst calendar month to the average load energy demand. Similarly, the storage capacity of the battery bank, C_s, is defined as the ratio of maximum storage energy to the average load energy demand [47].

Consequently, both C_A and C_s can be determined from:

$$C_A = \frac{\eta_{pvp} A_{pv} G_{av}}{L} \tag{7}$$

$$C_S = \frac{C}{L} \tag{8}$$

where A_{pvp} is the total area of the PV generators, η_{pv} is the PV module efficiency, G_{av} is the average daily global irradiation, L is the average daily energy consumption and C is the battery bank capacity. The output parameters of the system are the PV array area (A_{pvp}) as well as the useful battery capacity (C_{UP}).

Additional AI methods include an investigation that involves the use of fuzzy logic (FL) for optimizing the PV generator area and the size of the battery bank for an autonomous PV system located in Sfax, Tunisia [57]. Furthermore, hybrid optimisation techniques involving the merge of genetic algorithms (GA) and neural networks were also investigated by Mellit [58, 59]. A complete review of AI techniques for PV system sizing is provided by the same author in [56], as well as by the authors in [60].

4 Summary

Various AI techniques have been applied to three main PV applications, which include (1) Forecasting and modelling of meteorological data, (2) Basic modelling of solar cells and (3) Sizing of photovoltaic systems. In the case of solar radiation forecasting, it has been shown that AI can be used for the accurate prediction of solar radiation in the city of Beijing. Other scholars in the literature have successfully used various AI techniques for predicting the weather conditions of any geographic location. Consequently, AI techniques can then be extended to any geographic location. Furthermore, the estimation of energy production for a PV system has been shown using both conventional and AI techniques. Again, the results show the accuracy of AI techniques in comparison to analytical methods. Consequently, the chapter has illustrated how AI can be used to not only predict, but to accurately fill in the missing gaps of important information.

References

1. US Senate Committee on Energy and Natural Resources, Transcript of the Testimony of Richard E. Smalley to the US Senate Committee on Energy and Natural Resources, US Government Publishing Office, 27 April 2004. https://www.gpo.gov/fdsys/pkg/CHRG-108shrg95239/html/CHRG-108shrg95239.htm. Accessed 23 Sept 2018
2. Bar A, Feigenbaum E (1981) The handbook of artificial intelligence. Morgan Kaufmann, San Francisco
3. Samuel A (1959) Some studies in machine learning using the game of checkers. IBM J Res Dev 3(3):210–229
4. Friedman J, Hastie T, Tibshirani R (2001) The elements of statistical learning. Springer, New York
5. Haykin S (1994) Neural networks: a comprehensive foundation. Macmillan Publishing Company, New York
6. da Silva I, Spatti DH, Flauzino RA, Liboni L, dos Reis Alves S (2017) Artificial neural networks: a practical course. Springer, Switzerland
7. Yadav AK, Chandel SS (2014) Solar radiation prediction using artificial neural network techniques: a review. Renew Sustain Energy Rev 33:772–781
8. Khatib T, Mohamed A, Sopian K, Mahmoud M (2012) Assessment of artificial neural networks for hourly solar radiation prediction. Int J Photoenergy 2012:7 pp
9. Al-Daoud E (2009) A comparison between three neural network models for classification problems. J Artif Intell 2(2):56–64
10. Sutton R, Barto A (1998) Reinforcement learning: an introduction. MIT press, Cambridge
11. Nwigbo S, Madhu B (2016) Expert system: a catalyst in educational development in Nigeria. IOSR J Mob Comput Appl 3(2):8–11
12. Darlington K (2010) The essence of expert systems. Pearson Education, England
13. Holland J (1992) Adaptation in natural and artificial systems: an introductory analysis with applications to biology, control and artificial intelligence. MIT Press, Cambridge
14. Sivanandam S, Deepa S (2007) Introduction to genetic algorithms. Springer, Berlin
15. Dorigo M (1992) The metaphor of the ant colony and its application to combinatorial optimization. PhD thesis, Politecnico di Milano, Italy
16. Dorigo M, StÃijtzle T (2003) The ant colony optimization metaheuristic: algorithms, applications, and advances. Handbook of metaheuristics. International series in operations research and management science. Springer, Boston
17. Youssef A, El-Telbany M, Zekry A (2017) The role of artificial intelligence in photo-voltaic systems design and control: a review. Renew Sustain Energy Rev 78:72–79
18. Marini F, Walczak B (2015) Particle swarm optimization (PSO). A tutorial. Chemom Intell Lab Syst 149:153–165
19. Poli R (2008) Analysis of the publications on the applications of particle swarm optimisation. J Artif Evol Appl 2008:10 pp
20. Shi Y, Eberhart R (1998) A modified particle swarm optimizer. In: Proceedings of IEEE international conference on evolutionary computation
21. Zadeh L (1965) Fuzzy sets. Inf Control 8:338–353
22. Bose B (1994) Expert system, fuzzy logic, and neural network applications in power electronics and motion control. Proc IEEE 82(8):1303–1323
23. Metropolis N, Rosenbluth A, Rosenbluth M, Teller A, Teller E (1953) Equation of state calculation by fast computing machines. J Chem Phys 21:1087–1091
24. Kirkpatrick S, Gelatt C, Vecchi M (1983) Optimization by simulated annealing. Science 220(4598):671–680
25. Hu C, White R (1983) Solar cells: from basics to advanced systems. McGraw-Hill, New York
26. Meinel A, Meinel M (1976) Applied solar energy. Addison-Wesley, Reading
27. Kasap S (2018) Principles of electronic materials and devices. McGraw-Hill Education, New York

28. Mohandes M, Rehman S, Halawani T (1998) Estimation of global solar radiation using artificial neural networks. Renew Energy 14:179–184
29. Rehman S, Mohandes M (2009) Estimation of diffuse fraction of global solar radiation using artificial neural networks. Energy Sources 31:974–984
30. Lazzús J, Ponce A, Marín J (2011) Estimation of global solar radiation over the city of La Serena using a neural network. Appl Sol Energy 47(1):66–73
31. Khatib T, Mohamed A, Sopian K, Mahmoud M (2012) Solar energy prediction for Malaysia using artificial neural networks. Int J Photoenergy 2012:1–16
32. Mellit A, Pavan A (2010) A 24-h forecast of solar irradiance using artificial neural network: Application for performance prediction of a grid-connected PV plant at Trieste, Italy. Sol Energy 84:807–821
33. Mellit A (2008) Artificial intelligence techniques for modelling and forecasting of solar radiation data: a review. Int J Artif Intell Soft Comput 1:52–76
34. SUNDA, Beijing Sunda Solar Energy Technology Company, Ltd. http://www.sundasolar.com/. Accessed 18 June 2018
35. Tsai H, Tu C, Su Y (2008) Development of generalized photovoltaic model using MATLAB/SIMULINK. In: Proceedings of the World congress on engineering and computer science, WCECS 2008, San Francisco, USA, 22–24 October 2008
36. The German Energy Society (Deutsche Gesellshaft fur Sonnenenergie) (2008) Photovoltaic systems: a guide for installers, architects and engineers. Routledge, Abingdon
37. Easwarakhanthan T, Bottin J, Bouhouch I, Boutrit C (1986) Nonlinear minimization algorithm for determining the solar cell parameters with microcomputers. Int J Sol Energy 4:1–12
38. Al-Rashidi M, El-Naggar K, AlHajri M, Al-Othman A (2011) A new estimation approach for determining the I-V characteristics of solar cells. Sol Energy 85(7):1543–1550
39. Karatepe E, Boztepe M, Colak M (2006) Neural network based solar cell model. Energy Convers Manag 47:1159–1178
40. King D, Boyson W, Kratochvill J (2004) Photovoltaic array performance model. Sandia National Laboratories, Albuquerque, New Mexico
41. Townsend T (1989) A method for estimating the long-term performance of direct-coupled photovoltaic systems. Masters thesis, University of Wisconsin-Madison, Madison, WI, USA
42. Vergura S (2016) A complete and simplified datasheet-based model of PV cells in variable environmental conditions for circuit simulation. Energies 9:326
43. El-Naggar K, AlRashidi M, AlHajri M, Al-Othman A (2012) Simulated annealing algorithm for photovoltaic parameters identification. Sol Energy 86:266–274
44. Askarzadeh A, Rezazadeh A (2012) Parameter identification for solar cell models using harmony search-based algorithms. Sol Energy 86:3241–3249
45. Sharma V, Colangelo A, Spagna G (1995) Photovoltaic technology: basic concepts, sizing of a stand alone photovoltaic system for domestic applications and preliminary economic analysis. Energy Convers Manag 36(3):161–174
46. Tiwari G, Dubey S (2010) Fundamentals of photovoltaic modules and their applications. Royal Society of Chemistry Publishing, London
47. Lorenzo M, Egido E (1992) The sizing of stand alone PV-systems: a review and a proposed new method. Sol Energy Mater Sol Cells 26:51–69
48. Ahmad G (2002) Photovoltaic powered rural zone family house in Egypt. Renew Energy 6:379–390
49. Bhuiyan M, Asgar M (2003) Sizing of a stand-alone photovoltaic power system at Dhaka. Renew Energy 28:929–938
50. Committee National Electric Code (2017) Article 690: Solar Photovoltaic (PV) Systems, NFPA 70. National Electric Code, National Fire Protection Association (NFPA)
51. Nikhil P, Subhakar D (2013) Sizing and parametric analysis of a stand-alone photovoltaic power plant. IEEE J Photovolt 3:776–784
52. Hontoria L, Aguilera J, Zufiria P (2005) A new approach for sizing stand alone photovoltaic systems based in neural networks. Sol Energy 78:313–319

53. Almonacid F, Rus C, Pérez-Higueras P, Hontoria L (2011) Calculation of the energy provided by a PV generator. Comparative study: conventional methods versus artificial neural networks. Energy 36:375–384
54. Mellit A, Benghanem M, HadjArab A, Guessoume A (2005) An adaptive artificial neural network model for sizing stand-alone photovoltaic systems: application for isolated sites in Algeria. Renew Energy 30(10):1501–1524
55. Mellit A, Benghanem M, Arab AH, Guessoum A (2003) Modeling of sizing the photovoltaic system parameters using artificial neural network. In: Proceedings of IEEE, Conference on Control Application, vol 1, pp 353–357
56. Mellit A, Kalogirou S (2008) Artificial intelligence techniques for photovoltaic applications: a review. Prog Energy Combust Sci 34:574–632
57. Salah C, Lamamra K, Fatnassi A (2015) New optimally technical sizing procedure of domestic photovoltaic panel/battery system. J Renew Sustain Energy 7:1–14
58. Mellit A (2010) ANN-based GA for generating the sizing curve of stand-alone photovoltaic systems. Adv Eng Softw 41:687–693
59. Mellit A, Kalogirou S, Drif M (2010) Application of neural networks and genetic algorithms for sizing of photovoltaic systems. Renew Energy 35:2881–2893
60. Khatib T, Ibrahim I, Mohamed A (2016) A review on sizing methodologies of photovoltaic array and storage battery in a standalone photovoltaic system. Energy Conversion and Management 120:430–448

Applications of Improved Versions of Fuzzy Logic Based Maximum Power Point Tracking for Controlling Photovoltaic Systems

R. Boukenoui and A. Mellit

Abstract Many local and global fuzzy logic based MPPT methods have been proposed to seek the solar Photovoltaic (PV) system's MPP. In this chapter, various improved FL-MPPTs which have been proposed over the years are introduced, their advantages and disadvantages are also clarified. Local MPPTs (LMPPTs) based on improved versions of FL are presented in detail, including Modified Hill Climbing–FL Controller (HC–FLC) and Adaptive P&O–FLC. Then, based on simulation and experimental results, a comparative study is done by employing the main assessment criteria to figure out the relative merits and limitations of those MPPTs in tracking the maximum power . The last part of the chapter introduces advanced Global MPPT techniques based on FL to track the GMPP of complicated shading patterns. A performance comparison of those GMPPTs is done and useful information on how to choose a suitable MPPT depending on the application intended, are outlined.

Abbreviation

ABC :	Artificial Bee Colony
Adaptive :	P&O–FLC Adaptive Perturb Observe-Fuzzy Logic Controller
AI :	Artificial Intelligence
ANN :	Artificial Neural Network
COA :	Center Of Area
CDD :	Constraint Decision Defuzzification
CS :	Cuckoo Search

R. Boukenoui
Electronics Department, Blida 1 University, 09000 Blida, Algeria
e-mail: rachidboukenoui@gmail.com; rachid.boukenoui@ensea.fr

A. Mellit (✉)
Renewable Energy Laboratory, Jijel University, 18000 Jijel, Algeria
e-mail: adelmellit2013@gmail.com; amellit@ictp.it

A. Mellit
The International Centre for Theoretical Physics (ICTP), StaradaCostiera, 11-34151 Trieste, Italy

R.-E. Precup et al. (eds.), *Solar Photovoltaic Power Plants*, Power Systems,
https://doi.org/10.1007/978-981-13-6151-7_7

COG :	Center Of Gravity
ECOA :	Extended Center Of Area
FL :	Fuzzy Logic
FLC :	Fuzzy Logic Controller
GA :	Genetic Algorithm
HC-FLC :	Hill-Climbing Fuzzy Logic Controller
HC :	Hill-Climbing
InCond :	Incremental Conductance
LMPP :	Local Maximum Power Point
LMPPT:	Local Maximum Power Point Tracking
MPPT :	Maximum Power Point Tracking
PSCs :	Partial Shading Conditions
P&O :	Perturb and Observe
PV :	Photovoltaic
P–V :	Power-Voltage characteristic curve
PWM :	Pulse Width Modulation
SA :	Simulated Annealing

1 Introduction

Electricity extracted from green sources has received a great attention thanks to its sustainability and environmentally friendliness [1]. Photovoltaic (PV) systems have recently witnessed a big growth and concern as they are noiseless and almost no maintenance is required [2].

Considering Uniform Conditions (UCs), there exists on the Power–Voltage (P–V) characteristic curve of the PV Generator (PVG) only one maximum power point [3] Fig. 1a.

As confirmed by many studies [4], the non–uniformity of irradiance can reduce significantly the output power of a PV generator . Hence, bypass diodes should be connected in parallel within PV modules to avoid thermal destruction [5]. However, inserting bypass diodes will result in multimodal P–V curve that exhibits a single Global MPP (GMPP) and one or many Local MPPs (LMPPs), as shown in Fig. 1b.

Various solutions have been adopted in the literature, which can be classified mainly into two groups, hardware and software solutions. Hardware solutions comprise (1) Array connections [4–6], (2) PV system architectures [7] and (3) circuit topologies [8, 9]. Software solutions are related to MPPT methods that properly ensure tracking the real MPP and hence improving the overall efficiency of PV systems [10, 11]. This can be achieved by associating a MPPT controller to a DC–DC converter in order to adjust the duty cycle to much the load. The choice of one of the existing converters (type: Buck, Boost or Buck–Boost) depends on the voltage required by the load and the MPPT technique to be used with it. The control parameter provided by a MPPT controller is the duty cycle, it is used as an input of the Pulse

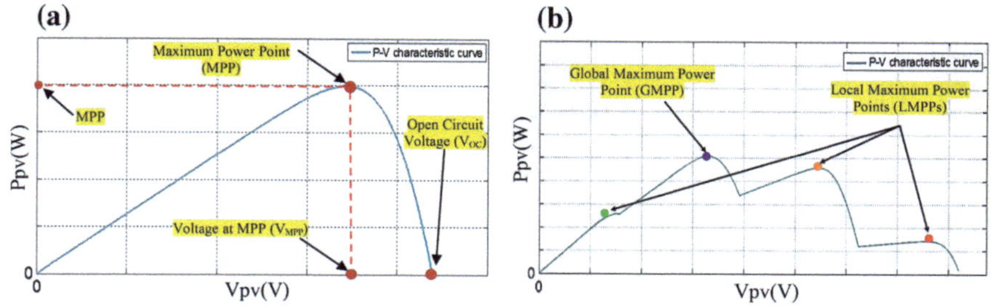

Fig. 1 P–V characteristics curve of a PVG **a** under UICs **b** under PSCs

Width Modulation (PWM) generator. The latter generates a PWM signal to control the switch of the DC–DC converter for maximum power extraction.

Conventional MPPTs such as Perturb and Observe (P&O), Incremental Conductance (InCond) and Hill Climbing (HC) are fundamentally based on the fact that the derivative of power with respect to either voltage or duty cycle is null at the MPP of PVG.

The new trend of research works is the integration of Artificial Intelligence (AI) techniques for modelling, optimization, prediction and control of renewable systems to harvest their maximum energy [12], especially, in the field of PV energy applications [13]. The fast convergence time and the high tracking accuracy are the principal advantages of MPPT controllers based on AI techniques. Conventional MPPTs and FL-MPPTs are both belong to LMPPTs. In this context, Many researchers have addressed the problem of Partial Shading by developing GMPPT techniques such as: soft computing and hybrid MPPTs. Some FL-MPPTs have been hybridized with other methods to escape LMPPs and hence properly identified the global peak under partial shading conditions (PSCs).

In this chapter, an attempt is made to present some local and global MPPTs based on improved versions of FL for enhancing the PV system performances. Clarifying their effectiveness in tracking the right MPP was the main focuses of this work. It is hoped that this chapter will help PV professionals and system designers.

2 LMPPTs Based on Improved FL

Fuzzy Logic has no need for an accurate mathematical model. Moreover, it can provide a high performance in controlling non–linear systems of arbitrary complexity by integrating the human experience into the control design process [13] see Fig. 2.

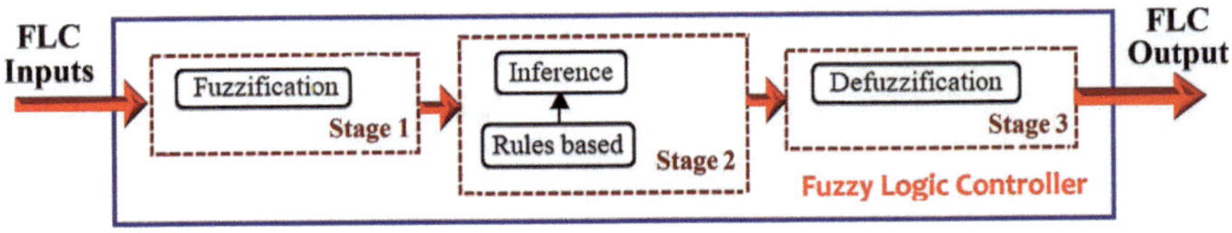

Fig. 2 Block diagram of FLC algorithm

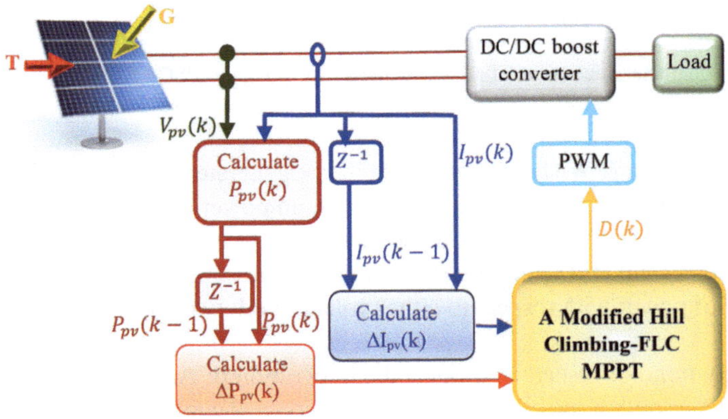

Fig. 3 A stand–alone PV system controlled by modified HC–FLC MPPT

2.1 *Modified HC–FLC [14]*

Modified HC–FLC MPPT shown in the Block diagram of Fig. 3 has been designed by Alajmi et al. [14] for standalone PV systems to address the main limitations of conventional HC.

The inputs of Modified HC–FLC are:

$$\Delta P_{pv}(n) = P_{pv}(n) - P_{pv}(n-1) \tag{1}$$

$$\Delta I_{pv}(n) = I_{pv}(n) - I_{pv}(n-1) \tag{2}$$

Equation 3 represents the FLC output which is the boost DC–DC converter

$$\Delta D(n) = D(n) - D(n-1) \tag{3}$$

The inputs (ΔP_{pv}, ΔI_{pv}) and the output (ΔD) are divided into 4 fuzzy logic subsets, which results in 16 fuzzy rules Fig. 4. The fuzzy logic inference employs Mamdani's Min–Max inference system which is shown in Table 1.

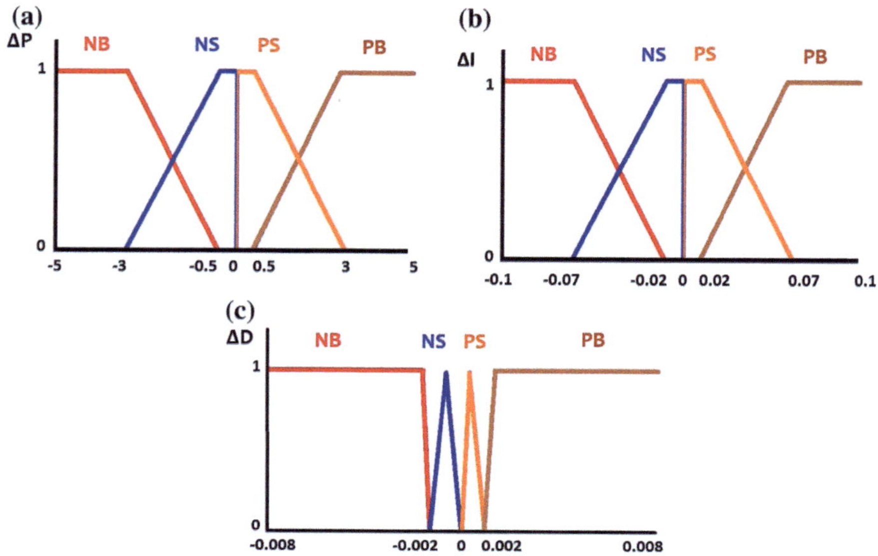

Fig. 4 Membership functions: inputs **a** ΔP_{pv}, **b** ΔI_{pv} and **c** the output ΔD [14]

Table 1 Fuzzy rules of modified HC–FLC [14]

$\Delta P_{pv} \Delta I_{pv}$	NB	NS	PS	PB
NB	PB	PB	NB	NB
NS	PS	PS	NS	NS
PS	NS	NS	PS	PS
PB	NB	NB	PB	PB

2.2 *Adaptive P&O–FLC [15]*

Zainuri et al. [15] designed an Adaptive P&O–FLC with 25 fuzzy rules that can operate only with a boost converter, to eliminate oscillations around the MPP and increase the PV system efficiency. As cab be seen from Fig. 5, the FLC inputs are the differential power and differential voltage, Eqs. 4 and 5, respectively;

$$\Delta P_{pv}(n) = P_{pv}(n) - P_{pv}(n-1) \tag{4}$$

$$\Delta V_{pv}(n) = V_{pv}(n) - V_{pv}(n-1) \tag{5}$$

$$\Delta D(n) = D(n) - D(n-1) \tag{6}$$

Figure 6 shows the membership functions (Table 2).

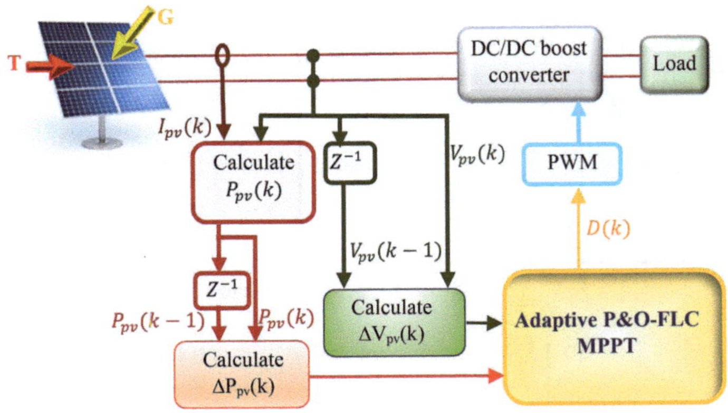

Fig. 5 Diagram of a SAPV controlled by Adaptive P&O–FLC MPPT

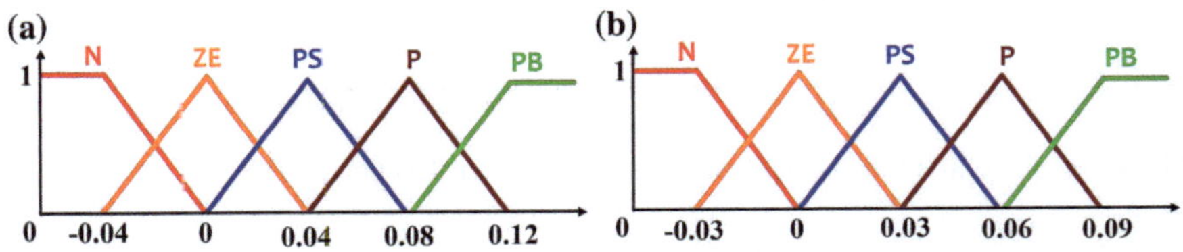

Fig. 6 MFs: **a** inputs ΔP_{pv}, and ΔV_{pv} **b** the output ΔD [15]

Table 2 Fuzzy rules of adaptive P&O–FLC [15]

$\Delta P_{\mathrm{pv}} \Delta V_{pv}$	N	ZE	PS	P	PB
N	ZE	PS	P	PB	PB
ZE	ZE	ZE	PS	P	PB
PS	N	ZE	ZE	PS	P
P	N	N	ZE	ZE	PS
PB	N	N	N	ZE	ZE

2.3 *Simulation Results and Comparison*

To evaluate the MPPT algorithms, one PV module type *KYOCERA KD210* of 210 W maximum power is used. Table 3 reports the electrical specifications.

Figure 7a shows waveforms of the extracted power by the three investigated MPPTs at $G = 1000\,\mathrm{W/m^2}$, and $T = 25\,^{o}C$.

Figure 7b shows waveforms of the extracted power by the three investigated MPPTs when the PVG undertakes variation of irradiance.

According to Fig. 7 and Table 4, it can be noticed that Modified HC–FLC performs better control than conventional P&O. The FL used by the Modified HC–FLC contributes in the complexity reduction (16 rules) compared to that of Adaptive

Table 3 Electrical characteristics of *KYOCERA KD210*

Designation	KYOCERA KD210
(P_{max})	210 W
(V_{MPP})	26.6 V
(I_{MPP})	7.9 A
(V_{oc})	33.2 V
(I_{sc})	8.58 A
NC	54 cells
Technology of cells	Multi–Crystalline silicon

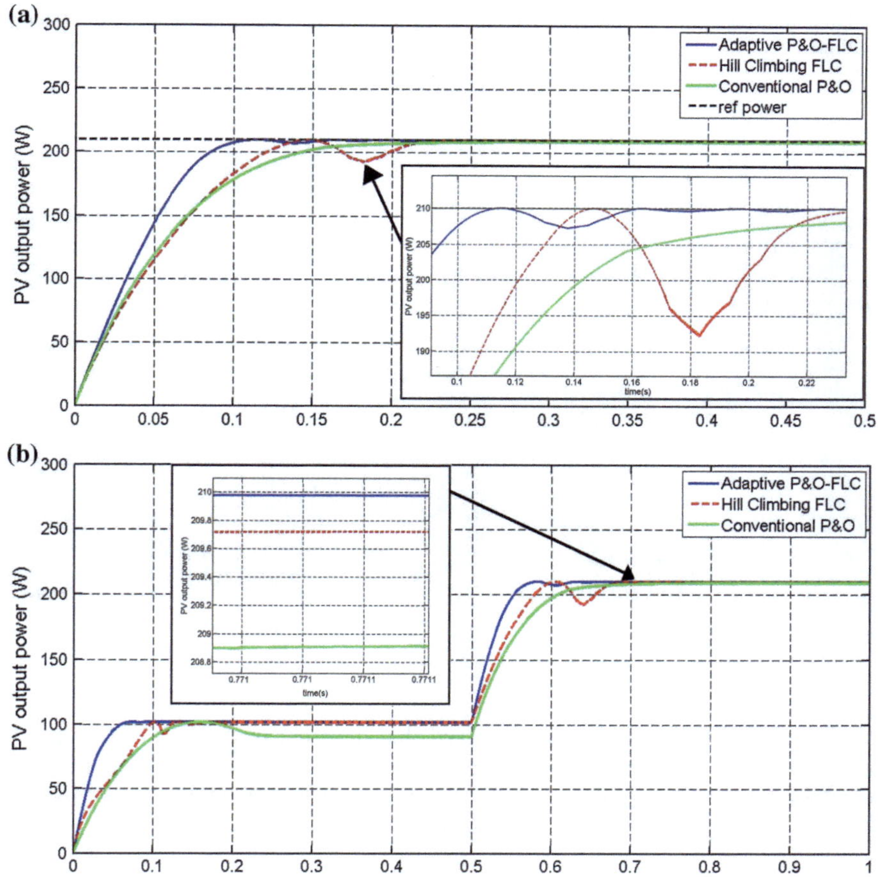

Fig. 7 The PV output power of the investigated MPPTs **a** at STC **b** under fast variations of irradiation

Table 4 Comparison of the modified HC–FLC, adaptive P&O–FLC and Conventional P&O

Evaluated indices	Conventional P&O	Modified HC–FLC [14]	Adaptive P&O–FLC [15]
Response time	Slow	Medium	Fast
Static error	1.1 W	0.3W	Negligible
Tracking technique	P&O	FLC–16 rules	FLC–25 rules
Tracking efficiency	99.47%	99.85%	99.99%
Complexity level	Low	Low to medium	Medium
Control variable	Voltage	Duty cycle	Duty cycle
Dependence on the PVG parameters	No	No	No
Initial parameter settings	1 parameter (The perturbation step size ΔVref)	3 parameters (3 scaling factors of the FLC inputs and output)	3 parameters (3 scaling factors of the FLC inputs and output)

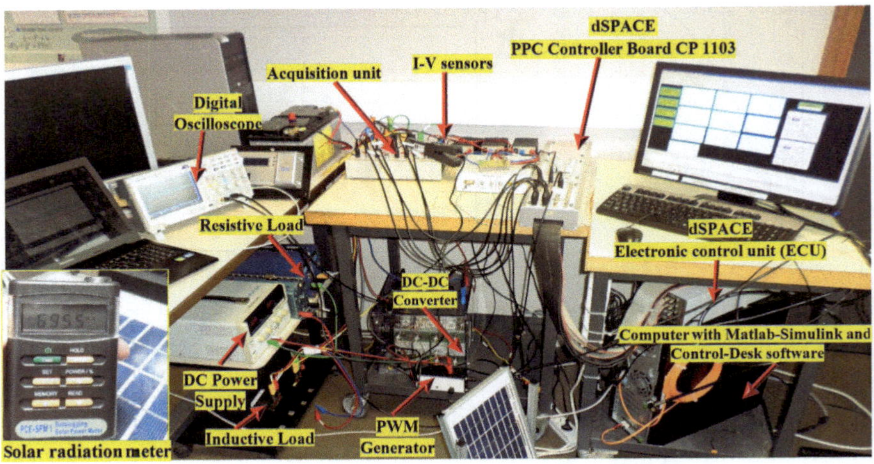

Fig. 8 Test facilities

P&O–FLC (25 rules). It can be observed that the improved P&O–FLC better to other MPPTs in terms of tracking efficiency and convergence time . It can be seen that three initial parameters need to be tuned by Adaptive P&O–FLC and Modified HC–FLC, whereas P&O needs the tuning of a single parameter.

2.4 Experimental Results and Comparison

Figure 8 shows the used test facility. A PV module of 10W receives light from a lamp as shown in Fig. 9. Its electrical specifications at STC are reported in Table 5.

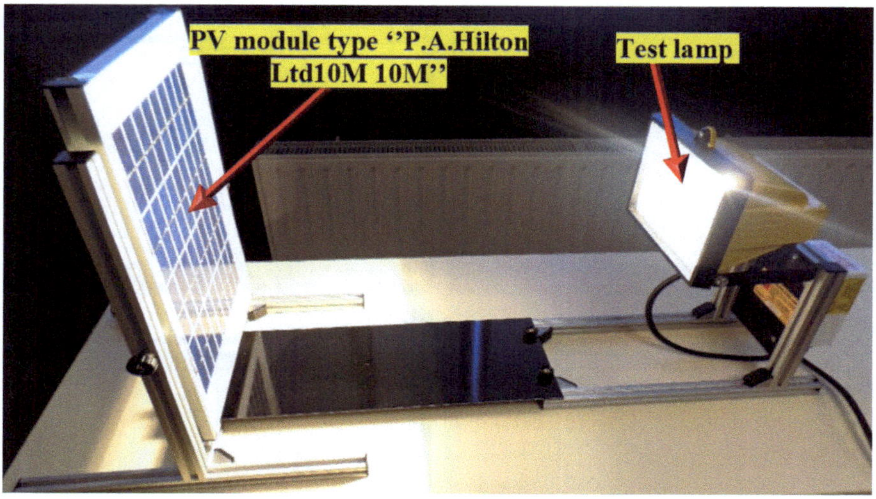

Fig. 9 The investigated PV module

Table 5 Electrical characteristics of *P.A.Hilton Ltd 10M*

Designation	P.A.Hilton Ltd 10 M
(P_{max})	10 W
(V_{MPP})	17.9 V
(I_{MPP})	0.56 A
(V_{oc})	22.1 V
(I_{sc})	0.58 A
NC	36 cells

As shown in Fig. 9, the investigated PV module is facing towards the lamp to get the maximum of light ($G = 858\,\mathrm{W/m^2}$, getting from solar irradiance meter measurement). The output Ipv, Vpv and Ppv are swept and the measured waveforms are shown in Fig. 10a. With reference to the last figure, the exact MPP is around 8.4W.

Measured Ipv, Vpv and Pv extracted by P&O, modified HC–FLC and adaptive P&O–FLC are illustrated in Fig. 10b, c and d, respectively. form the above figure, it can be observed that the investigated MPPTs achieve an operating point around the MPP (8.4 W). The improved P&O–FLC and Modified HC–FLC present negligible power ripples in steady–state condition. Therefore, the tracking efficiency is high. The power extracted by P&O is so affected by ripples, which in turn affects the accuracy in tracking the MPP.

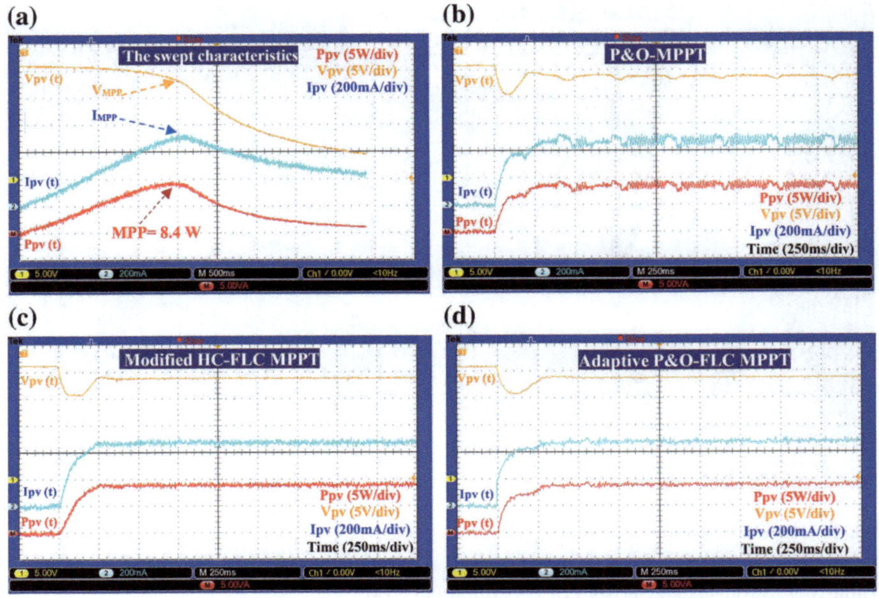

Fig. 10 Experimental waveforms of I_{pv} (cyan), V_{pv} (orange) and P_{pv} (red), **a** the swept characteristics **b** P&O **c** Modified HC–FLC MPPT **d** Adaptive P&O–FLC MPPT, under $G = 858\,\text{W/m}^2$

2.5 Other LMPPTs Based on Improved Versions of FL

In recent years, many MPPTs based on FL have been proposed to improve the PV system performance. Table 6 shows a summary of FL related research for MPPT applications.

3 GMPPTs Based on FL

A PVG may be constituted some PV modules connected together to form an array. when the PVG receives a uniform irradiance, the resulting P–V curve exhibits one MPP. However, the case in which different parts of it are partially shaded, so the P–V curve exhibits some power peaks. It is very likely that conventional MPPTs will be trapped on LMPP, resulting in extra power losses.

Due to the advantages of FL in the field of control, it has gained attention from researchers to be employed in the study of GMPPTs for partially shaded PVGs.

Table 6 Summary of MPPT techniques based on FLC

Reference	Inputs	Control variable	Number of rules	Power conditioning type	Remarks
Simoes et al. [16]	Two	$\Delta D(k)$ (the duty cycle change)	15 fuzzy rules	Boost converter	The convergence time has been improved. However, the system performs poorly in steady–state conditions
Won et al. [17]	Two	$\Delta D(k)$ (the duty cycle change)	25 fuzzy rules	Boost converter	The main limitations of this FLC are: poor performance in steady–state condition and divergence
Masoum et al. [18]	Three	$\Delta D(k)$ (the duty cycle change)	75 fuzzy rules	Buck converter	This FLC is designed by combining the FLC presented in [16] to that of [17]. The main limitation of this FLC lies in the huge number of rules
Kottas et al. [19]	Two	$\Delta D(k)$ (the duty cycle change)	25 fuzzy rules	Boost converter	This FLC aims to deal with the poor dynamic performance resulted in [17]. To this end, fuzzy cognitive network has been designed in order to be combined with the FLC of [17]. numerical simulations reveal that the convergence time is enhanced compared to [17]. However, This solution is not cost effective since its needs an additional current sensor and switch to be parallel connected with the PV system
Wu et al. [20]	Two	$\Delta D(k)$ (the duty cycle change)	25 fuzzy rules	Inverter	This FLC is designed to control inverters in both modes, standalone or grid–connection. A self–tuning algorithm has been proposed with this FLC to automatically tune the scaling factors of inputs and output. Based on experimental findings
Larbes et al. [21]	Two	D(k) (the instantaneous duty cycle)	25 fuzzy rules optimized by GA	Boost converter	By testing this FLC under varying weather conditions, GA–FLC outperforms P&O and conventional FLC in terms of convergence time and tracking efficiency

(continued)

Table 6 (continued)

Reference	Inputs	Control variable	Number of rules	Power conditioning type	Remarks
Messai et al. [22]	Two	D(k) (the instantaneous duty cycle)	25 fuzzy rules optimized by GA	Boost converter	GA has been used to optimize the membership functions and fuzzy rules of the FLC proposed in [23]. Simulation results reveal that both the dynamic response and the steady state performance of GA–FLC have been considerably enhanced compared to the original FLC [23] as well as conventional P&O
Al Nabulsi et al. [24]	Two	ΔVref (k) (the perturbation step size of P&O)	9 fuzzy rules	Buck converter	9 fuzzy rules FLC has been incorporated with P&O to get rid of oscillations around the MPP. Simulation findings demonstrate that this MPPT operates effectively. However, the oscillating behavior is not completely eliminated. Moreover, a PI controller is needed to provide the control signal to the Buck converter
Rajesh et al. [25]	Three	The reference current (I_{ref})	Unknown	Boost converter	This FLC intended to harvest maximum energy from PVGs. According to experimental results, it can be noticed that low fluctuations and fast transient response, are the main advantages of this FLC
Chao et al. [26]	Two	ΔD(k) (the duty cycle change)	25 fuzzy rules	Two–stage DC–DC converter	Experimental findings indicate that this FLC has a slow convergence to the MPP. Moreover, The tracking efficiency obtained by this FLC (90% to 96%) is much lower compared to that obtained by [24] (around 99%)

(continued)

Table 6 (continued)

Reference	Inputs	Control variable	Number of rules	Power conditioning type	Remarks
El Khateb et al. [27]	Four	D(n) (the instantaneous duty cycle)	50 fuzzy rules	SEPIC converter	This MPPT relies on two FLCs running simultaneously to control SEPI–Converter. However, sophisticated hardware equipment are necessary for implementing and processing two FLCs simultaneously
Guenounou et al. [28]	Two	ΔD(n) (the duty cycle change)	25 fuzzy rules	Boost converter	Test results reveal that this FLC operates very effectively compared to conventional FLC
Chen et al. [29]	Three	ΔD(n) (the duty cycle change)	77 fuzzy rules	Boost converter	This MPPT has been designed by combining conventional FLC and Auto–Scaling Variable Step–Size (ASVSS) algorithm. Simulation findings show that a high tracking accuracy is obtained under rapidly changing irradiance conditions. However, the high complexity level and the difficulty of implementation are the main drawbacks of this MPPT

3.1 GMPPT Proposed by Alajmi et al. [30]

Scanning, storing, perturbing and observing are the main procedures used by Alajmi et al. [30] MPPT to track the GMPP of partially shaded PVGs. Figure 11 shows the flowchart of the above method.

The inputs to the FLC are the power change ΔP_{pv}, the current change ΔI_{pv} and ΔP_m which represents the difference between the stored GMPP (P_M) and P_{pv},

$$\Delta P_{pv}(n) = P_{pv}(n) - P_{pv}(n-1) \tag{7}$$

$$\Delta I_{pv}(n) = I_{pv}(n) - I_{pv}(n-1) \tag{8}$$

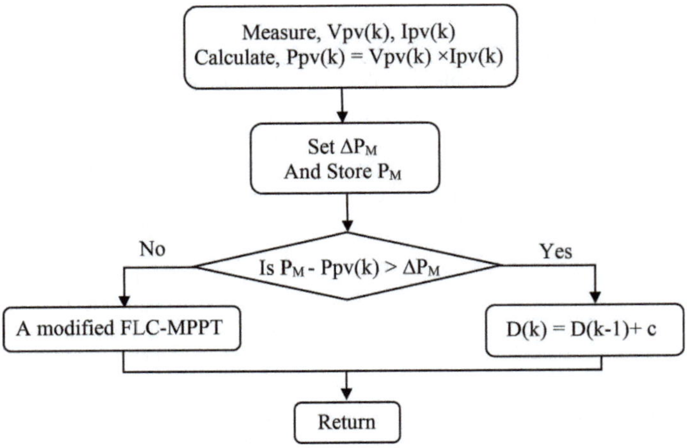

Fig. 11 Flowchart of the GMPPT proposed by Alajmi et al. [30]

Table 7 FLC rules proposed by Alajmi et al. [30]

$\Delta P_{pv}\Delta I_{pv}$	NB	NS	PS	PB	ΔPm
NB	PM	PM	NM	NM	
NS	PS	PS	NS	NS	PS
PS	NS	NS	PS	PS	
PB	NM	NM	PM	PM	
NB	PB	PB	PB	PB	
NS	PB	PB	PB	PB	PB
PS	PB	PB	PB	PB	
PB	PB	PB	PB	PB	

$$\Delta P_m(n) = P_M(n) - P_{pv}(n) \tag{9}$$

The FLC output is the duty cycle change ΔD for controlling the boost converter

$$\Delta D(n) = D(n) - D(n-1) \tag{10}$$

Four fuzzy subsets (NB, NS, PS, PB) are chosen for the variable inputs ΔP_{pv} and ΔI_{pv}, as well two subsets (PS and PB) for the variable input ΔP_m. The control output ΔD is obtained by processing six fuzzy subsets (NB, Negative Medium (NM), NS, Positive Medium (PM), PS, PB) Fig. 12.

As shown in Table 7, this FLC tracks the identified the GMPP by performing 32 FL rules.

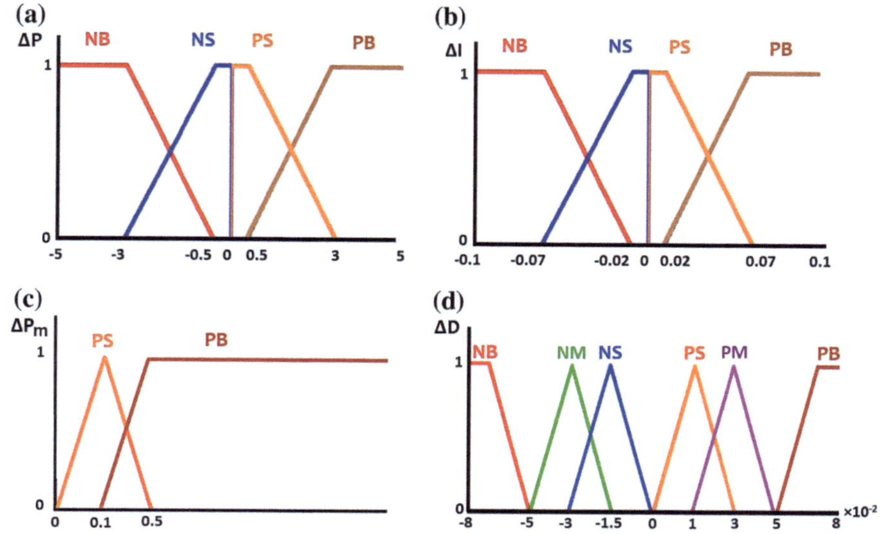

Fig. 12 Membership functions: inputs **a** $\varDelta P_{\mathrm{pv}}$, **b** $\varDelta I_{\mathrm{pv}}$, **c** $\varDelta P_{\mathrm{m}}$, and **d** the output $\varDelta$D [30]

3.2 GMPPT Proposed by Boukenoui et al. [31]

The GMPPT proposed by Boukenoui et al. [31] has been developed to identify and achieve the GMPP of PV systems operating under uniform and non–uniform irradiance conditions. It combines a scanning–storing loop with a tracking loop based on FLC–MPPT. The block diagram of a standalone PV system controlled by Boukenoui et al. [31] MPPT is depicted in Fig. 13.

The flowchart of the proposed GMPPP is given in Fig. 14. The pseudo algorithm reported below is applied to find the GMPP and D_{GMPP}. Subsequently, a tracking loop comprising FLC–MPPT takes over for accurately achieving the GMPP, regardless of irradiance variations, characteristics of the PVG and the shading pattern applied on it.

To identify the GMPP: the stored $P_{max}(n-1)$ should be compared to the instantaneous measured power $P_{PV}(n)$ and the GMPP is defined as the $P_{max}(n)$

$P_{max}(n) = \max\left(P_{PV}(n), P_{max}(n-1)\right)$

$\mathrm{GMPP} = \max(P_{max}(n))$

To identify the D_{GMPP}: $\varDelta P_{PV}(n)$ and the variation between the $P_{max}(n)$ and $P_{PV}(n)$ should be calculated,

$\varDelta P_{pv}(n) = P_{PV}(n) - P_{PV}(n-1).$

$\varDelta P_{max}(n) = P_{max}(n) - P_{PV}(n).$

The following conditions should be verified, knowing that D(n) is the instan-

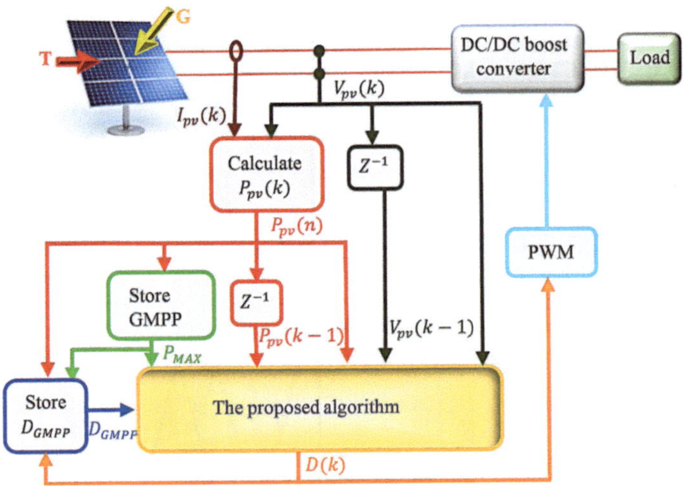

Fig. 13 Block diagram of a standalone PV system controlled by Boukenoui et al. [31] GMPPT

Table 8 Fuzzy rules proposed by Boukenoui et al. [31]

EΔE	NB	NS	ZE	PS	PB
NB	PB	PB	PS	PB	PB
NS	PB	PS	PS	PS	PB
ZE	NS	NS	ZE	PS	PS
PS	NB	NS	NS	NS	NB
PB	NB	NB	NS	NB	NB

taneous duty cycle.
If $\Delta P_{max}(n) \leq 0$ and $\Delta P_{PV}(n) \geq 0$ ***Then*** $d_{GMPP}(n) = D(n)$
Else $d_{GMPP}(n) = d_{GMPP}(n-1)$
End $D_{GMPP} = d_{GMPP}(n)$

The FLC proposed by Boukenoui et al. [31] uses as inputs, Eqs. 11 and 12

$$E(n) = \frac{P_{pv}(n) - P_{pv}(n-1)}{V_{pv}(n) - V_{pv}(n-1)} \tag{11}$$

$$\Delta E(n) = E(n) - E(n-1) \tag{12}$$

The fuzzy inference employs Mamdani's Min–Max inference system and 25 fuzzy rules shown in Table 8. Moreover, a Defuzzifier based on the COG technique is employed to generate the control signal (duty cycle).

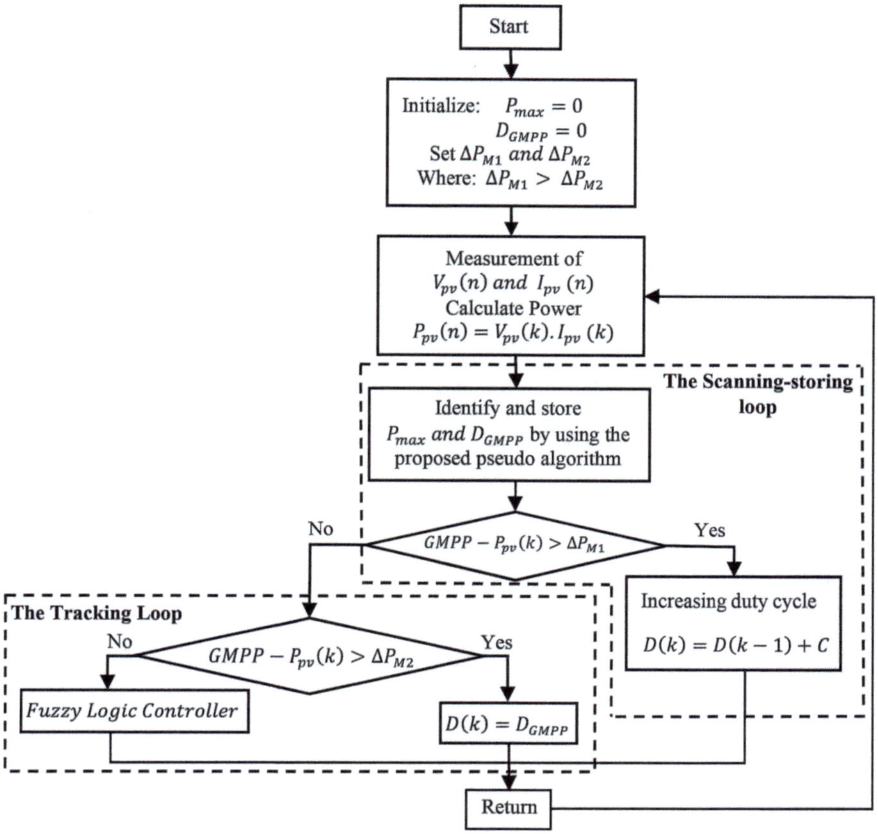

Fig. 14 Flowchart of the GMPPT proposed by Boukenoui et al. [31]

3.3 *Comparative Study*

To assess the robustness of Alajmi et al. [30] and Boukenoui et al. [31] in tracking the GMPP under PSCs, they have been compared to Ishaque et al. [32] and Ji et al. [33]. Each GMPPT has been associated to a stand–alone PV system to digitally simulate its performance using Matlab/Simulink software.

The GMPPT proposed by Ishaque et al. [32] used the PSO technique (it takes inspiration from flocking behavior of birds and fishes). In this work, PSO has been used as a direct duty cycle control MPPT in order to eliminate the PI control loop.

The GMPP proposed by Ji et al. [33] uses a LF that requires both Voc and Isc of the PVG to reveal the occurrence of PSCs as well as moving the operating point close to the GMPP. Then, a variable step size InCond is performed to achieve that GMPP.

The electrical characteristics of the employed PV module *Shell SP150–PC* are reported in Table 9. Each PV module has 72 PV cells connected in series, in addition

Table 9 Electrical characteristics of *Shell SP150–PC*

Designation	Shell SP150–PC
(P_{max})	150 W
(V_{MPP})	34 V
(I_{MPP})	4.40 A
(V_{oc})	43.40 V
(I_{sc})	4.80 A
NC	72 cells

Table 10 Performance comparison of the four investigated GMPPTs under PSCs

Evaluated indices	Alajmi et al. [30]	Ji et al. [33]	Ishaque et al. [32]	Boukenoui et al. [31]
Time	Around 113 ms	Around 106 ms	Around 138 ms	Around 94 ms
Error	2.3 W	10.6 W	4 W	1 W
Efficiency	99,22%	96,41%	98,64%	99,66%
Tracking technique	FLC–32 rules	InCond	PSO	FLC–25 rules
Algorithm's complexity	Medium to high	Low	High	High
Probability of successfully finding the GMPP for 100 shading patterns	High 91/100	Low 69/100	Medium to High 83/100	High 100/100
Sensors	(voltage, current)	(voltage, current)	(voltage, current)	(voltage, current)
Practical implementation	Yes	Yes	Yes	Yes
Dependence on the PVG parameters	No	Yes	No	No

to 3 bypass diodes. The PVG has been subjected to PSCs from the beginning. Two PV modules are fully illuminated (1 KW/m^2) and the other one receives an irradiance of 525 W/m^2.

Under the aforementioned shading pattern, the P–V curve exhibits two peaks with a global one of 295 W (see Fig. 15a). Under the same shading pattern, the four investigated GMPPT are tested, waveforms of the extracted power are depicted in Fig. 15b.

With reference to Fig. 15b and of Fig. 15a, it is clearly observed that all the investigated MPPTs have achieved the GMPP but with different performance indices, reported in Table 10.

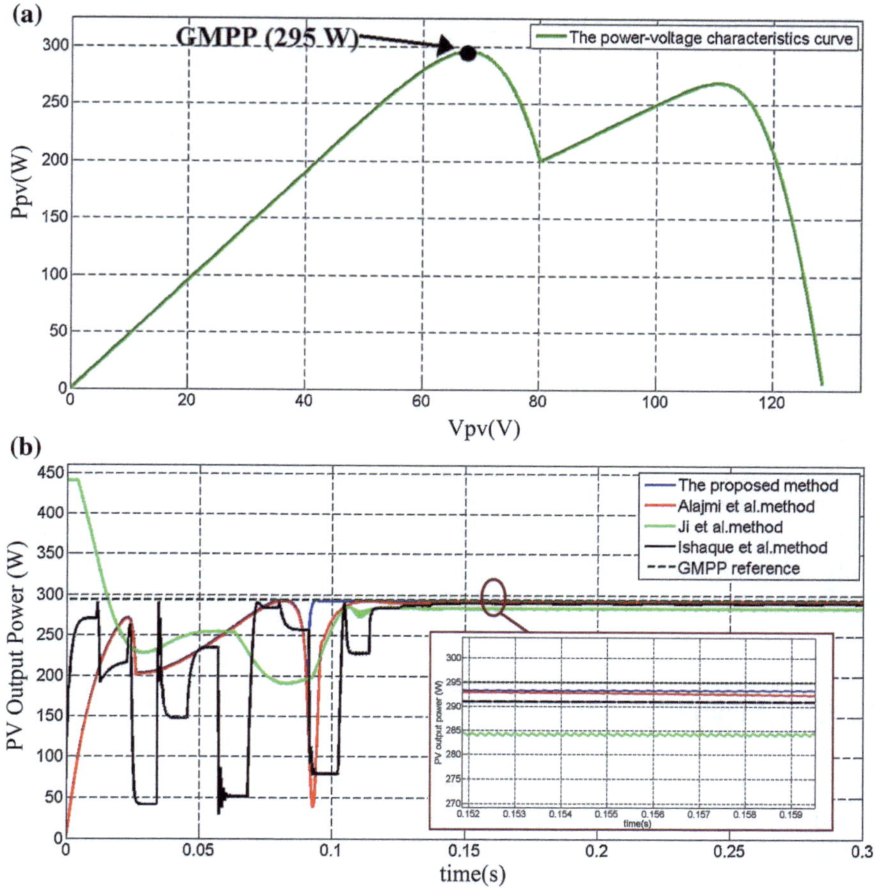

Fig. 15 **a** the P–V characteristic curve under PSCs (the shading pattern was 1, 1 KW/m^2 and 525 W/m^2) **b** waveforms of the power extracted by the investigated MPPTs under the aforementioned shading pattern

According to Table 10, it can be observed that Ishaque et al. [32] GMPPT takes a long time (138 ms) to converge to the GMPP. Moreover, it uses trial–error method to tune the four initial parameters of PSO (including the initial number of particles, population size, the inertia and learning factors). Therefore, determining their appropriate values is a demanding task.

Ji et al. [33] GMPPT has a low complexity. However, it exhibits the highest static error (10.6 W) and high power ripples when the GMPP is achieved. In addition to the aforementioned drawbacks, Ji et al. [33] GMPPT suffers from dependence on the PVG parameters which are Voc and Isc. Therefore, their values should to be readjusted when changing the PVG.

The FLC employed by Alajmi et al. [30] has 32 fuzzy rules and the one proposed by Boukenoui et al. [31] has only 25 rules. This leads to the high memory requirement during implementation. The FLC employed by Alajmi et al. [30] can fail to achieve the global power peak. Consequently, lower probability of successfully finding the GMPP (91/100) compared to the GMPPT proposed by Boukenoui et al. [31].

It should be noted that the choice of one of the existing MPPTs depends mainly on the application of the PVS. Therefore, there are some applications such as: powering household appliances or lighting for billboards, where the high accuracy of controller is not an important factor. Therefore, the PV system can operate by only a conventional MPPT. Other applications such as powering solar vehicles, satellites, industrial equipment, public–use facilities, lighting highway information signs, necessitate robust sensors to achieve a high accuracy of tracking and ensuring continuous supply of maximum electricity irrespective of the weather conditions.

Boukenoui et al. [31] appealing for PV energy conversion systems (grid connected or standalone PV systems applications) as a good candidate to be integrated in the existing PV applications. Firstly, to replace conventional MPPTs (P&O, InCond or conventional FLC, etc) where the high accuracy is necessary. Secondly, to effectively control PV systems installed in places where the non–uniformity of irradiance can be often found and difficult to be avoided, such as: building integrated PV systems.

4 Conclusions

Various improved FL–MPPTs applied either as LMPPT or GMPPT have been presented in this chapter. Those techniques show varying degrees of effectiveness. The number of Fuzzy rules varies from one MPPT to another (ranged from 9 to 77 rules), which leads to variations in the implementation complexity. Moreover, membership functions of some FLC–MPPT have been optimized by soft computing techniques (such as, GA) to achieve best results.

Adaptive P&O–FLC, Modified HC–FLC and P&O have been compared under variable insolation conditions. Simulation and experimental results demonstrated that the designed Adaptive P&O–FLC outperforms Modified HC–FLC and P&O. MPPTs based only on FLC are not able to track the GMPP under PSCs. To this end, researchers have developed novel mechanisms of GMPP identifying loop to be added to FLC–MPPTs, which enhance their effectiveness under PSCs.

Two GMPPTs based on FL for both UICs and PSCs have been presented, comparative study has been done and the obtained results reveal their relative merits and limitations in tracking the GMPP [31].

References

1. Bradai R, Boukenoui R, Kheldoun A, Salhi H, Ghanes M, Barbot J-P, Mellit A (2017) Experimental assessment of new fast MPPT algorithm for PV systems under non-uniform irradiance conditions. Appl Energy 199:416–429
2. Mekhilef S, Saidur R, Safari A (2011) A review on solar energy use in industries. Renew Sustain Energy Rev 15:1777–1790
3. Liu L, Meng X, Liu C (2016) A review of maximum power point tracking methods of PV power system at uniform and partial shading. Renew Sustain Energy Rev 53:1500–1507. https://doi.org/10.1016/j.rser.2015.09.065
4. Moballegh S, Jiang J (2014) Modeling, prediction, and experimental validations of power peaks of PV arrays under partial shading conditions. IEEE Trans Sustain Energy 5:293–300
5. Lijun G, Dougal RA, Liu S, Iotova AP (2009) Parallel-connected solar PV system to address partial and rapidly fluctuating shadow conditions. IEEE Trans Ind Electron 56:1548–1556. https://doi.org/10.1109/tie.2008.2011296
6. Boukenoui R, Bradai R, Salhi H, Mellit A (2015) Modeling and simulation of photovoltaic strings under partial shading conditions using Matlab/Simscape. In: 2015 International conference on clean electrical power (ICCEP). IEEE, pp 73–77
7. Gab-Su S, Jong-Won S, Bo-Hyung C, Kyu-Chan L (2014) Digitally controlled current sensorless photovoltaic micro-converter for DC distribution. IEEE Trans Ind Inform 10:117–126. https://doi.org/10.1109/tii.2013.2248015
8. Elserougi AA, Diab MS, Massoud AM, Abdel-Khalik AS, Ahmed S (2015) A switched PV approach for extracted maximum power enhancement of PV arrays during partial shading. IEEE Trans Sustain Energy 6:767–772
9. Bidram A, Davoudi A, Balog RS (2012) Control and circuit techniques to mitigate partial shading effects in photovoltaic arrays. IEEE J Photovolt 2:532–546
10. Kheldoun A, Bradai R, Boukenoui R, Mellit A (2016) A new Golden section method-based maximum power point tracking algorithm for photovoltaic systems. Energy Convers Manag 111:125–136. https://doi.org/10.1016/j.enconman.2015.12.039
11. Boukenoui R, Ghanes M, Barbot J-P, Bradai R, Mellit A, Salhi H (2017) Experimental assessment of maximum power point tracking methods for photovoltaic systems. Energy 132:324–340
12. Boukenoui R, Bradai R, Mellit A, Ghanes M, Salhi H (2015) Comparative analysis of P&O, modified hill climbing-FLC, and adaptive P&O-FLC MPPTs for microgrid standalone PV system. In: 2015 International conference on renewable energy research and applications (ICRERA). IEEE, pp 1095–1099
13. Mellit A, Kalogirou SA (2017) A survey on the application of artificial intelligence techniques for photovoltaic systems. McEvoy's handbook of photovoltaics, 3rd edn. Elsevier, Amsterdam, pp 735–761
14. Alajmi BN, Ahmed KH, Finney SJ, Williams BW (2011) Fuzzy-logic-control approach of a modified hill-climbing method for maximum power point in microgrid standalone photovoltaic system. IEEE Trans Power Electron 26:1022–1030
15. Zainuri MAAM, Radzi MAM, Soh AC, Rahim NA (2013) Development of adaptive perturb and observe-fuzzy control maximum power point tracking for photovoltaic boost DCDC converter. IET Renew Power Gener 8:183–194
16. Simoes MG, Franceschetti N (1999) Fuzzy optimisation-based control of a solar array system. IEEE Proc Electr Power Appl 146:552–558
17. Won C-Y, Kim D-H, Kim S-C, Kim W-S, Kim H-S (1994) A new maximum power point tracker of photovoltaic arrays using fuzzy controller. In: 25th Annual IEEE power electronics specialists conference, PESC'94 Record. IEEE, pp 396–403
18. Masoum M, Sarvi M (2002) Design, simulation and construction of a new fuzzy-based maximum power point tracker for photovoltaic applications. In: Australian universities power engineering conference, AUPEC. pp 1–6

19. Kottas TL, Boutalis YS, Karlis AD (2006) New maximum power point tracker for PV arrays using fuzzy controller in close cooperation with fuzzy cognitive networks. IEEE Trans Energy Convers 21:793–803
20. Wu T-F, Yang C-H, Chen Y-K, Liu Z-R (1999) Photovoltaic inverter systems with self-tuning fuzzy control based on an experimental planning method. In: Industry applications conference, 1999 thirty-fourth IAS annual meeting conference record of the 1999 IEEE. IEEE, pp 1887–1894
21. Larbes C, Cheikh SA, Obeidi T, Zerguerras A (2009) Genetic algorithms optimized fuzzy logic control for the maximum power point tracking in photovoltaic system. Renew Energy 34:2093–2100
22. Messai A, Mellit A, Massi Pavan A, Guessoum A, Mekki H (2011) FPGA-based implementation of a fuzzy controller (MPPT) for photovoltaic module. Energy Convers Manage 52:2695–2704. https://doi.org/10.1016/j.enconman.2011.01.021
23. Messai A, Mellit A, Guessoum A, Kalogirou SA (2011) Maximum power point tracking using a GA optimized fuzzy logic controller and its FPGA implementation. Sol Energy 85:265–277. https://doi.org/10.1016/j.solener.2010.12.004
24. Al Nabulsi A, Dhaouadi R (2012) Efficiency optimization of a DSP-based standalone PV system using fuzzy logic and dual-MPPT control. IEEE Trans Ind Inform 8:573–584
25. Rajesh R, Mabel MC (2016) Design and real time implementation of a novel rule compressed fuzzy logic method for the determination operating point in a photo voltaic system. Energy 116:140–153. https://doi.org/10.1016/j.energy.2016.09.068
26. Chao PC-P, Chen W-D, Chang C-K (2012) Maximum power tracking of a generic photovoltaic system via a fuzzy controller and a two-stage DCDC converter. Microsyst Technol 18:1267–1281
27. El Khateb A, Rahim NA, Selvaraj J, Uddin MN (2014) Fuzzy-logic-controller-based SEPIC converter for maximum power point tracking. IEEE Trans Ind Appl 50:2349–2358
28. Guenounou O, Dahhou B, Chabour F (2014) Adaptive fuzzy controller based MPPT for photovoltaic systems. Energy Convers Manag 78:843–850
29. Chen Y-T, Jhang Y-C, Liang R-H (2016) A fuzzy-logic based auto-scaling variable step-size MPPT method for PV systems. Sol Energy 126:53–63. https://doi.org/10.1016/j.solener.2016.01.007
30. Alajmi BN, Ahmed KH, Finney SJ, Williams BW (2013) A maximum power point tracking technique for partially shaded photovoltaic systems in microgrids. IEEE Trans Ind Electron 60:1596–1606
31. Boukenoui R, Salhi H, Bradai R, Mellit A (2016) A new intelligent MPPT method for standalone photovoltaic systems operating under fast transient variations of shading patterns. Sol Energy 124:124–142. https://doi.org/10.1016/j.solener.2015.11.023
32. Ishaque K, Salam Z, Shamsudin A, Amjad M (2012) A direct control based maximum power point tracking method for photovoltaic system under partial shading conditions using particle swarm optimization algorithm. Appl Energy 99:414–422. https://doi.org/10.1016/j.apenergy.2012.05.026
33. Ji Y-H, Jung D-Y, Kim J-G, Kim J-H, Lee T-W, Won C-Y (2011) A real maximum power point tracking method for mismatching compensation in PV array under partially shaded conditions. IEEE Trans Power Electron 26:1001–1009

A New Method for Generating Short-Term Power Forecasting Based on Artificial Neural Networks and Optimization Methods for Solar Photovoltaic Power Plants

Tugce Demirdelen, Inayet Ozge Aksu, Burak Esenboga, Kemal Aygul, Firat Ekinci and Mehmet Bilgili

Abstract In recent times, solar PV power plants have been used worldwide due to their high solar energy potential. Although the PV power plants are highly preferred, the main disadvantage of the system is that the output power characteristics of the system are unstable. As PV power plant system is connected to the grid side, unbalanced power flow effects all systems controls. In addition, the load capacitys is not exactly known. For this reason, it has become an important issue to be known correctly in PV output power and their time-dependent changes. The main aim of this work is to eliminate power plant instability due to the output power imbalance. For the short-term, power prediction is estimated by real-time data of 1 MW PV power plant in use. Estimation power data are compared with real-time data and precision of the proposed method is demonstrated. In the first phase, traditional artificial intelligence algorithms are used. Then, these algorithms are trained with swarm based optimization methods and the performance analyses are presented in detail. Among all the algorithms used, the algorithm with the lowest error is determined. Thus, this study provides useful information and techniques to help researchers who are interested in planning and modeling PV power plants.

T. Demirdelen (✉) · B. Esenboga
Electrical and Electronics Engineering Department, Adana Science and Technology University, Adana, Turkey
e-mail: tdemirdelen@adanabtu.edu.tr

I. Ozge Aksu
Computer Engineering Department, Adana Science and Technology University, Adana, Turkey

K. Aygul
Electrical and Electronics Engineering Department, Cukurova University, Adana, Turkey

F. Ekinci
Energy Systems Engineering Department, Adana Science and Technology University, Adana, Turkey

M. Bilgili
Mechanical Engineering Department, Cukurova University, Adana, Turkey

R.-E. Precup et al. (eds.), *Solar Photovoltaic Power Plants*, Power Systems,
https://doi.org/10.1007/978-981-13-6151-7_8

Keywords Neural networks · Optimization methods · Short-term · Power prediction · Photovoltaic plants · Firefly algorithm

1 Introduction

Energy is a crucial factor for a countrys economy and the ability of all living beings to survive. It is a primary source of sustainable economic growth and essential for the development of a country. Recently, energy demand has been increasing with industrialization and technological developments, rapid population growth, and competition among countries to reduce their external dependence. In order to meet energy demand, fossil fuels have been widely used all over the world for decades. However, they have lost their popularity due to global warming and a critical depletion of fossil fuels. Renewable energy presents effective solutions to the above issues of energy demand. These sources can be presented as an important alternative energy because they are natural such as sun, wind, and wave energy that time will not change in the foreseeable future. Hence, the use of alternative energy sources, as well as existing energy sources, is of great importance for countries and people in order to obtain energy with maximum efficiency. Solar energy has the greatest importance among renewable energy resources in terms of its available energy potential in most parts of the world. It is one of the most effective and easy ways to produce clean, sustainable, cheap and safe energy. PV panels are commonly known as a device or method for generating electric power from solar energy by using solar cells that catch sunlight and turn it directly into electricity. PV panels provide clean, cheap energy and there are no harmful greenhouse gas emissions during the generation of the electricity. The increasing popularity of this source leads to new studies in the solar energy area. It is used together in grid-connected systems or stand-alone in the meeting energy demands of the building, industries, and agriculture. One of the main areas of study is how to provide increased reliability and regular output power in PV panels, efficiently. This is because solar PV panels require direct access to solar energy but clouds and some environmental factors, such as dust, humidity, snow, and rain, prevent the constant production of power and maximizing the output power of PV panels. A PV panel output power strongly depends on solar radiation and but this solar radiation is not constant over time. Solar cells are used to produce electrical energy, but while they absorb 80% of solar radiation, they convert only a small portion to electricity. The remaining portion causes overheating of the photovoltaic cells. This causes decreasing the electrical efficiency of the panel. Each one degree increase in temperature results in a reduction of 0.4–0.5% in electrical efficiency [1]. Therefore, there is a necessity for exact modeling, estimation of solar radiation and temperature to increase control and decrease the negative impacts of PV power plants. In order to develop more reliable algorithms, modeling, optimization and the forecasting of hourly solar radiance and ambient temperature are required. Power forecasting from PV power plant generated, based on its forecasting time scale, is classified into four groups in Fig. 1.

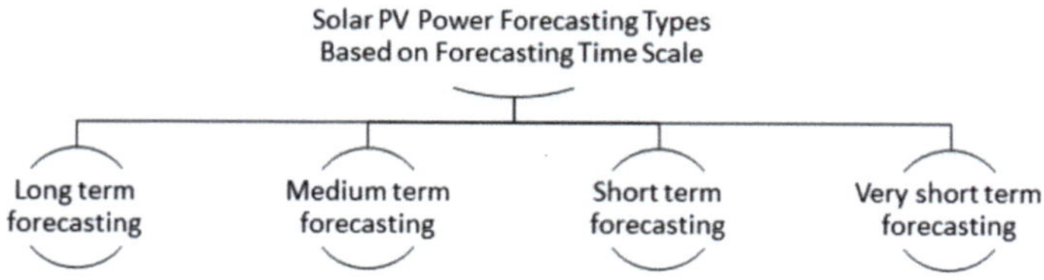

Fig. 1 Classification of solar PV generated power forecasting types based on forecasting time scale [2]

PV power forecasting is used to predict and balance energy generation and consumption for grid operators, thus it is notably important with respect to reduced penalties, grid stability, the reliability, the security of supply and lower maintenance. Forecasting time interval is classified into four groups.

Long-term forecasting ranges between one month and one year. In this type, it is generally used to operate operation security and energy planning.

Medium-term forecasting ranges from one week to one month. It regulates monthly maintenance of PV power by forecasting the availability of the generated electrical power.

Short-term forecasting timescales refer to one hour, one day or seven days. It is notably helpful for accurate planning of electrical transmission, distribution, and generation. Moreover, it has a critical issue in enhancing system stability, reliability and operation.

Very short-term forecasting timescales refer to a few seconds, one minute or several minutes (less than1hour). It uses small-scale data to obtain optimum forecasting, but it is not preferred because it is not necessary and is challenging for the estimation process in PV power plant. Accurate power forecasting from PV power plant prevents the effect of PV panel output power uncertainty and improves the system reliability. It also becomes more important and interesting because of the increasing penetration of PV power in many areas. Instantaneous fluctuating in temperature and solar radiance take place, mainly arising from the existence of clouds. Consequently, short-term, real solar parameters, such as solar radiation and temperature make it possible to obtain more accurate results so that stability, reliability and security of the PV system are increased by using real short-term data. In this study, the real-data of 1 MW PV power plant in Turkey is used for forecasting and the short-term generated power for PV power plant is estimated by Artificial Neural Network (ANN). The primary contribution of this paper is to provide useful information for academics and professionals who are interested in modeling and planning of PV power plants. Following a literature survey, Table 1 presents the forecasting and optimization methods used for solar parameters such as PV generated power, solar radiation, and temperature that benefit directly from solar energy. Different methods are used to forecast solar energy parameters over different time-scales. These methods are classified in Table 1

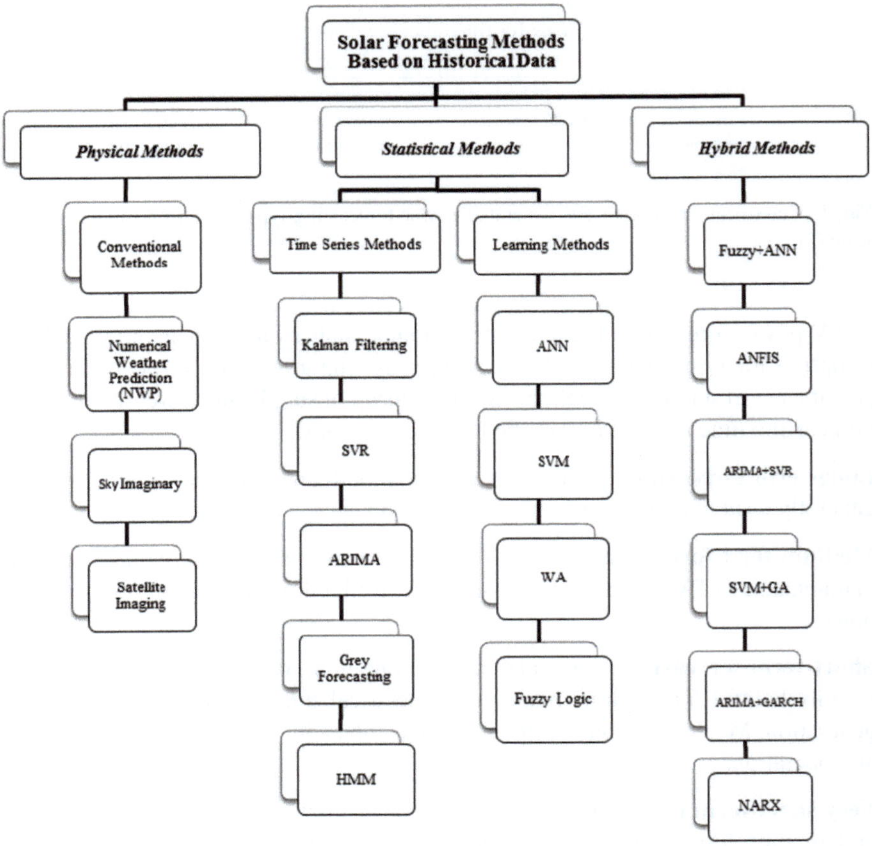

Fig. 2 Classification of solar forecasting methods and models based on historical data [54]

and discussed in detail. Finally, ANN, ANN-PSO and ANN-FA are selected for performance analysis does these methods are discussed in detail. Also, in particular, this method will present simulation results.

2 Solar Forecasting Methods

Researchers have work on solar power forecasting with different methods in order to obtain more accuracy. These methods shown in Fig. 2 will be presented in detail. Forecasting methods used to forecast PV parameters or meteorological data presented in Fig. 2 can be categorized into three groups based on historical data and meteorological variables.

Table 1 Various solar forecasting methods

References	Forecast horizon	Input variable	Forecast variable	Forecast method
[3]	24 h ahead	Solar radiance air temperature	Solar radiance	Artificial neural network multilayer perceptron architecture
[4]	72 h	Solar radiance cloud cover air temperature	Solar power	Artificial neural network
[5]	24 h ahead	Meteorological data	Solar radiance	Nonlinear autoregressive recurrent neural networks with exogenous inputs (NARX)
[6]	60 min ahead	Meteorological data	Solar radiance	Auto-regressive moving average model (ARMA) generalized auto regressive conditional heteroscedasticity (GARCH)
[7]	60 min ahead	Solar power air temperature solar radiation	Solar power	Artificial neural network back propagation algorithm
[8]	Intra-day and days ahead	Solar radiation	Solar PV power	Artificial neural network autoregressive integrated moving average
[9]	6 h ahead	Meteorological data	Solar power	Machine learning algorithm
[10]	A month ahead	Solar radiation solar PV power	Solar radiation	Auto-regressive models with exogenous inputs (ARX)
[11]	Hourly	Solar radiation Air temperature Sky condition (rainy, cloudy, sunny)	Solar radiation	Fuzzy and neural network methods
[12]	24 h ahead	Climatic conditions (humidity, temperature) Solar radiation Air temperature	Solar PV power	Artificial neural network back propagation algorithm

(continued)

Table 1 (continued)

References	Forecast horizon	Input variable	Forecast variable	Forecast method
[13]	6 h ahead	Meteorological data	Solar radiation	Wavelet neural network Levenberg–Marquardt (LM) training
[14]	Seasonal	Meteorological data (Air temperature, solar radiation)	Solar PV power	Forward neural network particle swarm optimization architecture (PSO)
[15]	One-hour measurement intervals for one year	Energy percentage of total Energy	Residential demand	FeedForward artificial neural network (FFANN)
[16]	Hourly,ranging from 07:00 to 16:00	Solar radiance ambient temperature	Solar radiation	Deep recurrent neural networks
[17]	Monthly	Meteorological data	Solar radiation	Artificial neural network back propagation algorithm
[18]	24 h ahead	Climatic condition	Solar radiation	Artificial neural network back propagation algorithm
[19]	A day ahead	Solar radiation air temperature humidity	Solar radiation	Artificial neural network
[20]	Hourly	Solar radiation weather data	Solar power	Artificial neural network Back propagation algorithm
[21]	30 min ahead	Weather condition air temperature solar radiance humidity	Solar power	Extreme learning machine
[22]	24 h ahead	Meteorological data	Solar PV power	Fuzzy logic system
[23]	30 min ahead	Meteorological data	Solar radiation	Fuzzy time series
[24]	24 h ahead	Solar radiation air temperature	Solar radiation	Artificial neural networks autoregressive models

(continued)

Table 1 (continued)

References	Forecast horizon	Input variable	Forecast variable	Forecast method
[25]	Intra-day and days ahead	Meteorological data (Air temperature, solar radiation, humidity)	Solar PV power	Stationary wavelet transform (SWT) artificial neural networks
[26]	One day ahead	Historical-past meteorological data	Solar radiation	Artificial neural networks
[27]	One day ahead	Meteorological data	Solar power	Deep learning artificial neural networks
[28]	24 h ahead	Clearness index	Solar radiation	Artificial neural networks support vector regression (SVR)
[29]	24 h ahead	Historical-past meteorological data	Solar power	Gaussian process regression (GPR)
[30]	Minutes ahead	Cloud movement ambient temperature Humidity wind speed	Solar radiation	Artificial neural networks fuzzy logic preprocessing
[31]	7 days ahead	Meteorological data	Solar radiation	Artificial neural networks back propagation algorithm
[32]	Hourly	Climatic conditions	Solar power	Extreme learning machine
[33]	24 h ahead	Air temperature cloud cover wind speed humidity	Solar power	Artificial neural networks
[34]	24 h ahead	Meteorological data	Solar radiation	Nonlinear autoregressive neural networks (NAR)
[35]	One day ahead	Meteorological data	Solar radiation	Hybrid neural fuzzy inference system (ANFIS)
[36]	One hour ahead	Clearness index Sky conditions	Solar PV power	Machine learning support vector regression
[37]	15 days ahead	Air temperature cloud cover wind speed	Solar PV power	Artificial neural networks
[38]	Hourly, ranging from 08:00 to 17:00 h	Meteorological data	Solar radiation	Support vector regression

(continued)

Table 1 (continued)

References	Forecast horizon	Input variable	Forecast variable	Forecast method
[39]	24 h ahead	Meteorological data	Solar radiation	Self-organization map (SOM)
[40]	Hourly	Solar radiation Air temperature	Solar PV power	Artificial neural networks multilayer perceptron architecture
[41]	10 min ahead	Zenith angle Solar radiation	Solar radiation	Artificial neural networks multilayer perceptron architecture
[42]	24 h ahead	Weather condition (humidity, temperature, radiance)	Solar PV power	Fuzzy theory radial basis neural network
[43]	One day ahead	Solar radiation Ambient temperature	Solar radiation	Artificial neural networks multilayer perceptron architecture
[44]	One hour ahead	Solar PV power ambient temperature	Solar PV power	Artificial neural networks nonlinear autoregressive neural network
[45]	(5 min step data)	Zenith angle Horizon angle solar radiation	Solar radiation	Artificial neural networks multilayer Perceptron Architecture
[46]	Hourly	Meteorological data	Solar radiation	Artificial neural networks
[47]	Last 50 h	Solar radiation Ambient temperature	Solar radiation	Neural network Levenberg–Marquardt (LM) training
[48]	Not specified	Solar radiation Ambient temperature	Solar PV power	Artificial neural networks ELM algorithm
[49]	Hourly	Cloud cover Solar radiation	Solar radiation	Satellite image analysis artificial neural networks
[50]	One day ahead	Meteorological data	Solar PV power	Artificial neural networks
[51]	Hourly	Solar radiation air temperature	Solar PV power	ARIMA SVR
[52]	Hourly	Weather conditions	Solar PV power	Artificial neural networks

2.1 Physical Methods

Numerical Weather Prediction (NWP) is a method of weather forecasting. It is the process of determining a future state by making mathematical solutions of the equations that express changes of the variables indicating the state of the atmosphere (temperature, wind, humidity, and pressure) [53].

Sky Imaginary Forecasting Method (SIFM) is used for the detection of clouds, estimating the behavior and attitude cloud. This method generates superior resolution image of the sky from horizon to horizon. The sky imaginary method is generally used as a short-term power estimation for the power output of the PV plant. Satellite imaging is a method relatively similar to the sky imaginary method [53].

Satellite Image Methods (SIM) allow cloud motion to be traced in order to forecast future cloud movement. This forecasting method presents an effective way to forecast very short-term radiance. However, it delivers less effective performance when clouds are rapidly forming or dissipating [53].

2.2 Statistical Methods

Statistical methods are time series analyses that deal with time series and historical data. Time series data is in a series of particular time periods or intervals. Statistical methods are a useful data-driven approach that is able to forecast the future behavior of a power plant. This section will give brief information on widely used methods [7].

Support Vector Machine (SVM) method is used as a classification, regression and anomaly detection. It is found in a statistical and mathematical theory to achieve accurate forecasting. It also deals with nonlinear problems and solves complex computational problems. This method is generally used in forecasting, regression analysis, and classification.

Wavelet Analysis (WA) method is a useful way to satisfy noise in real-time input datasets before the forecasting method is applied. Thus, it provides improvements to the reliability of estimation. It is relatively effective for the analyzing of frequency and time-dependent data because of its capability of eliminating non-periodic and transient signals.

Fuzzy logic is a data-driven algorithm method based on a human-like way of thinking. This method can be relatively useful compared to other methods if there are a large number of input variables. It is used for forecasting solar radiation and temperature, or as an optimal clustering process. It categorizes several sets of temperature, cloud conditions and meteorological data.

ANN method is one of the major tools used in machine learning. It is a brain-inspired system based on a learning/training method. An ANN consists of the input layer, hidden layer(s), output layers, neurons and connections. Every layer includes neurons as part of the network structure and each neuron links to another neuron located in the next layer. ANN is widely used in solving various classifications and forecasting problems because of non-linearity in meteorological data. It is particularly suitable when compared to other statistical methods when the data is non-linear.

ARIMA can be thought of as an improved model of traditional linear regression. It is a popular forecasting model that utilizes historical information to make predictions. This type of model is a basic forecasting technique that can be used as a foundation for more complex models.

Kalman Filtering is actually an estimator. It is a method of estimating the state of many different field systems. It mathematically estimates the states of linear systems (linear equation first order equations in computational equations). This method uses real-time statistical data and provides real-time forecasting of power generation.

Support Vector Regression (SVR) is quite different than SVM. Because the SVR method uses real-time outputs, it becomes very difficult to predict the information. The main idea is to minimize error and provide accurate forecasting.

Grey forecasting method is the others forecasting method to predict for the behavior of non-linear time series. This method is especially effective as the data is insufficient. Its prediction results may be inaccurate sometimes, so other statistical methods can be more useful than this method.

Hidden-Markov models (HMMs) are a common tool for modeling time series data. This model seeks to recover the sequence of states from the observed data. It has been used for forecasting applications in recent years because of its flexibility and computational efficiency.

2.3 Hybrid Methods

These methods are the combination of two or more forecasting methods such as Fuzzy-ANN, ANFIS, ARIMA and SVR, ARIMA and GARCH, NARX, and SVM and GA. The main aim is to improve forecasting accuracy and reduce the forecasting error. These hybrid methods can provide better forecasting performance compared with each forecasting method [55].

3 Methodology for Solar Power Forecasting

In this study, a Multi-Layer Feed Forward (MLFF) neural network structure, which is a type of Artificial Neural Network (ANN), was used. The MLFF network structure consists of the input layer, the hidden layer, and the output layer. Numbers of the neuron for the input and output layer depend on problem structure. The neuron number at the hidden layer is defined with a trial and error method. Network training is the process of identifying weight values for the nerve element connections in ANN. Initially, these values are determined randomly. Then, network parameters are updated in order to get the best yield from the network. In this study, the Firefly Algorithm (FA) and Particle Swarm Optimization (PSO) are applied to train the network coefficients.

3.1 Particle Swarm Optimization

Particle Swarm Optimization (PSO) algorithm is a swarm-based heuristic algorithm widely used all over the world. First studies on the PSO algorithm were done by Kennedy and Eberhart in 1995. This method is a simulation of the food search behaviors of flocks of birds and shoals of fish [56]. Each possible solution in the PSO algorithm is called a particle. The algorithm begins to work with randomly distributed particles in the solution space. In this algorithm, particles have speed and location. The distance of each particle to the solution (food) is expressed by a function determined by the current position and speed of the particle. After each iteration, the particles update their speeds for themselves and the swarm, and they move accordingly. When the best solution for ith particle up to that point is expressed byp_{ibest}, and the best solution of all particles up to that point is expressed byg_{best}, the expression used to update the speed values (v_i) of the particles at each iteration is:

$$v_i(t+1) = v_i(t) + c_1 \times r_1 \times (p_{ibest} - x_i(t)) + c_2 \times r_2 \times (g_{best} - x_i(t)) \quad (1)$$

where t is the iteration number, i is the index of the corresponding particle, x_i is the location of the particle r_1 and r_2 values are random numbers generated in the interval [0, 1], c_1 and c_2 are acceleration coefficients, which in general are chosen in the interval [0, 2].

Where the c_1 coefficient takes the particles to the local best, the c_2 coefficient takes the particles to the global best. The term w is the inertia weight and provides the balance between the local and global best. The term w is calculated as follows:

$$w_k = (w_{\min} - w_{\max}) \frac{(K-k)}{K} + w_{\max} \quad (2)$$

where $w_{\max}$ and $w_{\min}$ are the maximum and minimum values initially determined for the inertia value, k is the number of the iteration, and K is the maximum number of iterations. Each iteration step, the position update of the particles is done:

$$x_i(t+1) = x_i(t) + v_i(t+1) \quad (3)$$

The flowchart for the PSO algorithm is presented in Fig. 3. In addition, pseudo-code for PSO algorithm with ANN expressed as 5 steps are seen in Fig. 4.

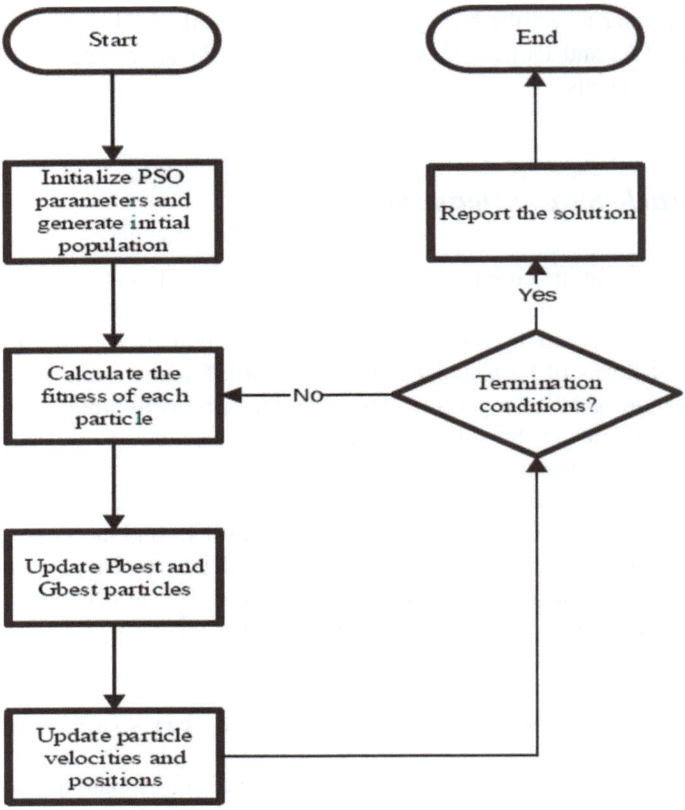

Fig. 3 Flowchart of the conventional PSO algorithm [57]

Fig. 4 Pseudo-code for the particle swarm optimization algorithm with ANN

Step 1: Initialize the neural network
Define the structure of the neural network
Step 2: Determine the parameters of the neural network
Initialize the bias and the weight coefficients
Step 3: Optimization step
Generate the initial population
Initialize the PSO parameters
Calculate the fitness value of each particle
While (reach terminate conditions ?)
Update Pbest and Gbest particles
Update particle velocities and positions
End while
Step 4: Validate the neural network
Step 5: Examine the testing data with neural network

3.2 *Firefly Algorithm*

Firefly Algorithm (FA) is based on the swarm method that is inspired by the natural life behavior of fireflies and their communication through brightness. Fireflies use their brightness to protect themselves from predators and to attract their prey. FA algorithm is selected to optimize ANN parameters for this study because of few parameters, it is easily adaptable to the problem being worked on and it does not have a complex structure. In the FA method, there are two important criteria: the change of light intensity and the attractiveness of the firefly (β). The attractiveness value of a firefly (β) is changed according to the distance to other fireflies. Let us assume that x_i and x_j values are the positions of the ith and jth fireflies. The distance between these two fireflies (r_{ij}) is calculated as follows:

$$r_{ij} = \left\| x_i - x_j \right\| \tag{4}$$

The attractiveness of the firefly is proportional to the light it emits and it decreases accordingly as the intensity of light decreases. The attractiveness of the firefly (β) is calculated as follows:

$$\beta(r) = \beta_0 e^{-\gamma r^2} \tag{5}$$

where the value β_0 is the maximum attractiveness parameter at $r = 0$ and γ is the light emitting coefficient. The ith firefly will move to the jth firefly, which is more attractive than itself. This position change is as follows:

$$x_i = x_i + \beta_0 e^{-\gamma r_{ij}^2} (x_j - x_i) + \alpha \left(rand - {}^1\!/_2 \right) \tag{6}$$

where the $rand$ value is a random real number in the interval $[0,\ 1]$ and α is a random selection parameter, respectively. This equation consists of three components: the first component represents the current position of the firefly, the second component shows the brightness of Firefly, and the third component represents the random movement of the Firefly when there is no brighter firefly around itself. The convergence speed of the algorithm and its ability to find the local/global best solution depends on the α, β and γ parameters used in the working steps of the algorithm. When the study investigated this, $\beta_0 = 1$ and $\alpha \in [0,\ 1]$ values were used. The γ parameter, which has a big impact on the working speed of the algorithm, has a value $[0.1,\ 10]$ for practical applications [58]. The flowchart is presented in Fig. 5. In addition, pseudo-code for FA algorithm with ANN expressed as 5 steps are shown in Fig. 6.

4 Data Representation and Pre-processing

The behavior of solar radiation can be easily interpreted for daily and seasonal, where in the range of [0, 1000] indicates stronger radiation. In winter, dawn to dusk period

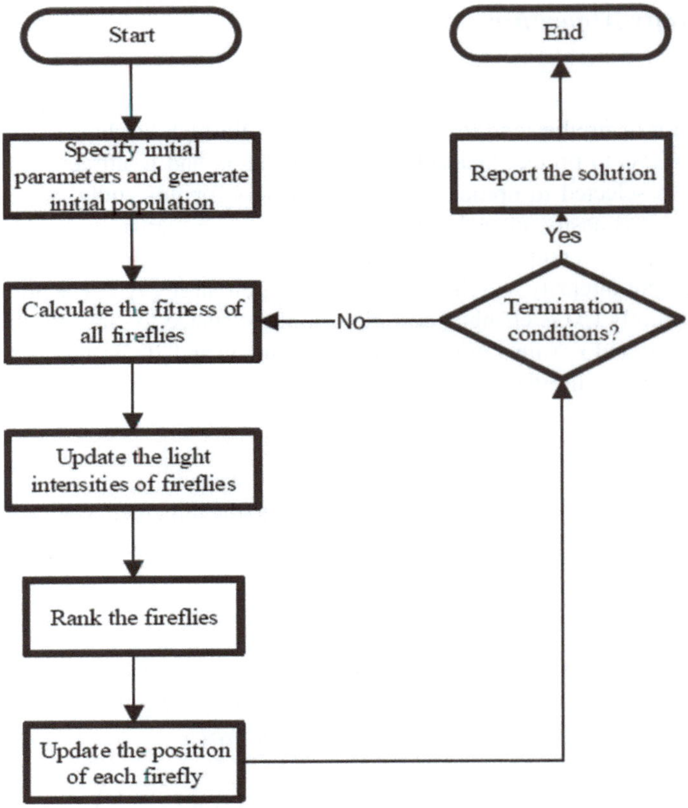

Fig. 5 Flowchart of FA algorithm

Fig. 6 Pseudo-code for the Firefly algorithm with ANN

Step 1: Initialize the neural network
Define the structure of the neural network
Step 2: Determine the parameters of the neural network
Initialize the bias and the weight coefficients
Step 3: Optimization step
Generate the initial population
Initialize the FA parameters
Calculate the fitness of each firefly
While (reach terminate conditions ?)
Update the light intensities of fireflies
Rank the fireflies
Update the position of each firefly
Find the current best
End while
Step 4: Validate the neural network
Step 5: Examine the testing data with neural network

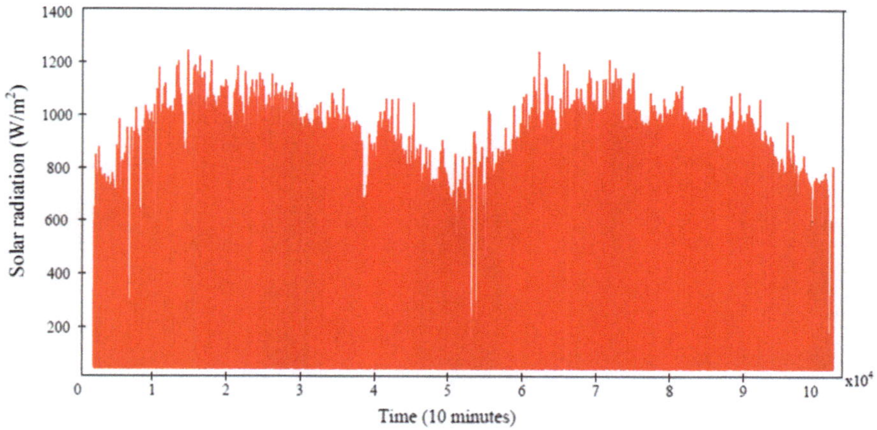

Fig. 7 Series of PV panels in a solar power plant

is shorter than that of summer, while in summer, radiation at noon is the strongest of the whole year for Turkey. Such a 1D representation shows a significant insight into the solar radiation pattern as a function of time, which is represented in Fig. 7.

4.1 Correlation Analysis

The embedded dimension of the input for the prediction model, i.e., the number of previous data samples used as the input, is determined by the auto-correlation coefficients of the samples:

$$r_k = \frac{1}{(N-k)s^2}\sum_{i=k}^{N}(x_i-\mu)(x_{i-k}-\mu) \tag{7}$$

where μ and s are the mean and variance of the samples, respectively, r_k is the sample autocorrelation coefficient, k is *a* delay, x is data set and N is the number of samples of the series. Figure 8 shows a 1D view of the autocorrelation coefficients of the solar radiation in 2016 and 2017.

An important observation in Figs. 8 and 9 is that there are strong correlations between the solar radiation, not only in consecutive hours but also during some hours of consecutive days.

The correlation between two consecutive days for the same hour is stronger than that between the current hour and two hours ahead in the same day. Therefore, when constructing a prediction model, the data from the previous day at the time of prediction must be used with a higher priority than the data from the previous two

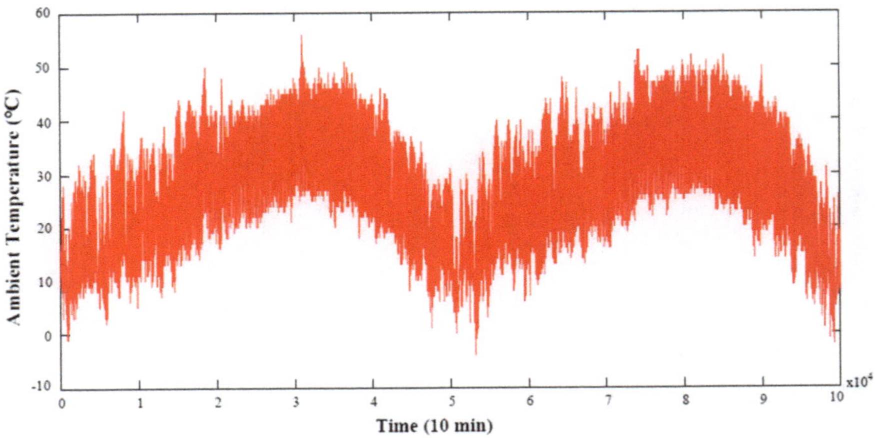

Fig. 8 Correlation between ambient temperature and time (Daily hours)

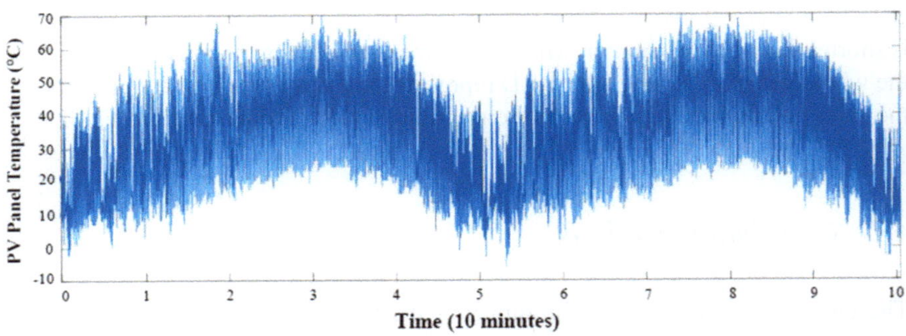

Fig. 9 Correlation between PV panel temperature and time

hours. In this study, the former two days' solar radiation data at the time of prediction, and the data at the current time, are used as the input to the prediction model.

5 Simulation Results

A PV plant consists of one on-grid inverter and 38 series PV panels that have the same characteristics in order to meet the energy demands of an industry. Table 2 shows the parameters of PV panels and inverter. Selected PV panels are shown in Fig. 10. In this study, real-time data were obtained from one of the inverters set in PV power plant, thus, simulation studies are done effectively by processing data with three different methods: ANN, ANN-FA, and ANN-PSO. The mean absolute error (MAE), mean absolute percentage error (MAPE), the coefficient of determination (R^2), and correlation coefficient (ρ) are used to evaluate the performance of the

Table 2 Inverter and PV panel datasheet

PV panel	Parameter	On-grid inverter	Parameter
Power rating	265 Wp	Output power	10 kW
Open circuit voltage (Voc)	38 V	MPPT range	260–850 V
Voltage at power point	31.8 V	Max. input voltage	1000 V
Short circuit current (Isc)	8.7 A	Max. input DC current	22 A
Current at the power point	8.1 A	Operating temperature	−40 °C.....85 °C
Cell efficiency	17.8%	Max. efficiency	98%
Power tolerance	±2.7%	Protection class	IP65

Fig. 10 Series of PV panels in a solar power plant

Solar Power Prediction (SPP) models. MAPE and regression criteria are analyzed in particular using ANN, ANN-PSO and ANN-FA methods in the simulation results section below.

In the study, two different swarm-based methods were used for network training. Swarm-based methods first work with a randomly distributed population in the search space. The success of that particle is then calculated using an objective function whose fitness value is determined for each particle in the search space. In the next step, the information of the particles is updated according to the structure of the algorithm (according to the relevant equations in the algorithm) and a new generation is created. These steps are continued until the termination criterion is reached. Once the training

Table 3 Results of the test phase

Method	MAPE	Regression
ANN	22.2098	0.97669
ANN-PSO	19.5525	0.97671
ANN-FA	19.4149	0.97993

is complete, ANN is generated using the optimum values from the best particle and passed to the test phase.The data that were used for training and testing stages of the artificial neural network is recorded at 2015 and 2016 in Turkey. By using ANN the data for 2017 is estimated, and the results were compared with the real data. The output layer of the neural network structure is an instantaneous PV plant power. The input layer of the network is fed with as follows:

- Ambient temperature [°C]
- Solar radiation [W/m^2]
- PV Panel temperature [°C]

For each method used, 8 neurons are used for hidden layer in the network structure. In addition, 1 neuron is used in the input layer and 3 neurons are used in the output layers. The training coefficients during network training are as follows:

- Weights for the 24 interconnections between the input and the hidden layer
- The bias value for the 8 neurons in hidden layers
- Weights for the 8 interconnections between the hidden and the output layer
- The bias value for the single output layer

During network training, 41 parameters are trained. The algorithms are run for 300 iterations, and during the training, 20 individuals are used for each optimization method. A total of 100479 data sets are sent to the network. After the training phase is finished, the test phase is complete.

Each method was used in the neural network test phase, which is created with the best network parameters obtained from 25 iterations. Estimation results obtained at the end of the optimization studies are analyzed according to MAPE criteria. The results obtained from the three different methods are given in Table 3 according to the MAPE criterion.

The best result is obtained in ANN-FA, as shown in Table 3. Real-time results and test results obtained with ANN-FA are shown graphically in Figs. 11, 12, 13 and 14 over a 3-month period:

- January–February–March results are shown in Fig. 11
- April–May–June results are shown in Fig. 12
- July–August–September results are shown in Fig. 13
- October–November–December results are shown in Fig. 14.

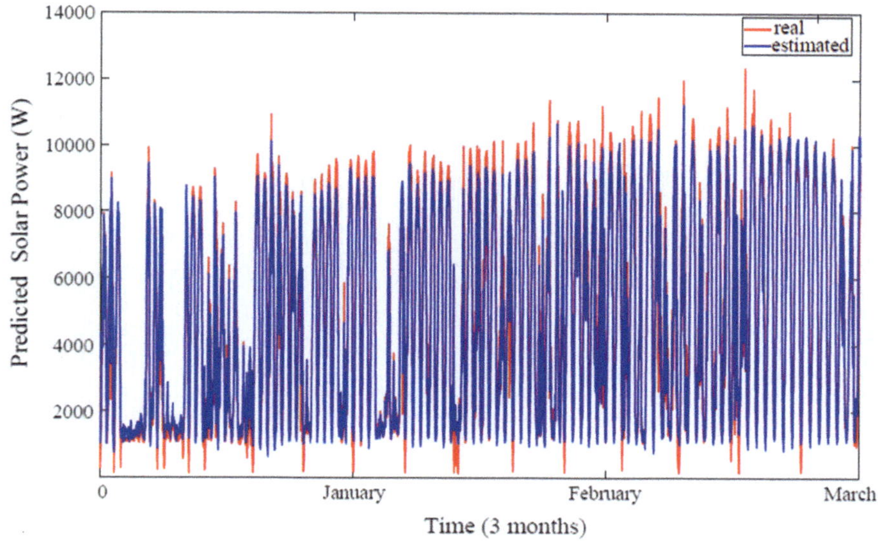

Fig. 11 Solar-generated PV power forecasting result (January–February–March)

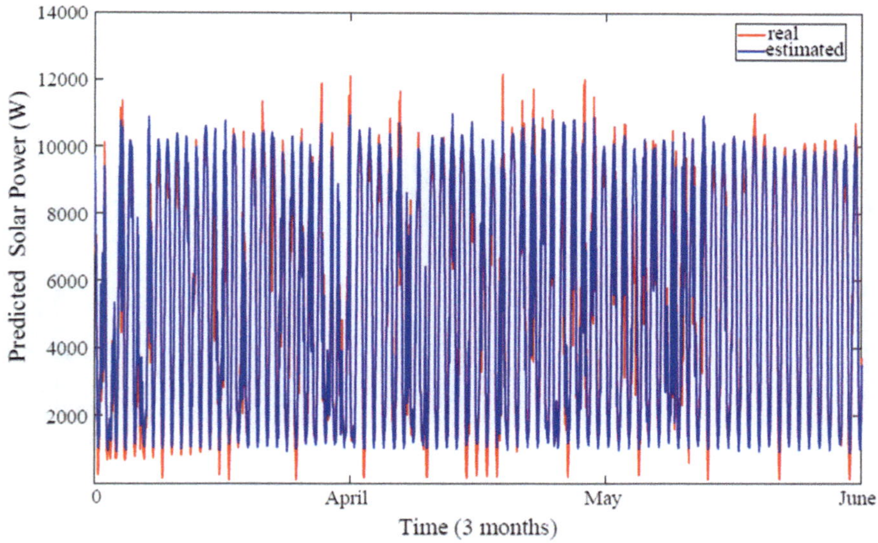

Fig. 12 Solar-generated PV power forecasting result (April–May–June)

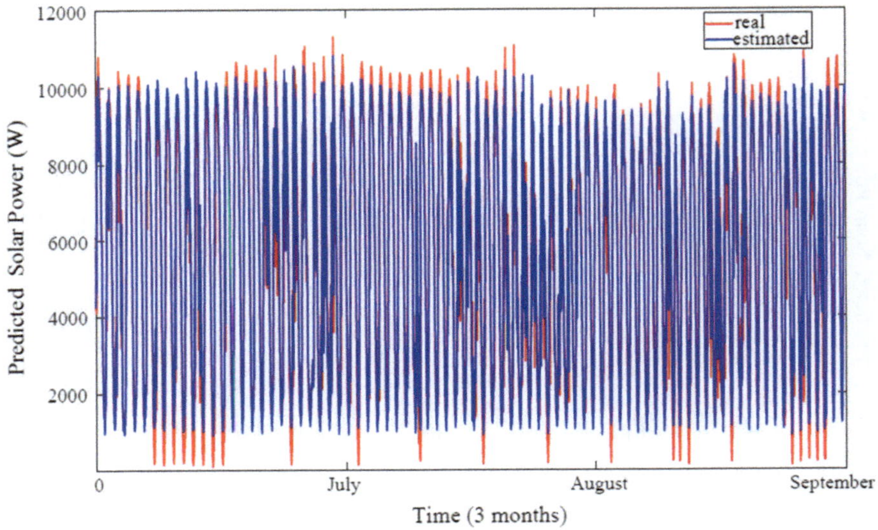

Fig. 13 Solar-generated PV power forecasting result (July–August–September)

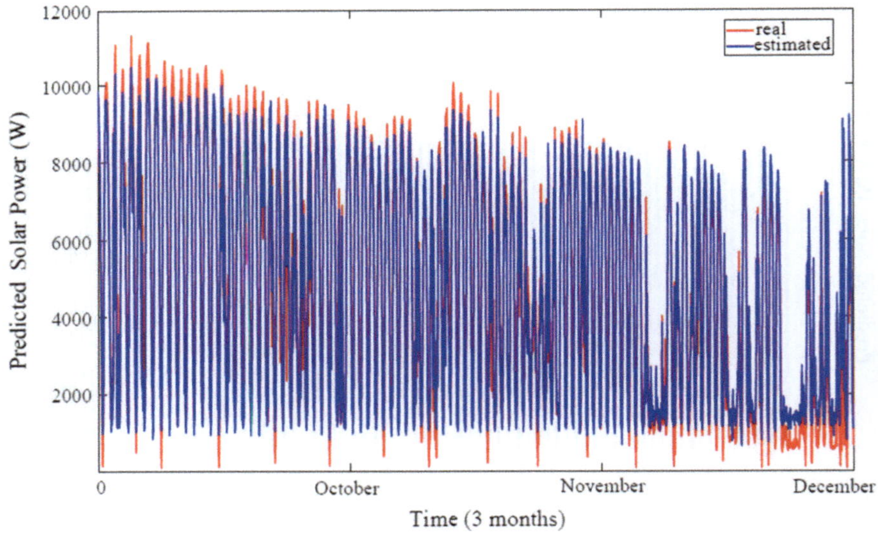

Fig. 14 Solar-generated PV power forecasting result (October–November–December)

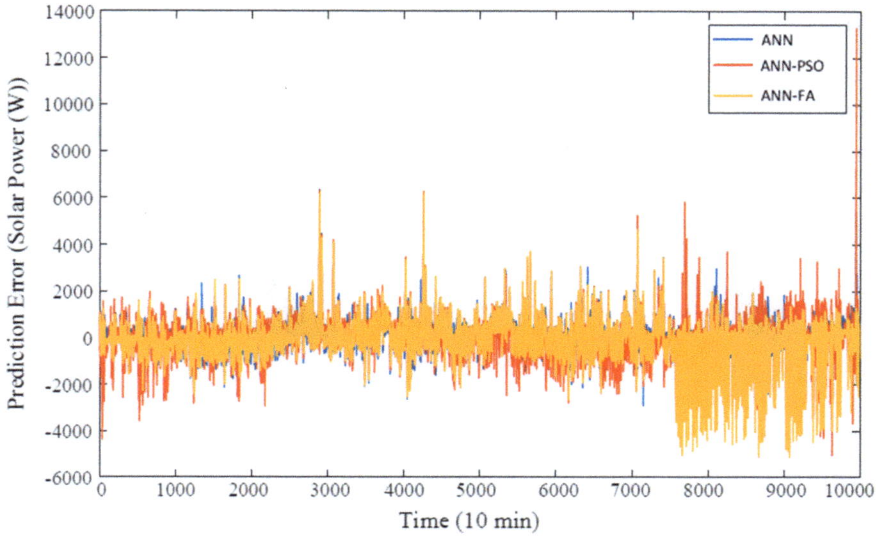

Fig. 15 The prediction error values for the proposed methods

Figure 15 shows the error graphs obtained from the results of the three methods. As clearly shown in this figure, the least error is obtained with the ANN-FA method.

The regression graphs obtained at the end of the test phase are shown in Fig. 16. It can be seen that the best data distribution is shown in Fig. 16 shows that the least error obtained at the end of the test phase was achieved with ANN-FA. When examined against the numerical results given in Table 3, it is seen that the best method for PV power forecasting is obtained by using ANN-FA method.

6 Conclusion

The use of solar energy has increased significantly in recent years in meeting energy demand. Many studies have been carried out in order to make maximum use of these resources. Solar energy is one of the renewable energy sources in meeting the energy supply. Although solar energy is abundant in nature, this energy is not benefited at certain times of the day. There are many factors such as cloudiness, dust, solar radiation and average sunrise time changes during the day for solar power prediction. Therefore, it has no stable behavior structure. Power estimation is important for energy planning so that this study is carried out. The data of 1 MW PV power plant in Turkey is used to estimate output power by real-time data mining for short time prediction. The data of solar radiation, ambient temperature, and panel temperature were used for input parameters. Firstly, estimation was performed by traditional ANN. However, the expected performance has not been reached. It was seen that three input parameters obtained are insufficient to estimate in traditional ANN. It was decided

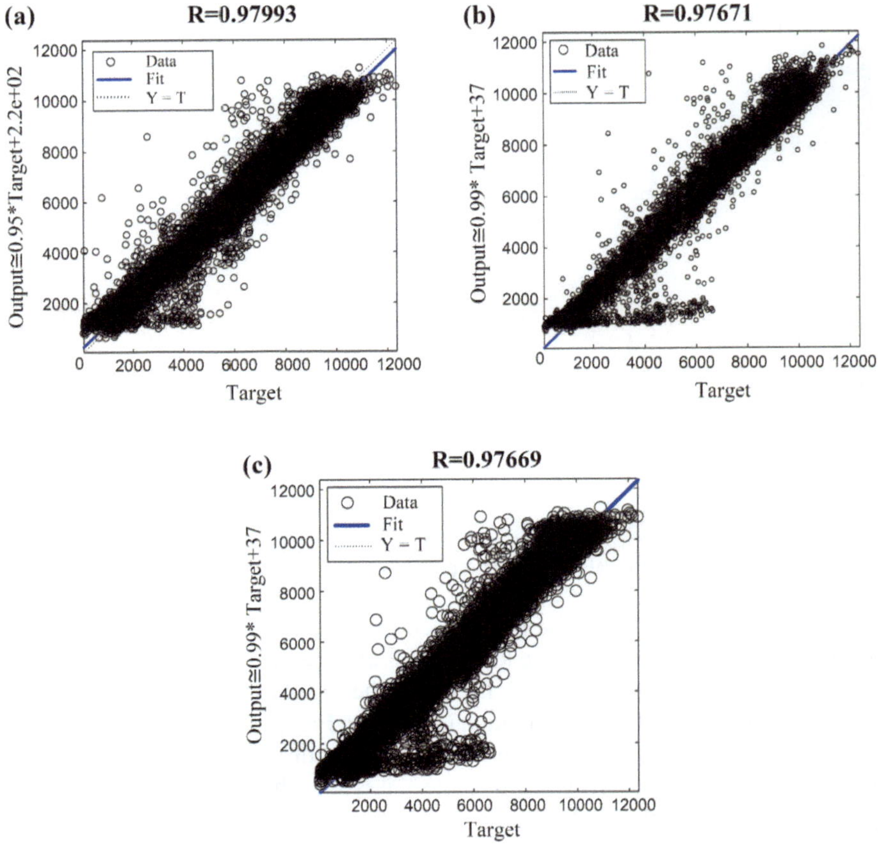

Fig. 16 Regression graph **a** ANN-FA, **b** ANN-PSO, **c** ANN

that ANN should be used with optimization method to increase performance. Then, input data was optimized using different optimization methods. Optimized input data was processed in conventional ANN and the short-term prediction was performed. ANN, ANN-PSO and ANN-FA results used for estimation were compared in detail. In the performance analysis, it was seen that the artificial neural network algorithm was not able to predict short time output power by using a few input data alone. In the literature, it has been tried to obtain the optimum prediction by Particle Swarm Optimization (PSO) method which is one of the most used optimization management. The desired performance could not be reached due to continuous insertion of the local minimum values in the PSO method. Firefly Algorithm (FA), a new optimization method, has been shown to be the most efficient algorithm for short-term solar estimates by removing the problem of sticking to local minimum values.

Acknowledgements The authors would like to acknowledge the Scientific Project Unit of Adana Science and Technology University (Project Number: 18103015 and 18103016) for full financial support.

References

1. Razak A, Irwan YM, Leow WZ, Irwanto M, Safwati I, Zhafarina M (2016) Investigation of the effect temperature on photovoltaic (PV) panel output performance. Int J Adv Sci Eng Inf Technol 6(5):682–688
2. Chang W-Y (2014) A literature review of wind forecasting methods. J Power Energy Eng 2(4)
3. Mellit A, Pavan AM (2010) A 24-h forecast of solar radiance using artificial neural network: application for performance prediction of a grid-connected PV plant at Trieste, Italy. Sol Energy 84(5):807–821
4. Sperati S, Alessandrini S, Delle Monache L (2016) An application of the ECMWF ensemble prediction system for short-term solar power forecasting. Sol Energy 133:437–450
5. Ahmad A, Anderson TN, Lie TT (2015) Hourly global solar radiation forecasting for New Zealand. Sol Energy 122:1398–1408
6. David M, Ramahatana F, Trombe P-J, Lauret P (2016) Probabilistic forecasting of the solar irradiance with recursive ARMA and GARCH models. Solar Energy 133:55–72
7. Izgi E, Öztopal A, Yerli B, Kaymak MK, Şahin AD (2012) Short-mid-term solar power prediction by using artificial neural networks. Sol Energy 86(2):725–733
8. Kardakos EG, Alexiadis MC, Vagropoulos SI, Simoglou CK, Biskas PN, Bakirtzis AG (2013) Application of time series and artificial neural network models in short-term forecasting of PV power generation, In: 2013 48th international universities power engineering conference (UPEC), pp 1–6
9. Persson C, Bacher P, Shiga T, Madsen H (2017) Multi-site solar power forecasting using gradient boosted regression trees. Sol Energy 150:423–436
10. e Silva RA, Brito MC (2018) Impact of network layout and time resolution on spatio-temporal solar forecasting. Sol Energy 163:329–337
11. Chen SX, Gooi HB, Wang MQ (2013) Solar radiation forecast based on fuzzy logic and neural networks. Renew Energy 60:195–201
12. Ramaswamy S, Sadhu PK (2017) Forecasting PV power from solar irradiance and temperature using neural networks. In: 2017 international conference on infocom technologies and unmanned systems (trends and future directions)(ICTUS), pp 244–248
13. Dewangan CL, Singh SN, Chakrabarti S (2017) Solar irradiance forecasting using wavelet neural network. In: Asia-Pacific power and energy engineering conference (APPEEC). IEEE PES, pp 1–6
14. Raza MQ, Nadarajah M, Ekanayake C (2017) A multivariate ensemble framework for short term solar photovoltaic output power forecast. In: Power and energy society general meeting. IEEE, pp 1–5
15. Percy SD, Aldeen M, Berry AM (2018) Residential demand forecasting with solar-battery systems: a survey-less approach. IEEE Trans Sustain Energy
16. Alzahrani A, Shamsi P, Ferdowsi M, Dagli C (2017) Solar irradiance forecasting using deep recurrent neural networks. In: 2017 IEEE 6th international conference on renewable energy research and applications (ICRERA), pp 988–994
17. Wang Y, Shen Y, Mao S, Cao G, Nelms RM (2018) Adaptive learning hybrid model for solar intensity forecasting. IEEE Trans Ind Inform 14(4):1635–1645
18. Praynlin E, Jensona JI (2017) Solar radiation forecasting using artificial neural network. In: 2017 innovations in power and advanced computing technologies (i-PACT), pp 1–7
19. Çevik S, Çakmak R, Altaş İH (2017) A day ahead hourly solar radiation forecasting by artificial neural networks: a case study for Trabzon province. In: 2017 international artificial intelligence and data processing symposium (IDAP), pp 1–6
20. Sahoo AK, Sahoo SK (2016) Energy forecasting for grid connected MW range solar PV system. In: 2016 7th India international conference on power electronics (IICPE), pp 1–6
21. Majumder I, Behera MK, Nayak N (2017) Solar power forecasting using a hybrid EMD-ELM method. In: 2017 international conference on circuit, power and computing technologies (ICCPCT). pp 1–6

22. Hassan S, Khanesar MA, Hajizadeh A, Khosravi A (2017) Hybrid multi-objective forecasting of solar photovoltaic output using Kalman filter based interval type-2 fuzzy logic system. In: 2017 IEEE international conference on fuzzy systems (FUZZ-IEEE), pp 1–6
23. Severiano CA, Silva PC, Sadaei HJ, Guimarães FG (2017) Very short-term solar forecasting using fuzzy time series. In: 2017 IEEE international conference on fuzzy systems (FUZZ-IEEE), pp 1–6
24. Orjuela-Cañón AD, Hernández J, Rivero CR (2017) Very short term forecasting in global solar irradiance using linear and nonlinear models. In: 2017 IEEE workshop on power electronics and power quality applications (PEPQA), pp 1–5
25. Chiang P-H, Chiluvuri SPV, Dey S, Nguyen TQ (2017) Forecasting of solar photovoltaic system power generation using wavelet decomposition and bias-compensated random forest. In: Ninth annual IEEE green technologies conference (GreenTech), pp 260–266
26. Moustris KP, Kavvadias KA, Kokkosis AI, Paliatsos AG (2016) One day-ahead forecasting of mean hourly global solar radiation for energy management systems purposes using artificial neural network modeling
27. Gensler A, Henze J, Sick B, Raabe N (2016) Deep learning for solar power forecasting–an approach using AutoEncoder and LSTM neural networks. In: 2016 IEEE international conference on systems, man, and cybernetics (SMC), pp 002858–002865
28. Huang C, Zhang Z, Bensoussan A (2016) Forecasting of daily global solar radiation using wavelet transform-coupled Gaussian process regression: case study in Spain. In: Innovative smart grid technologies-Asia (ISGT-Asia). IEEE 2016, pp 799–804
29. Sheng H, Xiao J, Cheng Y, Ni Q, Wang S (2018) Short-term solar power forecasting based on weighted Gaussian process regression. IEEE Trans Ind Inform 65(1):300–308
30. Sivaneasan B, Yu CY, Goh KP (2017) Solar forecasting using ANN with fuzzy logic pre-processing. Energy Procedia 143:727–732
31. Watetakarn S, Premrudeepreechacharn S (2015) Forecasting of solar irradiance for solar power plants by artificial neural network. In: Innovative smart grid technologies-Asia (ISGT ASIA). IEEE, 2015, pp 1–5
32. Behera MK, Majumder I, Nayak N (2018) Solar photovoltaic power forecasting using optimized modified extreme learning machine technique. Eng Sci Technol Int J
33. Verma T, Tiwana APS, Reddy CC, Arora V, Devanand P (2016) Data analysis to generate models based on neural network and regression for solar power generation forecasting. In: 2016 7th international conference on intelligent systems, modelling and simulation (ISMS), pp 97–100
34. Gairaa K, Chellali F, Benkaciali S, Messlem Y, Abdallah K (2015) Daily global solar radiation forecasting over a desert area using NAR neural networks comparison with conventional methods. In: 2015 international conference on renewable energy research and applications (ICRERA), pp 567–571
35. Silva F, Teixeira B, Teixeira N, Pinto T, Praça I, Vale Z (2016) Application of a hybrid neural fuzzy inference system to forecast solar intensity. In: 2016 27th international workshop on database and expert systems applications (DEXA), pp 161–165
36. Alfadda A, Adhikari R, Kuzlu M, Rahman S (2017) Hour-ahead solar PV power forecasting using SVR based approach. In: Power and energy society innovative smart grid technologies conference (ISGT). IEEE, 2017, pp 1–5
37. Abuella M, Chowdhury B (2015) Solar power forecasting using artificial neural networks. In: North American power symposium (NAPS). 2015, pp 1–5
38. Sreekumar S, Sharma KC, Bhakar R (2016) Optimized support vector regression models for short term solar radiation forecasting in smart environment. In: Region 10 conference (TENCON). IEEE, 2016, pp 1929–1932
39. Ghayekhloo M, Ghofrani M, Menhaj MB, Azimi R (2015) A novel clustering approach for short-term solar radiation forecasting. Sol Energy 122:1371–1383
40. Mellit A, Sağlam S, Kalogirou SA (2013) Artificial neural network-based model for estimating the produced power of a photovoltaic. Renew Energy 60:71–78

41. Notton G, Paoli C, Ivanova L, Vasileva S, Nivet ML (2013) Neural network approach to estimate 10-min solar global irradiation values on tilted planes. Renew Energy 50:576–584
42. Yona A, Senjyu T, Funabashi T, Kim C-H (2013) Determination method of insolation prediction with fuzzy and applying neural network for long-term ahead PV power output correction. IEEE Trans Sustain Energy 4(2):527–533
43. Amrouche B, Le Pivert X (2014) Artificial neural network based daily local forecasting for global solar radiation. Appl Energy 130:333–341
44. Almonacid F, Pérez-Higueras PJ, Fernández EF, Hontoria L (2014) A methodology based on dynamic artificial neural network for short-term forecasting of the power output of a PV generator. Energy Convers Manag 85:389–398
45. Dahmani K, Dizene R, Notton G, Paoli C, Voyant C, Nivet ML (2014) Estimation of 5-min time-step data of tilted solar global irradiation using ANN (Artificial Neural Network) model. Energy 70:374–381
46. Kaushika ND, Tomar RK, Kaushik SC (2014) Artificial neural network model based on interrelationship of direct, diffuse and global solar radiations. Sol Energy 103:327–342
47. Kashyap Y, Bansal A, Sao AK (2015) Solar radiation forecasting with multiple parameters neural networks. Renew Sustain Energy Rev 49:825–835
48. Teo TT, Logenthiran T, Woo WL (2015) Forecasting of photovoltaic power using extreme learning machine. In: Innovative smart grid technologies-Asia (ISGT ASIA). IEEE, 2015, pp 1–6
49. Dong Z, Yang D, Reindl T, Walsh WM (2014) Satellite image analysis and a hybrid ESSS/ANN model to forecast solar irradiance in the tropics. Energy Convers Manag 79:66–73
50. Quan DM, Ogliari E, Grimaccia F, Leva S, Mussetta M (2013) Hybrid model for hourly forecast of photovoltaic and wind power. In: 2013 IEEE international conference on fuzzy systems (FUZZ), pp 1–6
51. Bouzerdoum M, Mellit A, Pavan AM (2013) A hybrid model (SARIMA-SVM) for short-term power forecasting of a small-scale grid-connected photovoltaic plant. Sol Energy 98:226–235
52. Gandelli A, Grimaccia F, Leva S, Mussetta M, Ogliari E (2014) Hybrid model analysis and validation for PV energy production forecasting. In: 2014 international joint conference on neural networks (IJCNN), pp 1957–1962
53. Tuohy A et al (2015) Solar forecasting: methods, challenges, and performance. IEEE Power Energy Mag 13(6):50–59
54. Das UK et al (2018) Forecasting of photovoltaic power generation and model optimization: a review. Renew Sustain Energy Rev 81:912–928
55. Debnath KB, Mourshed M (2018) Forecasting methods in energy planning models. Renew Sustain Energy Rev 88:297–325
56. Yano F, Shohdohji T, Toyoda Y (2007) An improvement of particle swarm optimization with a neighborhood search algorithm. Ind Eng Manag Syst 6(1):64–71
57. Ab Wahab MN, Nefti-Meziani S, Atyabi A (2015) A comprehensive review of swarm optimization algorithms. PLoS One 10(5):e0122827
58. Dada EG, Ramlan EI (2015) Primal-dual interior point method particle swarm optimization (pdipmPSO) algorithm. In: 3rd international conference on advances in engineering sciences and applied mathematics (ICAESAM'2015), London (UK), 23–24 March 2015. https://doi.org/10.15242/iie.e0315072,2015

Evaluation on Training Algorithms of Back Propagation Neural Network for a Solar Photovoltaic Based DSTATCOM System

Nor Hanisah Baharudin, Tunku Muhammad Nizar Tunku Mansur, Rosnazri Ali and Muhammad Irwanto Misrun

Abstract This chapter discusses evaluation on the Back Propagation Neural Network (BPNN) control algorithm based on Fast Fourier Transform (FFT) control algorithm with different BPNN training algorithms for Distribution Static Compensator (DSTATCOM) with integrated solar photovoltaic system. Furthermore, the comparison is performed with different weight or bias training functions such as supervised and unsupervised. Each training algorithms have been utilized to investigate its performance in generating the target pattern for harmonic elimination in term of accuracy, learning epochs and training time. The performance of the BPNN training algorithms is determined by calculating the error between the target and output pattern using Mean Squared Error (MSE). The lower value of the MSE shows the higher accuracy of the output pattern according to the target pattern given. Number of iterations (epochs) and training time are evaluated to investigate the performance of different BPNN training algorithms on DSTATCOM for harmonic reduction under nonlinear load condition.

1 Introduction

Due to the fossil fuel depletion and threats of current global warming issues, the renewable energy resources are gaining momentum rapidly in recent energy development industry. The development of micro-grid from interconnected Distributed Generations (DGs) with renewable energies are beneficial for both sides between

---

N. H. Baharudin (✉) · T. M. N. T. Mansur · R. Ali · M. I. Misrun
School of Electrical System Engineering, Universiti Malaysia Perlis (UniMAP), Arau, Perlis, Malaysia
e-mail: norhanisahbaharudin@gmail.com

T. M. N. T. Mansur
e-mail: tunkunizar@unimap.edu.my

R. Ali
e-mail: rosnazri@unimap.edu.my

M. I. Misrun
e-mail: irwanto@unimap.edu.my

R.-E. Precup et al. (eds.), *Solar Photovoltaic Power Plants*, Power Systems,
https://doi.org/10.1007/978-981-13-6151-7_9

consumers and power utilities since the electricity can be produced locally [1]. However, the tremendous growth of individual grid-connected solar photovoltaic systems at the distribution systems have introduced the disturbances of power quality such as harmonics, voltage profile and power losses due to its uncontrollable natural characteristic of electricity generation based on meteorological conditions to the Point of Common Coupling (PCC) [2–7]. There are available technical assessments and international standards on grid-connected Photovoltaic (PV) systems such as IEC 61727 (1995-06) and IEEE 929-2000 to provide proper guidance on compatible operation of grid connected renewable energy resources in terms of power system and power quality issues [8, 9]. In addition, numerous usages of nonlinear loads at the distribution system have deteriorated the power quality due to the harmonic distortion. Thus, more researchers are concentrating on power quality improvement features of PV system to improve power quality distortion issues as well as generating power to ensure the availability of electricity supply [10–13]. The works that have proposed by using different control algorithms such as Leaky Least Mean Fourth (LLMF) [13], Synchronous Reference Frame Theory (SRF) [12, 14, 15], SRF with optimized controller using Differential Evolution (DL) [16], Fuzzy Logic Controller (FLC), Improved Linear Sinusoidal Tracer (ILST) [17], offset reduction Second Order Generalized Integrator (SOGI) [18], Predictive Direct Power Control (P-DPC) [19], direct current control method with PV/Wind based DSTATCOM [20], sliding mode control with PV/micro turbine based DSTATCOM [21], adaptive neural learning algorithm [22], fuzzy PI controller with Constant Switching Frequency Sub-Harmonic Pulse Width Modulation (CSFSHPWM) [23], adaptive FLC [24, 25], decoupled current controller [26], Least Mean Eight (LME) adaptive algorithm [27], Adaptive Modified Least Mean Fourth (MLMF) [28], dual-sign error control algorithm [29] with PV based DSTATCOM to provide electricity supply to the connected loads, put the excessive PV energy onto the utility grid and solve power quality issues such as harmonic elimination, reactive power compensation, load balancing as well as controlling the system to maintain its stability even in rapid change [30].

This research proposes different training algorithms of BPNN since BPNN have been reported widely in certain areas such as pattern identifications and function approximations, however a limited literature is available related to the usage of BPNN control algorithm with different training algorithms for PV based DSTATCOM in detecting harmonic current [31–33].

2 Power Quality Issues

Fast industrialization has increased the numbers of power electronic loads tremendously in electric power distribution system. These nonlinear loads inject harmonic distortions which cause non-sinusoidal current in the distribution network. These nonlinear loads associated with transformers, High-Voltage DC (HVDC) systems, renewable energy sources, Switch Mode Power Supplies (SMPSs), Uninterruptible Power Supplies (UPSs), arc furnaces, welding systems, telecommunication equip-

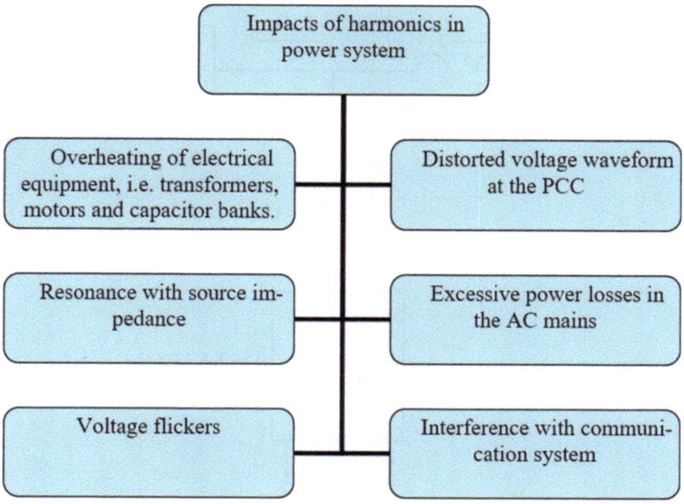

Fig. 1 Impact of harmonics in the power system [42–45]

ment and many others create power quality issues at the PCC in the distribution network which led to waveform distortion. Poor power quality will affect both utilities and consumers such as malfunction of sensitive equipment in industries, loss of revenue, increase power losses and resonance with source impedance [33–38]. Figure 1 illustrates the impact of harmonics in the power system. There are stringent international standards had been established to curb the power quality issues at the PCC, for instance, IEEE Standard 1531-2003, IEEE Standard 519-2014 and IEC 61000-3-2:2018 [39–41]. In this chapter, the compliance to IEEE Standard 519-2014 has been emphasized to improve harmonic current distortions.

3 Active Power Line Conditioners

Active Power Line Conditioners (APLCs) are capable to provide remedial measures with a more cost effective solution to mitigate power quality problems in the distribution network [10]. These compensating devices are capable to compensate necessary reactive power and harmonic component from connected nonlinear loads [46]. The APLCs can be classified into three different topologies of active power filters such as DSTATCOM, Dynamic Voltage Restorer (DVR) and Unified Power Quality Conditioner (UPQC) as shown in Figs. 2, 3 and 4. These compensating devices are used to solve different problems of power quality in the distribution network which the DSTATCOM is capable to solve current based power quality problems, the DVR is proposed to suppress the voltage based power quality problems while the UPQC is used to solve both types of power quality problems [47]. These active power filters

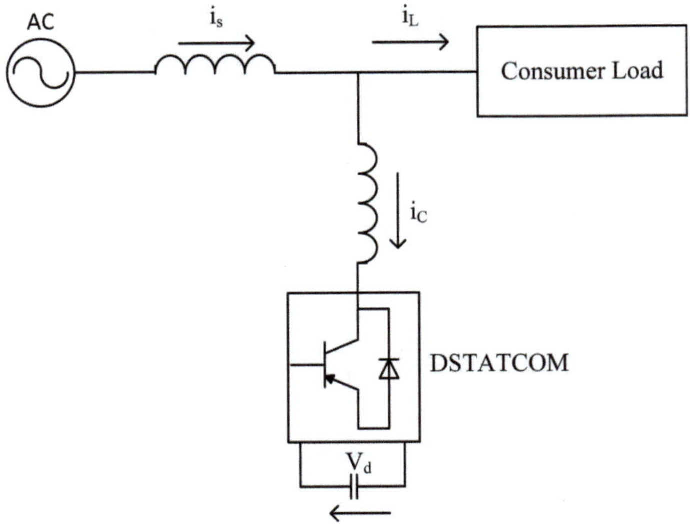

Fig. 2 The equivalent circuit diagram Dynamic Voltage Restorers (DVRs)

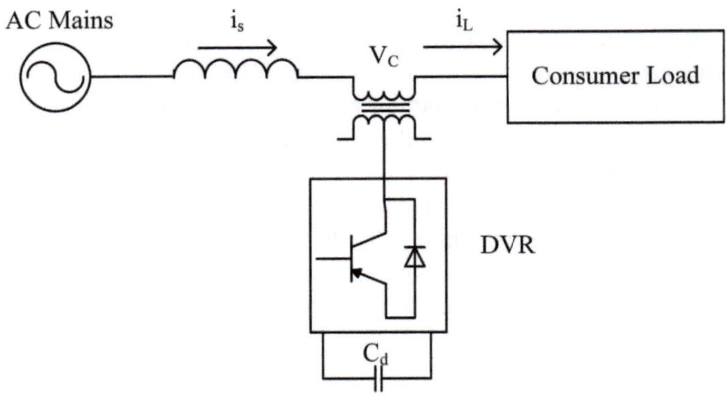

Fig. 3 The equivalent circuit diagram Distribution Static Compensator (DSTATCOM)

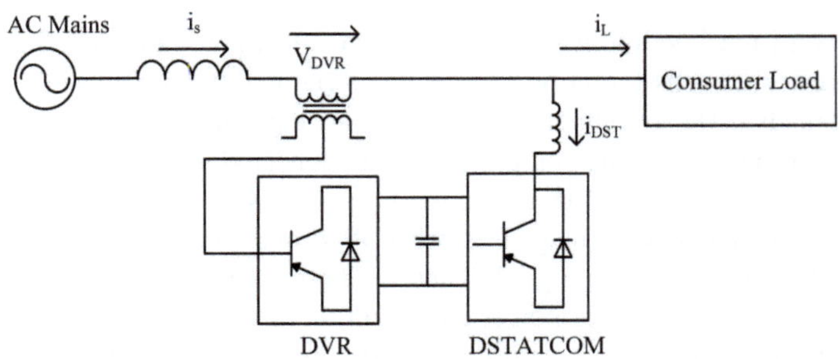

Fig. 4 The equivalent circuit diagram Unified Power Quality Conditioner (UPQC)

are realized based on Voltage Source Inverter (VSI) since it has better performance than Current Source Inverter (CSI) due to its light weight, cheaper, expandable to multi-step and multi-level versions as well as its capability to deal with lower switching frequency [48]. Among these, DSTATCOM has been considered as one of the most effective compensating device with a very fast response for reactive power compensation [49–51]. The DSTATCOM's compensation current injection through the interface inductor at the Point of Common Coupling (PCC) can eliminate harmonic currents due to the distorted nonlinear load and provide reactive power compensation [10]. In recent years, the increasing concern of power quality issues has led to the tremendous applications of DSTATCOM and it has become the most widely used filter in the industrial processes [43, 52–54]. Owing to the advancement in power electronics control of DSTATCOM, its primary benefit is the voltage at the PCC can be effectively regulated. In addition, the DSTATCOM can also provide power factor correction, harmonics current elimination, voltage regulation during sag, swell and others with exceptional features of fast response as well as enhancement of transient overload capability under lower voltage level [55].

4 Design of Back Propagation Neural Network Control Algorithm

This research is suggested solar PV based DSTATCOM with BPNN control algorithm since limited application of neural network in Renewable Energy Sources (RES) with DSTATCOM in the previous literatures for power quality improvement features [56, 57]. In the available literature, BPNN has become the most popular controller in various applications with advanced theoretical development. However, its implementation are still lacking in electrical power engineering for three-phase system in hardware implementation for industrial applications [33, 58]. The investigation on the BPNN control algorithm is done based on Voltage Reference Configuration (VRC) based FFT control algorithm for different training algorithms of BPNN such as Levenverg–Marquadt, Bayesian regulation, Broyden–Fletcher–Goldfarb–Shanno (BFGS) Quasi-Newton, conjugate-Gradient (Powell–Beale restart, Fletcher–Reeves and Polak–Ribiere updates), Gradient descent (adaptive learning rate BPNN), one step secant BPNN, scaled conjugate Gradient BPNN and resilient BPNN (RPROP). Further, comparisons are also done with different weight/bias training functions such as supervised and unsupervised. The training algorithms that have been used for this investigation are summarized in Fig. 5. Through the training process of the BPNN, the steady state line current from 0.3 to 0.4 s peak amplitude of 5 cycles with 10,000 samples has been set as target data whereas the input pattern of nonlinear load current has been used with similar time frame and number of samples. The 10,000 samples are divided into three sets of sample which 7001 samples are used for training set, 1500 samples for validation set and another 1500 for testing set. The training set samples are used to calculate the Gradient for the network weights and biases com-

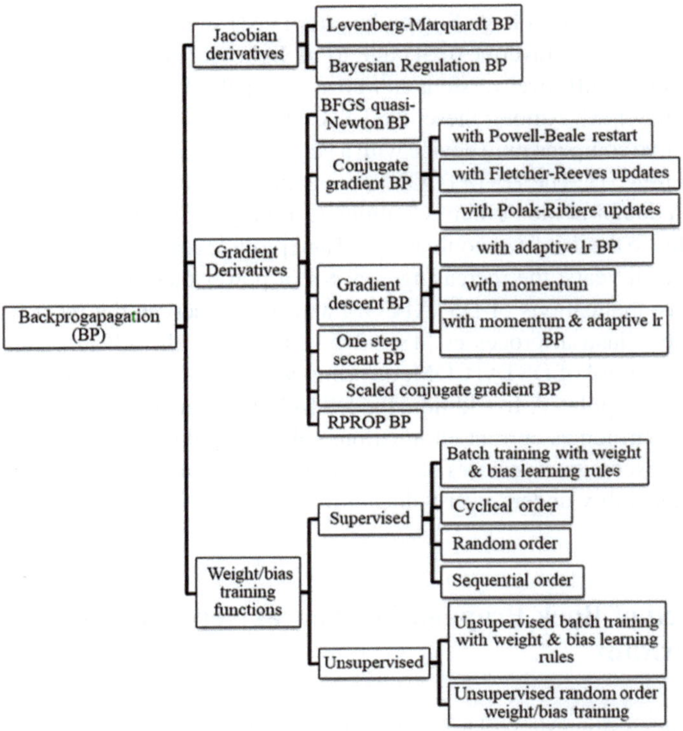

Fig. 5 Summary of the proposed BPNN learning algorithms

putation to follow the target pattern. The validation set samples are used to validate the training set samples and its validation error is observed throughout the training process to prevent over fitting of the training set samples. Then, the testing set samples will be used for independent test for network generalization. Each training algorithms as stated in Fig. 5 have been utilized to investigate its performance in generating the target pattern for harmonic elimination in term of accuracy, learning epochs and training time. Training algorithm accuracy is determined by calculating the error between the target and output pattern using MSE. The lower value of the MSE shows the higher accuracy of the output pattern according to the target pattern given. Number of iterations (epochs) and their training time are compared to investigate the performance of different BPNN training algorithms on DSTATCOM.

5 Results and Discussions

The preliminary model of DSTATCOM system are simulated based on VRC control algorithm and the simulation results are presented in Figs. 6, 7, 8 and 9 [59]. The distorted nonlinear load current is observed in Fig. 6 which can be compensated

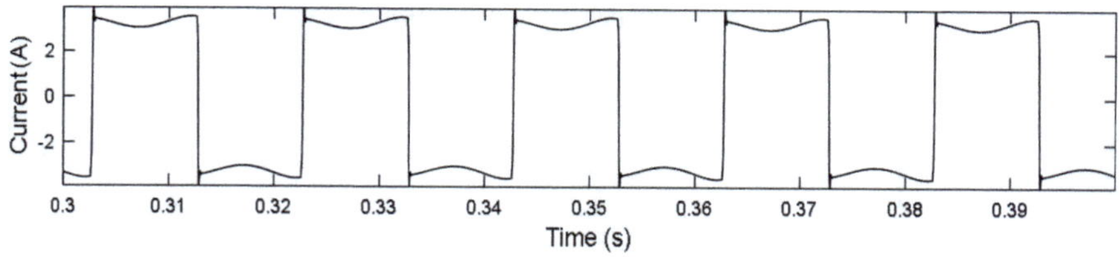

Fig. 6 Phase A load currents waveform (i_{La})

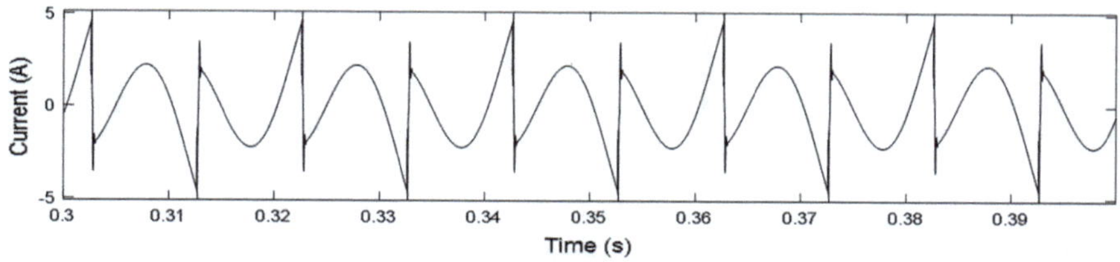

Fig. 7 Phase A compensating currents waveform (i_{Ca})

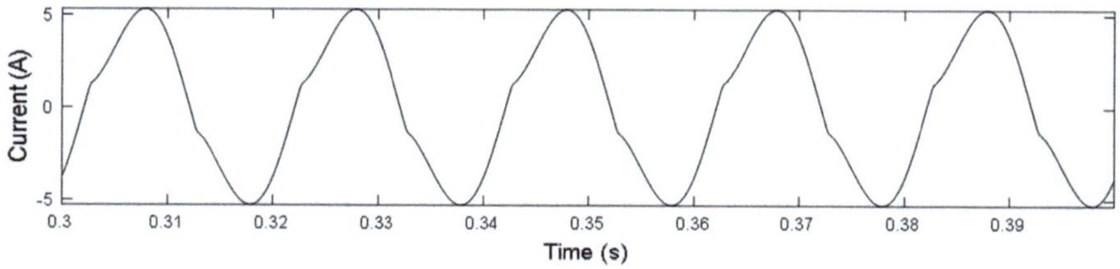

Fig. 8 Phase A line currents waveform after compensation (i_{Sa})

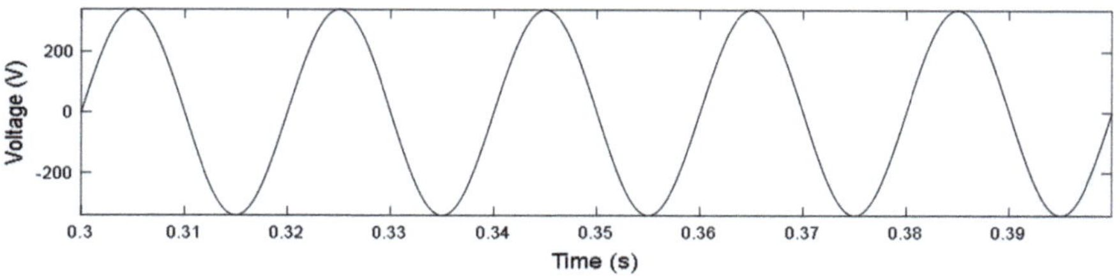

Fig. 9 Phase A line voltage waveform after compensation (v_{Sa})

by DSTATCOM through compensating current injection at the PCC as shown in Fig. 7. The DSTATCOM is capable to improve the harmonic currents distortion at the PCC as the line current and voltage are sinusoidal and in phase by referring to Figs. 8 and 9. Then, the steady state line current from 0.3 to 0.4 s peak amplitude of 5 cycles with 10,000 samples has been set as target data for the BPNN training. This offline training use load current of the same time frame and number of samples as the input data samples.

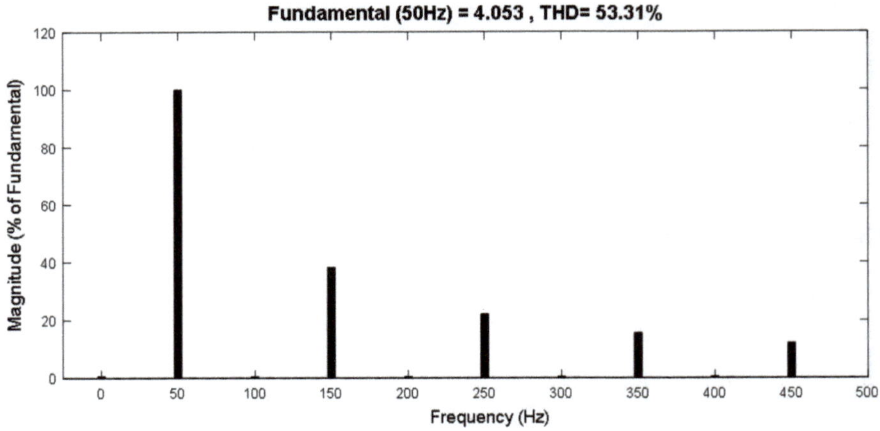

Fig. 10 Total Harmonic Distortion (THD) of the load current

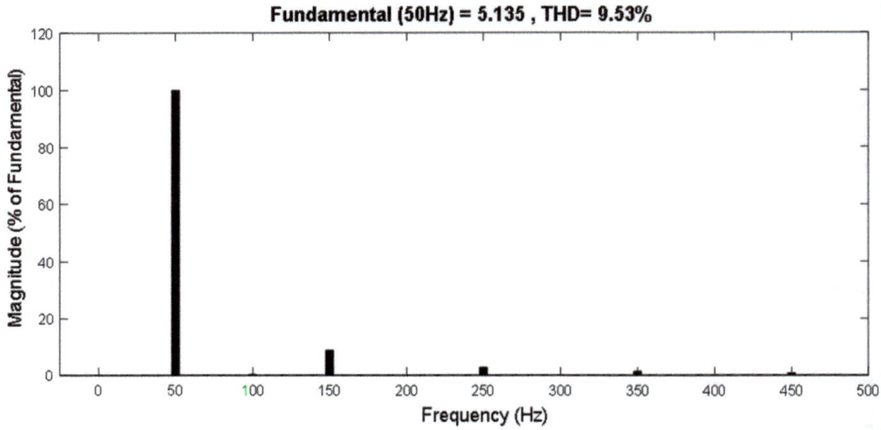

Fig. 11 Total Harmonic Distortion (THD) of the line current after compensation

After DSTATCOM compensation, the voltage at PCC is regulated and the load current can be compensated accordingly. The Total Harmonic Distortion (THD) of the nonlinear load current is 53.31% which can be reduced to 9.53% as represented in Figs. 10 and 11. The odd harmonics of 3rd, 5th, 7th and 9th are also dampened after the compensation. Thus, the DSTATCOM is capable to regulate the line voltage at PCC and reduce the power quality problem in term of harmonic current reduction. Thus, these simulation results can be used as the base data for BPNN input training for the preliminary analysis by using different training algorithms as summarized in Fig. 5.

The purpose of this study is to compare the training time as well as its accuracy for different type of BPNN training algorithms and RBF. Since previous work on power quality improvement by using DSTATCOM with BPNN control algorithm

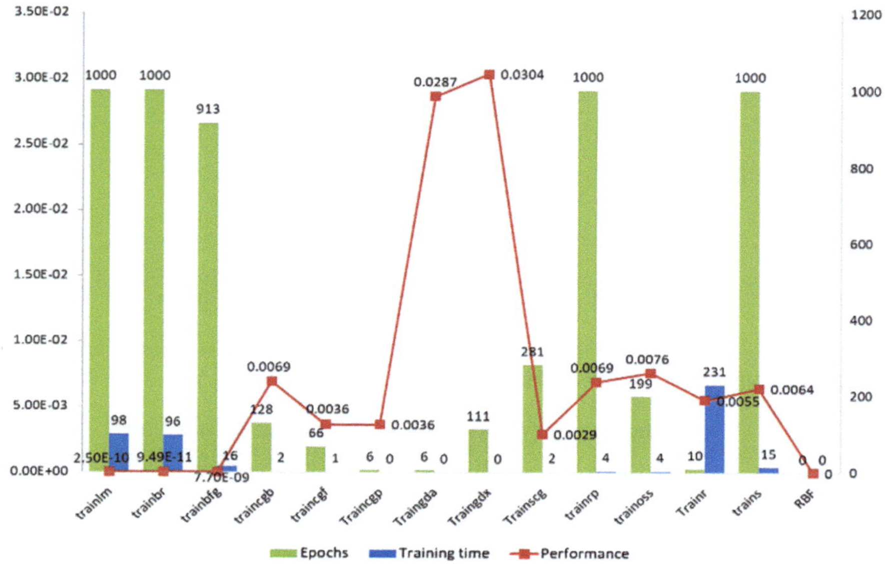

Fig. 12 Comparison performance of different learning algorithms based on its accuracy, training time and no. of epochs

have stated the disadvantages of its large training time in the application of the complex system and the selection of the number of the hidden layers [33] which is based on Levenberq-Marquadt BP [60] and exclusive connecting network [61] whereas, RBF based DSTATCOM has been proven suitable to reduce THD less than 5% with reduced computation time of the DSP [31].

There are lots of other BP training algorithms that are available that can be used for DSTATCOM control which yet to be confirmed its performance on DSTATCOM. As can be seen in Fig. 12, different training algorithms of Back Propagation (BP) have shown that training time is proportional to number of iterations (epochs). However, there is a trade-off between performance in terms of degree of accuracy and training time. The most accurate training algorithms of BP are Levenverg–Marquadt (trainlm), Bayesian regulation (trainbr) and BFGS Quasi-Newton (trainbfg) with accuracy higher than 7.7×10^{-9} as illustrated in Figs. 13 and 14. However, the training time is between 16 s and 1 min 38 s which is quite a long time and unsuitable for dynamic system such as DSTATCOM.

Then, the specified accuracy for DSTATCOM performance limits to 0.01 which lots of training algorithms can perform better with better training time which is less than 4 s such as conjugate-Gradient with Powell–Beale restart (traincgb), conjugate-Gradient with Fletcher–Reeves updates (traincgf), conjugate-Gradient with Polak–Ribiere updates (traincgp), scaled conjugate Gradient BP (trainscg), RPROP resilient BP (trainrp) and one step secant BP (trainoss) which are demonstrated in Fig. 15. Several BPNN training algorithms that diverged during the experiment such as Gradient descent (traingd), Gradient descent with momentum (traingdm), Gradient descent

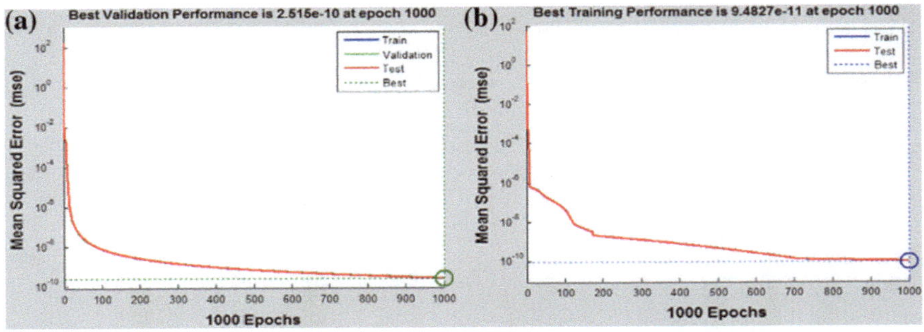

Fig. 13 Performance of BPNN training algorithm with Jacobian derivatives **a** Levenberg-Marquardt (trainlm) **b** Bayesian Regulation (trainbr)

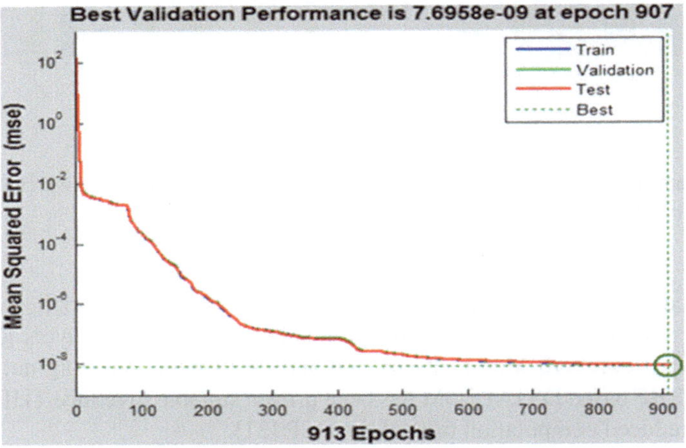

Fig. 14 Performance of BPNN training algorithm with Gradient derivatives using Broyden, Fletcher, Goldfarb and Shanno (BFGS) Quasi-Newton method (train-bfg)

with adaptive learning rate (LR), Gradient descent with momentum and adaptive learning rate, supervised weight/bias training functions with batch training(trainb), cyclical order (trainc), random order (trainr) and sequential order (trains). These performances are presented in Figs. 16 and 17. This divergence may occur due to the lagging of training process which increases the possibility of falling at undesirable local minima caused by multiple layers of neural network structure and the BPNN algorithm's greedy nature [62].

These BPNN training algorithms such as conjugate-Gradient with Powell–Beale restart (traincgb), conjugate-Gradient with Fletcher–Reeves updates (traincgf), conjugate Gradient with Polak–Ribiere updates (traincgp), scaled conjugate gradient BP (trainscg) and resilient BP (trainrp) and one step secant BP (trainoss) are amongst the fastest response with accurate performance. The disadvantages of BPNN is arbitrarily long training time for complex system applications as well as tedious selection

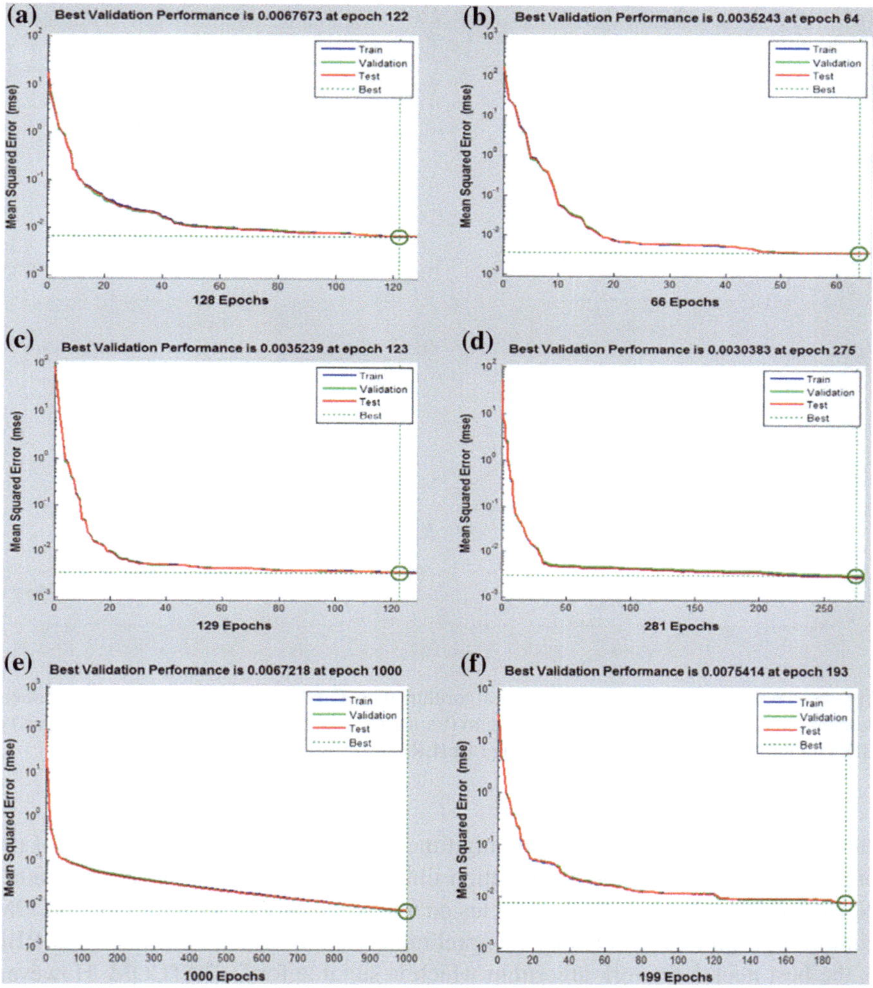

Fig. 15 Performance of BPNN training algorithm with Gradient derivatives using conjugate gradient **a** Powell–Beale restart (traincgb) **b** Fletcher–Reeves up-dates (traincgf) **c** Polak–Ribiere updates (traincgp) **d** scaled conjugate gradient (trainscg) **e** RPROP resilient back propagation (trainrp) **f** one step secant (trainoss)

of hidden layers for neural networks [33]. As compared with Levenverg–Marquadt (trainlm), Bayesian regulation (trainbr) and BFGS Quasi-Newton (trainbfg), it has poor performance on dynamic system because of large computation time but it has proven to be the most accurate learning algorithm. Levenverg–Marquadt learning algorithm is typically used for BP control algorithm in the previous literatures. On the other hand, RBF is also proven to be the fastest response with the most accurate performance. As mentioned earlier, the architecture of RBF is simple, accurate and its dynamic response is improved since its process to update algorithm parameters

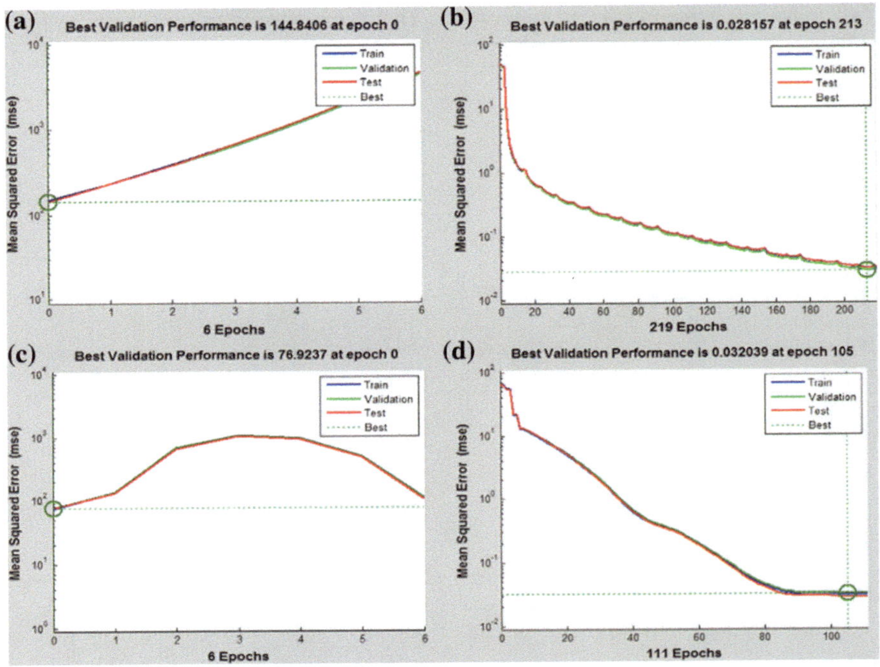

Fig. 16 Performance of BPNN training algorithm with Gradient derivatives using **a** Gradient descent (GD) (traingd) **b** GD with momentum (traingdm) **c** GD with adaptive learning rate (LR) **d** GD with momentum and adaptive learning rate (LR)

can occur concurrently while avoiding time delay to complete computation of the previous values. Thus, the computation time of RBF is much reduced than other NN algorithms [32, 33]. As the conclusion, the preference of RBF for DSTATCOM control algorithm is justified by this preliminary study which concludes that RBF is the best neural network algorithm which is suitable for DSTATCOM. However, the output of the DSTATCOM as presented previously by using control algorithm based on VRC control algorithm can reduce supply current harmonics from 53.31 to 9.53% which is still higher than the specified THD limit by IEEE Standard 519-2014. Thus, this research based on previous works based on DSTATCOM based RBF can further decrease the THD of the line current up to 4.7% which is implemented with self-supporting dc-link compensation [31]. Moreover, this analysis is also based on previous works based on BPNN DSTATCOM which the THD of the line current can be improved up to 2.9% which is implemented with self-supporting dc-link compensation [33]. Thus, these learning algorithms can further reduce the THD according to the IEEE Standard 519-2014 guidelines with less training time. In summary, the application of neural network control algorithm can be further developed in the area of power electronics and electrical machine drives especially on distribution system [56, 57]. Its implementation for electrical power engineering in three-phase system through hardware implementation and industrial applications can be further improved

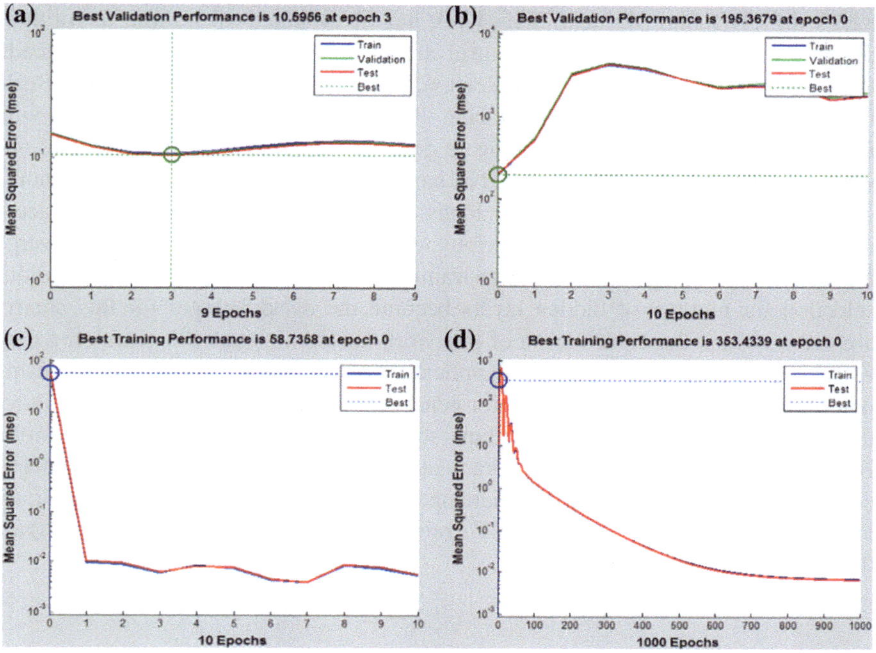

Fig. 17 Performance of BPNN training algorithm with supervised weight/bias training functions **a** Batch training with weight abd bias learning rules (trainb) **b** Cyclical order (trainc) **c** Random order (trainr) **d** Sequential order (trains)

with the advancement of theoretical neural network research works [33, 58]. Thus, it can be concluded that neural network is the recent research area in power engineering especially DSTATCOM with lots of improvement that have been done by researchers theoretically in order to enhance its performance. The most widely used NN control algorithms in the previous literatures for DSTATCOM are Adaptive Linear Neuron (Adaline) and BPNN. However, the drawback of BPNN is the large computation time which is not suitable for dynamic and complex system such as DSTATCOM. Thus, this work has proposed other learning algorithms of back propagation that can reduce its large training time such as conjugate-gradient with Powell–Beale restart, conjugate-Gradient with Fletcher–Reeves updates, conjugate-Gradient with Polak–Ribiere updates, scaled conjugate-Gradient BP and resilient BP and one step secant BP with accurate performance as well as RBFNN for faster response.

6 Conclusions

The investigation on BPNN control algorithm for DSTATCOM in order to improve the harmonic distortions demonstrates successful performance in harmonic compensation under nonlinear loads which has improved from 53.31 to 9.53% with

82.1% of THD improvement. This work has investigated 18 training algorithms of BPNN for the DSTATCOM control algorithm such as Levenverg–Marquadt, Bayesian regulation, BFGS Quasi-Newton, conjugate-Gradient with Powell–Beale restart, conjugate-Gradient with Fletcher–Reeves updates, conjugate-gradient with Polak–Ribiere updates, scaled conjugate gradient BP, RPROP resilient BP and one step secant BP. It is concluded that BPNN based DSTATCOM is suitable for harmonic current elimination under nonlinear loads and improve its training time for accurate DSTATCOM performance. Previous work shows that BPNN with Levenverg–Marquadt training algorithm has large training time for complex system and tedious selection for number of hidden layers become the disadvantages for this control algorithm. Thus, the contribution of this work has demonstrated less training time with high accuracy in the complex application of PV based DSTATCOM by using other training algorithms of BPNN for generating compensating currents was introduced and suitable training algorithms such as conjugate-gradient with Powell–Beale restart (traincgb), conjugate-Gradient with Fletcher–Reeves updates (traincgf), conjugate-Gradient with Polak–Ribiere updates (train-cgp), scaled conjugate Gradient BP (trainscg), RPROP resilient BP (trainrp) and one step secant BP (trainoss) are demonstrated.

References

1. Xue Y, Chang L, Kjær SB, Bordonau J, Shimizu T (2004) Topologies of single-phase inverters for small distributed power generators: an overview. IEEE Trans Power Electron 19(5):1305–1314
2. Shahnia F, Ghosh A, Ledwich G, Zare F (2014) Voltage unbalance improvement in low voltage residential feeders with rooftop PVs using custom power devices. Int J Electr Power Energy Syst 55:362–377
3. Trichakis P, Taylor PC, Lyons PF, Hair R (2008) Predicting the technical impacts of high levels of small-scale embedded generators on low-voltage networks. IET Renew Power Gener 2(4):249–262
4. Hussin MZ, Hasliza N, Yaacob A, Zain ZM, Omar AM, Shaari S (2012) A development and challenges of grid-connected photovoltaic system in Malaysia. In: 2012 IEEE control and system graduate research colloquium (ICSGRC 2012). pp 191–196
5. Papathanassiou SA (2007) A technical evaluation framework for the connection of DG to the distribution network. Electr Power Syst Res 77:24–34
6. Eltawil MA, Zhao Z (2010) Grid-connected photovoltaic power systems: technical and potential problems a review. Renew Sustain Energy Rev 14:112–129
7. Karimi M, Mokhlis H, Naidu K, Uddin S, Bakar AHA (2016) Photovoltaic penetration issues and impacts in distribution network a review. Renew Sustain Energy Rev 53:594–605
8. Hernández JC, Ortega MJ, De Cruz J, Vera D (2011) Guidelines for the technical assessment of harmonic, flicker and unbalance emission limits for PV-distributed generation. Electr Power Syst Res 81:1247–1257
9. IEEE recommended practice for utility interface of photovoltaic (PV) systems: IEEE Std 929-2000. Technical report (2000)
10. Kannan VK, Rengarajan N (2014) Investigating the performance of photovoltaic based DSTATCOM using I cosΦ algorithm. Int J Electr Power Energy Syst 54:376–386
11. Mahela OP, Shaik AG (2015) A review of distribution static compensator. Renew Sustain Energy Rev 50:531–546

12. Kannan VK, Rengarajan N (2012) Photovoltaic based distribution static compensator for power quality improvement. Int J Electr Power Energy Syst 42(1):685–692
13. Agarwal RK, Hussain I, Singh B (2017) Implementation of LLMF control algorithm for three-phase grid tied SPV- DSTATCOM system. IEEE Trans Ind Electron PP(99):1–10
14. Mahela OM, Shaik AG (2016) Power quality improvement in distribution network using DSTATCOM with battery energy storage system. Int J Electr Power Energy Syst 83:229–240
15. Chidurala A, Saha TK, Mithulananthan N (2013) Power quality enhancement in unbalanced distribution network using Solar-DSTATCOM. In: Australasian universities power engineering conference, AUPEC 2013. Hobart, pp 1–6
16. Mallick RK, Sinha S, Mohanty S, Kumar S (2016) Design of optimal controller for DSTATCOM using differential evolution technique. In: International conference on electrical, electronics, and optimization techniques (ICEEOT). pp 1432–1437
17. Singh B, Jain C, Goel S (2014) ILST control algorithm of single-stage dual purpose grid connected solar PV system. IEEE Trans Power Electron 29(10):5347–5357
18. Jain C, Singh B (2015) An offset reduction second order generalized integrator based control algorithm for single-phase S-DSTATCOM. In: 2015 39th National systems conference (NSC). pp 1–6
19. Ouchen S, Betka A, Abdeddaim S, Menadi A (2016) Fuzzy-predictive direct power control implementation of a grid connected photovoltaic system, associated with an active power filter. Energy Convers Manag 122:515–525
20. Menniti D, Pinnarelli A, Sorrentino N (2010) An hybrid PV-wind supply system with D-Statcom interface for a water-lift station. In: 2010 international symposium on power electronics electrical drives automation and motion (SPEEDAM). pp 1387–1392
21. Samuel P, Gupta R, Chandra D (2009) Grid interface of photovoltaic-micro turbine hybrid based power for voltage support and control using VSI in rural applications. In: 2009 IEEE Power & energy society general meeting. pp 1–6
22. Senthilkumar A, Poongothai K, Selvakumar S, Silambarasan M (2015) Mitigation of harmonic distortion in microgrid system using adaptive neural learning algorithm based shunt active power filter. Procedia Technol 21:147–154
23. Sangeetha B, Geetha K (2014) Performance of multilevel shunt active filter for smart grid applications. Int J Electr Power Energy Syst 63:927–932
24. Kumar TP, Venkateshwarlu S (2017) Investigating the performance of PV interfaced PBT based DSTATCOM by adaptive fuzzy logic controller for reactive power management. In: 2017 11th international conference on intelligent systems and control (ISCO). Coimbatore, pp 160–165
25. Mishra S, Ray PK (2016) Power quality improvement using photovoltaic fed DSTATCOM based on JAYA optimization. IEEE Trans Sustain Energy 7(4):1672–1680
26. Hussain I, Kandpal M, Singh B (2017) Grid integration of single stage SPV-STATCOM using cascaded 7-level VSC. Int J Electr Power Energy Syst 93:238–252
27. Shivam S, Hussain I, Singh B (2016) Real-time implementation of SPV system with DSTATCOM capabilities in three-phase four-wire distribution system. IET Gener Transm Distrib 11(2):495–503
28. Beniwal N, Hussain I, Singh B, Chandra A, Al-Haddad K (2016) Adaptive control scheme for three-phase four wire grid tied SPV system with DSTATCOM capabilities. In: 2016 National power systems conference (NPSC). Bhubaneswar, pp 1–6
29. Shivam S, Hussain I, Singh B (2016) Dual-sign error based adaptive control for three phase grid tied SECS with DSTATCOM capabilities. In: 2016 IEEE international conference on power electronics, drives and energy systems (PEDES). Trivandrum, pp 1–6
30. Ullah W, Mekhilef S, Seyedmahmoudian M, Horan B (2017) Active power filter (APF) for mitigation of power quality issues in grid integration of wind and photovoltaic energy conversion system. Renew Sustain Energy Rev 70:635–655
31. Singh B, Arya SR, Dube SK, Chandra A, Al-Haddad K (2013) Implementation of DSTATCOM using neural network based radial basis function. In: 2013 IEEE industry applications society annual meeting. Lake Buena Vista, pp 1–8

32. Chang GW, Chen C-I, Teng Y-F (2010) Radial-basis-function-based neural network for harmonic detection. IEEE Trans Ind Electron 57(6):2171–2179
33. Singh B, Arya SR (2014) Back-propagation control algorithm for power quality improvement using DSTATCOM. IEEE Trans Ind Electron 61(3):1204–1212
34. Arya SR, Singh B (2013) Performance of DSTATCOM using leaky LMS control algorithm. IEEE J Emerg Sel Top Power Electron 1(2):104–113
35. Emadi A, Nasiri A, Bekiarov SB (2005) Uninterruptible power supplies and active filters. CRC Press LLC, Boca Raton
36. Ewald FF, Mohammad ASM (2008) Power quality in power systems and electrical machines. Elsevier Academic Press, London
37. Singh B, Bhuvaneswari G, Arya SR (2012) Review on power quality solution technology. Asian Power Electron J 6(2):19–27
38. Singh B, Arya SR, Chandra A, Al-Haddad K (2012) Implementation of adaptive filter based control algorithm for distribution static compensator. In: 2012 IEEE industry applications society annual meeting. Las Vegas, pp 1–8
39. IEEE. IEEE Std 1531–2003 IEEE. Technical report (2003)
40. IEEE. IEC 61000-3-2:2018. Technical report (2018)
41. IEEE. IEEE Std 519–2014 (Revision of IEEE Std 519-1992). Technical report (2014)
42. Arya SR, Singh B (2014) Power quality improvement under nonideal AC mains in distribution system. Electr Power Syst Res 106:86–94
43. Padiyar KR (2007) FACTS controllers in power transmission and distribution. New Age International (P) Limited, Publishers, New Delhi
44. Akagi H, Watanabe EH, Aredes M (2007) Instantaneous power theory and applications to power conditioning. Wiley, Hoboken
45. Rechka S, Ngandui E, Jianhong X, Sicard P (2003) Analysis of harmonic detection algorithms and their application to active power filters for harmonics compensation and resonance damping. Can J Electr Comput Eng 28(1):41–51
46. Moreno-Munoz A (2007) Power quality mitigation technologies in a distributed environment. Springer, Berlin
47. Singh B, Jayaprakash P, Somayajulu TR, Kothari DP (2009) Reduced rating VSC with a zigzag transformer for current compensation in a three-phase four-wire distribution system. IEEE Trans Power Deliv 24(1):249–259
48. Singh B, Chandra A, Al-Haddad K (2015) Power quality: problems and mitigation techniques. Wiley, West Sussex
49. Chen C-S, Lin C, Hsieh W-L, Hsu C-T, Te-tien K (2013) Enhancement of PV penetration with DSTATCOM in Taipower distribution system. IEEE Trans Power Syst 28(2):1560–1567
50. Da Costa Teixeira C (2004) Power quality solutions for low and medium voltage critical loads. In: 2004 IEEE/PES transmision and distribution conference and exposition: Latin America (IEEE Cat. No. 04EX956). Sao Paulo, pp 326–331
51. Masand D, Jain S, Agnihotri G (2006) Control algorithms for distribution compensator. In: IEEE ISIE. Montreal, pp 1830–1834
52. El-Habrouk M, Darwish MK, Mehta P (2000) Active power filters: a review. IEE Proc Electr Power Appl 147(5):403–413
53. Singh B, Al-Haddad K, Chandra A (1999) A review of active filters for power quality improvement. IEEE Trans Ind Electron 46(5):960–971
54. Arya SR, Singh B, Chandra A, Al-Haddad K (2012). Control of shunt custom power device based on anti-hebbian learning algorithm. In: IECON proceedings (Industrial electronics conference). pp 1246–1251
55. Masand D, Jain S, Agnihotri G (2008) Control strategies for distribution static compensator for power quality improvement. IETE J Res 54(6):421–428
56. Arya SR, Singh B (2014) Neural network based conductance estimation control algorithm for shunt compensation. IEEE Trans Ind Inform 10(1):569–577
57. Bose BK (1994) Expert system, fuzzy logic, and neural network applications in power electronics and motion control. Proc IEEE 82(8):1303–1323

58. Ahn H-S, Chen YQ, Moore KL (2007) Iterative learning control: brief survey and categorization. IEEE Trans Syst Man Cybern Part C: Appl Rev 37(6):1099–1121
59. Baharudin NH, Mansur TMNT, Hassan SIS, Saad P, Ali R, Lada MY (2015) A comparison of distribution static synchronous compensator (DSTATCOM) control algorithms for harmonic elimination. ARPN J Eng Appl Sci 10(22)
60. Ying C, Qingsheng L (2009) New research on harmonic detection based on neural network for power system. In: 2009 Third international symposium on intelligent information technology application. Shanghai, pp 113–116
61. Jung I, Wang G (2007) Pattern classification of back-propagation algorithm using exclusive connecting network. World Acad Sci Eng Technol 1(12):180–184
62. Lu Y, Sundararajan N, Saratchandran P (1998) Performance evaluation of a sequential minimal radial basis function (RBF) neural network learning algorithm. IEEE Trans Neural Netw 9(2):308–18

Power Extraction from PV Module Using Hybrid ANFIS Controller

Tata Venkat Dixit, Anamika Yadav, S. Gupta and Almoataz Y. Abdelaziz

Abstract The characteristic of PV module is nonlinear, complex in nature and its performance depends on different environmental factors. In order to enhance the efficiency of photovoltaic power system, selection of a suitable power converters and control strategies are essential. In this chapter, the performance of soft-computing techniques of MPPT such as ANN and Hybrid-ANFIS are compared with well established Modified Incremental Conductance method under load and solar irradiance change. The ANFIS is able to exploit both data and knowledge to formulate more efficient hybrid intelligent system. It learns the information from experimental data and automatically determines the best membership parameters and rule bases associated to Fuzzy Inference System (FIS) to map given input output data. In this chapter, the parameters of FIS are tuned by Back-Propagation (BP) or hybrid (combination of Least Square Estimation and BP) method. Also, the effect of load impedance and converter topologies on ANFIS controller design has been investigated. Further, the detailed description of hardware implementation of ANFIS controller on DSP/FPGA platform has been presented.

T. V. Dixit
Department of Electrical Engineering, Goverment Polytechnic Kondagaon (C.G.), Jagdalpur, India
e-mail: tvdixit@gmail.com

A. Yadav · S. Gupta
Department of Electrical Engineering, National Institute of Technology, Raipur, India
e-mail: ayadav.ele@nitrr.ac.in

S. Gupta
e-mail: sgupta.ele@nitrr.ac.in

A. Y. Abdelaziz (✉)
Faculty of Engineering, Electrical Power and Machines Department, Ain Shams University, Cairo, Egypt
e-mail: almoatazabdelaziz@hotmail.com

R.-E. Precup et al. (eds.), *Solar Photovoltaic Power Plants*, Power Systems,
https://doi.org/10.1007/978-981-13-6151-7_10

List of Abbreviations/Nomenclature

I_{PV} :	Solar cell load current
I_{SC} :	Short-circuit current
I_{ph} :	Photocurrent ($I_{pv} \approx I_{sc}$)
V_{PV} :	Terminal voltage of solar panel
V_{OC} :	Open circuit voltage of panel
I_O :	Reverse saturation current
q :	Electron charge (1.602E-19 C)
STC/stc :	Standard Test Condition
k :	Boltzmann's Constant (1.381E-23 J/K)
T_{stc} :	Junction temperature (in Kelvin) at STC
β_v :	Temperature coefficient of open circuit voltage
AM :	Air-Mass
G :	Solar irradiance (W/m^2)
R_O :	Load impedance
d :	Duty ratio of converter
R_{in} :	Input resistance of converter
V_{mp} :	Voltage of panel at MPP
I_{mp} :	Current at MPP
A :	Thermal voltage timing completion factor
P_{mp} :	Power at maximum power point
d_{mp} :	Duty ratio at MPP
$\alpha_I(mA/°C)$:	Temperature coefficient of short-circuit current

1 Introduction

The Solar power plants are playing a vital role in the energy market due to its eco-friendly nature and low operational and maintenance cost [1]. In fact, the energy supplied from the solar power is increasing by 20–25% per annum over past three decades [2, 3]. Noteworthy, the efficiency of Photovoltaic module/system is poor due to its non-linear characteristic, cell temperature, and solar irradiation dependent nature [4]. Hence, extracting the maximum power from PV system for a given load is an indispensable task. The efficiency of these systems depends mainly on load impedance, converter topology, and control strategies. In order to enhance the power extraction efficiency of PV system, efficient algorithms are integrated with PV module. Presently, several MPPT algorithms have been proposed by researchers to extract the maximum power from PV module. These algorithms differ in many aspects, such as cost and complexity, sensor requirement, converter speed, topology and complexity, robustness, tracking efficiency under environment variation and hardware requirement. The objective of these algorithms is to track and extract the maximum power for photovoltaic module. This is achieved by matching the load

Table 1 Comparison of Expert System (ES), Fuzzy Set (FS), Neural Network (NN) and swarm intelligence [8]

Features	Expert system	Fuzzy set	Neural network	Evolutionary computation
Knowledge representation	Good	Moderate	Poor	Modest
Uncertainty tolerance	Good	Moderate	Moderate	Moderate
Imprecision tolerance	Poor	Moderate	Moderate	Moderate
Adaptability	Poor	Moderate	Moderate	Moderate
Learning ability	Poor	Poor	Moderate	Moderate
Explanation ability	Moderate	Moderate	Poor	Modest
Knowledge discovery and data mining	Poor	Modest	Moderate	Good
Maintainability	Poor	Modest	Moderate	Good

impedance of the circuit with the source impedance of the photovoltaic panel at MPP by varying the duty ratio of the converter.

Nowadays, popular artificial intelligence technique such as fuzzy logic and ANN are widely adopted for solving complex nonlinear uncertain problems [5]. The Fuzzy Inference System (FIS) can model qualitative aspect of several uncertain problems without employing precise quantitative analysis. Although, the FIS has many practical limitations such as, lack of standard methods for: (a) transformation into rule-base (b) effective tuning of membership functions for better performance index. Further, ANN model is a block-box approach of controller design and it needs huge experimental data for effective design [6, 7]. Thus, ANFIS acts as a tool for constructing the fuzzy rules based on IF-THEN rules with proper membership functions to produce stipulated relationship between input-output. In general, hybrid learning system are preferred due to the problems associated with basic learning rules (gradient descent and chain rule) such as, slowness and convergence at local minima. The hybrid-ANFIS has two modes of adaptive learning such as Batch Learning and On-line learning are adopted. The comparative analysis of various intelligence systems on the basis of uncertainty and imprecision tolerance, adaptability and learning ability etc. are reported in Table 1. In this chapter, the performance of soft-computing based Maximum Power Extraction (MPE) technique such as Hybrid-ANFIS is compared with well-established Incremental Conductance method under load and solar irradiance variation while effect of temperature variation are ignored during this study. Noteworthy, the effect of load impedance and converter topology on ANFIS controller design has been investigated. Also, the detailed description of hardware implementation of ANFIS controller on DSP/FPGA platform has been presented.

2 Empirical Model of PV Module

In literature, several mathematical models are reported that illustrates the operation and behavior of photovoltaic panel. These models differ in computational burden, accuracy and numbers of parameters needed to estimate the voltage-current characteristics [9–11]. In this section, an empirical model of photovoltaic module for RTDS (OPAL-RT) platform has been discussed. In fact, approximate equivalent circuit modeling of PV cell is mandatory for real time simulation. The five parameter model reported in the literature involves iterative solutions due to an algebraic loop. Noteworthy, the successful results of these models on SIMULINK/RTDS platform could be achieved provided the algebraic loop solver converges to a definite answer. Therefore, the implementation of these models on Real Time Digital Simulator or HIL simulation platform (OPAL-RT) results compilation error. This limitation can be overcome by empirical model of solar cell presented in the next subsection.

2.1 *Improved Model of PV for RTDS Implementation*

Although, this model has no physical interpretation but it can be characterized by simple resolution which requires only four parameters V_{oc}, I_{sc}, I_{mp} and V_{mp} [12]. Afore-mentioned, the form of this model is essentially based on five parameter model of solar cell.

$$I_{pv} = I_{sc}\left\{1 - C_1\left(e^{\frac{V_{pv}}{C_2 V_{oc}}} - 1\right)\right\} \tag{1}$$

where, the above equation exactly fits to short circuit point. Further, the open circuit and maximum power point can be approximated as

$$C_2 = \frac{\left(\frac{V_{mp}}{V_{oc}} - 1\right)}{\ln\left(1 - \frac{I_{mp}}{I_{sc}}\right)} \tag{2}$$

$$C_1 = \left(1 - \frac{I_{mp}}{I_{sc}} \exp\left(-\frac{V_{mp}}{C_2 V_{oc}}\right)\right) \tag{3}$$

2.2 *Effect of Solar Irradiance and Temperature Variation*

In this section, the effect of temperature and solar irradiance variation on the characteristics of solar panel has been investigated. Usually, the solar panel specifications such as V_{oc}, I_{sc}, I_{mp} and V_{mp} are always reported at STC (at cell temperature 25 °C, AM 1.5 and 1 kW/m^2 solar irradiance) by manufacturer. Therefore, the Eqs. (1)–(3) are valid characteristics equation at STC. In order to include the influence of solar irradiance and temperature variation on cell parameter V_{oc}, I_{sc}, I_{mp} and V_{mp} several methods are discussed in the literature.

The Ref. [13] has considered normalized parameters of V_{oc}, I_{sc}, I_{mp} and V_{mp} per unit change in the temperature. The influence of solar irradiance on temperature coefficient of V_{oc} and V_{mp} is assumed to be insignificant. Whereas the temperature coefficient of short-circuit and maximum power point current are scaled by the ratio of the actual irradiance level to irradiance level at STC. The set of equation to estimate the performance of panel is given by Eq. (4). References [12, 14–16] have proposed a model that shows modest variation in open circuit voltage due to its logarithmic relationship with solar irradiance variation while the short circuit current is directly proportional to per unit solar irradiance.

$$\left.\begin{aligned} I_{SC} &= I_{SC_{stc}}[1 - \alpha_{Isc}(T_{stc} - T)] \\ I_{mp} &= I_{mp_{stc}}[1 - \alpha_{I_{mp}}(T_{stc} - T)] \\ V_{OC} &= V_{OC_{stc}} - \beta_{VOC} V_{OC_{stc}}(T_{stc} - T) \\ V_{mp} &= V_{mp_{stc}} - \beta_{Vmp} V_{mp_{stc}}(T_{stc} - T) \end{aligned}\right\} \text{Where, } \beta_{VOC} \text{ and } \beta_{Vmp} < 0; \text{ and } \alpha_{Isc} > 0 \quad (4)$$

It is noteworthy that the simultaneous effect of solar irradiance and temperature variation on can be included in equation by using superposition principle as given by Eqs. (5) and (6). In Eq. (5) $\alpha_I(mA/°C)$ shows positive temperature coefficient of short-circuit or maximum power point current. Similarly, in Eq. (6) $\beta_V(V/°C)$ is the negative temperature coefficient of voltage corresponding to V_{oc} and V_{mp}

$$I(G, T)_x = I_{x_{stc}} \frac{G}{G_{stc}} + [\alpha_I(T - T_{stc})];\ Where, \alpha_I > 0 \quad (5)$$

$$V(G, T)_x = V_{y_{stc}} + \left(\frac{AKT}{q}\right) \frac{G}{G_{stc}} + [\beta_V(T - T_{stc})];\ Where, \beta_V < 0 \quad (6)$$

In the present work, Eqs. (1)–(6) are considered to implement the solar module in HIL platform (RT-LAB) and the limitation of algebraic loop has been overcome by adding memory block in load voltage sensing path. The current-voltage and power-voltage characteristics of solar panel during change in solar irradiance and temperature are shown in Figs. 1 and 2 respectively.

3 Maximum Power Extraction Constraints of PV Module

As the efficiency of solar system is low, it requires an efficient converter and algorithm to extract the maximum power from the solar module at given solar irradiance and temperature. Noteworthy, the solar module has a single operating point at which

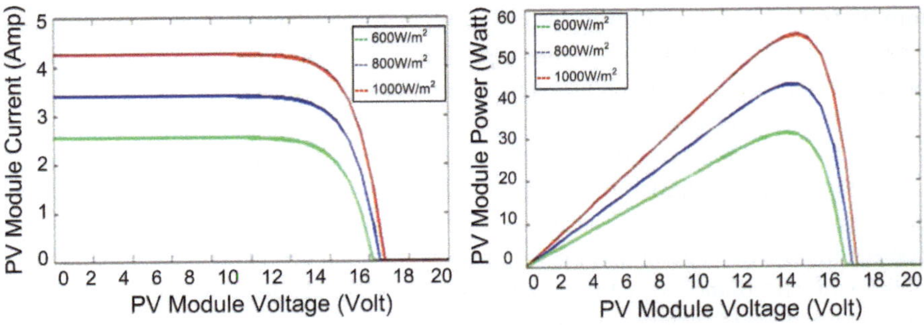

Fig. 1 Current-voltage and power -voltage characteristics during change in solar irradiance

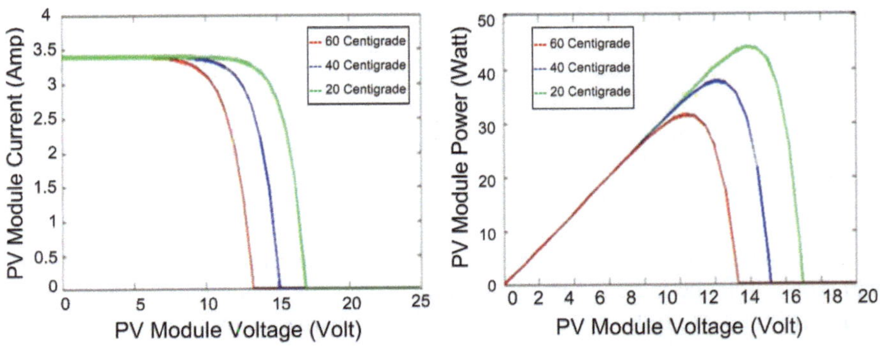

Fig. 2 Current-voltage and power -voltage characteristics during change in temperature

maximum available power can be extracted at particular solar irradiance and temperature as shown in Fig. 1. Hence, the purpose of maximum power point tracking technique is to track the MPP under given environmental condition [17, 18]. As per impedance matching principle, in order to transfer maximum power, the load impedance should match with the source impedance of the solar module. This can be achieved by DC–DC converter that acts as agent to match the load impedance with the source impedance. In [19, 20], several non-isolated DC–DC converters such as, buck, boost, buck-boost, Cuk, [21] and SEPIC are reviewed for photovoltaic applications. The comparative study of these converter based on number of component required, load constraint on MPPT, output voltage polarity, input and output current ripple, cost, control effort, performance under temperature and irradiance variation are summarized in [19, 20, 22]. The selection of right configuration of DC–DC converter and effect of load line on the performance has not been investigated widely [23]. In this work, basic boost topologies such as conventional Boost Converter (CBC) and SEPIC converters with resistive load have been considered. In this context the performance of advanced MPPT algorithms such as ANN and ANFIS are compared with well-established methods such as P&O [24] and Incremental Conductance [25] for all above configurations.

Table 2 Comparison of proposed and existing MPPT methods [26–28]

Performance parameters	Modified P&O	Modified IC	Fuzzy	ANN	ANFIS
Type	Direct	Direct	Direct	Hybrid	Hybrid
Performance under irradiance change	Good	Good	Good	Depends on training data	Good
Complexity	Low	Low	High	High	High
PV array dependent	No	No	Yes	Yes	Yes
Speed	Slow	Medium	Fast	Fast	Fast
Reliability	Medium	Medium	Medium	Medium	Medium
Implementation	Digital/Analog	Digital	Digital	Digital	Digital
Memory requirement	No	No	Yes	Yes	Yes
Parameter tuning	Not required	Not required	Required	Required	Required
Sensor used	Voltage, current	Voltage, current	Voltage, current	Voltage, current	Voltage, current
% Overshoot	No	No	No	No	Moderate
Power extraction efficiency % for CBC	~94.96–97.72	~97.57–99.81	~87–90.5*	~97.61–98.93	~98.63–99.63
Cost	Inexpensive	Inexpensive	Expensive	Expensive	Expensive

3.1 *Maximum Power Extraction Techniques*

Aforementioned, to extract the maximum power from the PV module, load impedance must match with the source impedance. Therefore, several algorithms are proposed in the literature for achieving MPPT by varying the duty ratio of the converter. In the present work, the performance of soft-computing MPPT techniques such as ANN, ANFIS [26] are compared with well-established incremental conductance methods. A comparison among these MPPT methods based on type, efficiency, transient performance, complexity, speed, reliability, implementation constraint, memory and sensor requirement, necessity of tuning, percentage overshoot and cost are tabulated in Table 2.

3.2 *Modified Incremental Conductance method*

This MPPT algorithm is based on hill climbing philosophy of power curve. If the slope of power curve of photovoltaic module is positive $\left(\frac{dP_{pv}}{dV_{pv}} > 0\right)$ then with increase in terminal voltage $V_{pv(n+1)-V_{pv(n)}}$ of the module, the duty ratio will increase [29]. Similarly, if slope of power curve is negative, the increase in terminal voltage will be the cause of reduction of duty ratio. The complete operation of incremental conductance method is briefed as follows:

1. If slope of power curve is positive $\left(\frac{dP_{pv}}{dV_{pv}} > 0\right)$ and change in the terminal voltage of panel is also positive ($\Delta V > 0$) then the Operating Point (OP) lies left side of the MPP.In order to achieve MPP the control variable V_{pv} should increase as a result $\Delta V > 0$ becomes more positive.
2. If $\left(\frac{dP_{pv}}{dV_{pv}} > 0\right)$ but $\Delta V < 0$, i.e. the OP lies on the left side of MPP. Hence, to reach MPP the control variable V_{pv} should increase for ensuring positive change in the terminal voltage of panel ($\Delta V > 0$).
3. If $\left(\frac{dP_{pv}}{dV_{pv}} < 0\right)$ and $\Delta V > 0$ i.e. the OP lies on the right side of MPP. Therefore, for reaching MPP the control variable V_{pv} should decrease that results positive change in terminal voltage of solar panel ($\Delta V < 0$).
4. Similarly, if $\left(\frac{dP_{pv}}{dV_{pv}} < 0\right)$ and $\Delta V < 0$ i.e. the OP lies on the right side of MPP. Thus, for achieving MPP, the control variable V_{pv} should decrease which results more negative ($\Delta V < 0$).
5. For $\Delta P = 0$ the OP lies on MPP and control variable should remain same.
6. If $\Delta V = 0$; $\Delta P \neq 0$ and irradiance level has changed then

   a. Control variable V_{pv} should increase if $\Delta P > 0$ and irradiance increased hence ΔV will be positive.
   b. Control variable V_{pv} should decrease if $\Delta P < 0$ and irradiance level decreased results positive.

Where, $\Delta V = V_{pv(n+1)} - V_{pv(n)}$ and $\Delta P = P_{pv(n+1)} - P_{pv(n)}$; and 'n' is the sampled instant.

The advantage of this method is that it does not oscillate around MPP unlike classic P&O method as well as it can track the MPP under sudden solar isolation change. The flowchart of modified incremental conductance method is shown in Fig. 3.

3.3 *ANN Based MPPT Controller*

Basically, ANN is a black box data driven modeling approach to track the MPP. This model does not require any information about the solar panel. The block diagram for implementing ANN-based MPPT scheme is depicted in Fig. 4.The input to proposed controller is voltage and current of PV panel and output is the duty ratio. Where, this duty ratio is used to operate the DC–DC converters (CBC and SEPIC). The selection of transfer function and neuron in each layer are obtained by trial and error. The most popular Levenberg–Marquardt algorithm has been adopted to train the neural network for all three converter topologies. To train the ANN model data is obtained

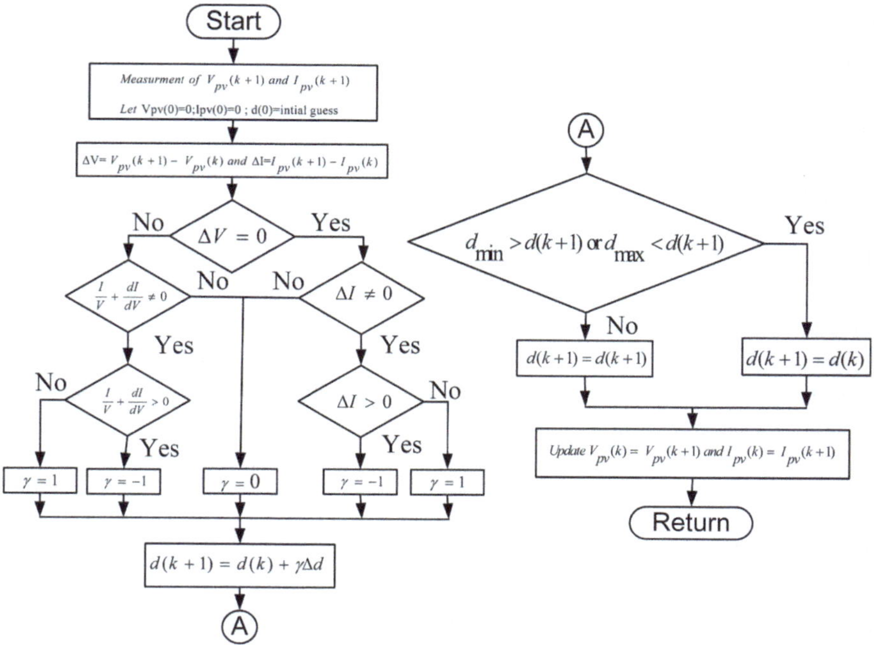

Fig. 3 Flow chart of modified IC MPPT algorithm

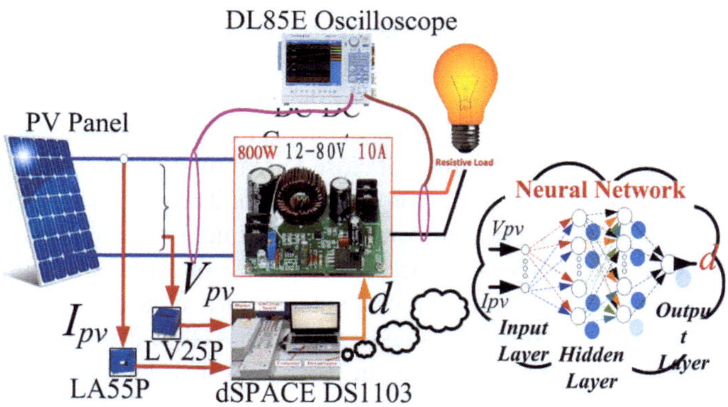

Fig. 4 Implementation of ANN-based MPPT control scheme on dSPACE DS1103 controller

by P&O MPPT algorithm for different converter topologies under load variation and isolation level change. In order to train, test and validate the neural network for CBC and SEPIC converter 69904 patterns are fed to neural networks. The details of MLP network for each NN (for CBC and SEPIC) are tabulated in Table 3. Further, the training and testing results of ANN for CBC and SEPIC converters are depicted in Figs. 5 and 6 respectively.

Table 3 Details of MLP structure for different converter topologies

Parameters of MLP structure		CBC	SEPIC
No of nodes in input layer		2	2
In first hidden layer	Neurons	20	36
	Type	TANSIG	LOGSIG
In second hidden layer	Neurons	10	20
	Type	LOGSIG	TANSIG
In third hidden layer	Neurons	...	10
	Type	...	TANSIG
In output layer	Neurons	1	1
	Type	PURLINE	PURLINE
Training goal (MSE)		8.71E-6 @200 epochs	2.7E-4@200 epochs
% error in output prediction		<0.27	<9

3.4 ANFIS Based Controller

The popular artificial intelligence technique such as fuzzy logic and ANN are found effective for solving engineering problems such economic load dispatch, load forecasting, security assessment of power system while the conventional techniques failed to provide accurate solution. The advancement in digital signal processors and control strategies made artificial intelligence an attractive solution to industrial applications such as automobile, missile technology, robotics and energy harvesting etc. Moreover, the fuzzy logic has ability to convert the linguistic variables into single valued number using Fuzzy Inference System (FIS) [16]. Another advantage of FIS is that it allows setting the rule quit close to real time process and they can be easily interpretable [30]. The limitation of FIS is that it requires vast expert knowledge to define fuzzy rules base and computational time for tuning the membership parameters. Although, ANN is a block-box approach of artificial intelligence that acquires the knowledge by training process but it requires huge data-set and the analysis of trained network is difficult to understand. An ANFIS is the hybridized form of ANN ad FIS earlier proposed by Jang in 1990s [31]. This intelligent controller do not require exact mathematical model of the system. The ANFIS can be applicable for all the applications where intelligent controllers such as FIS and neural network are applicable for modeling and control of complex system [32]. The ANFIS is able to exploit both data and knowledge to formulate more efficient hybrid intelligent system. The ANFIS controller automatically determines the best membership parameters and rule bases associated to FIS for mapping given input output data-set. The parameters of FIS are tuned by Back-Propagation (BP) or hybrid (combination of Least Square Estimation and BP) method.

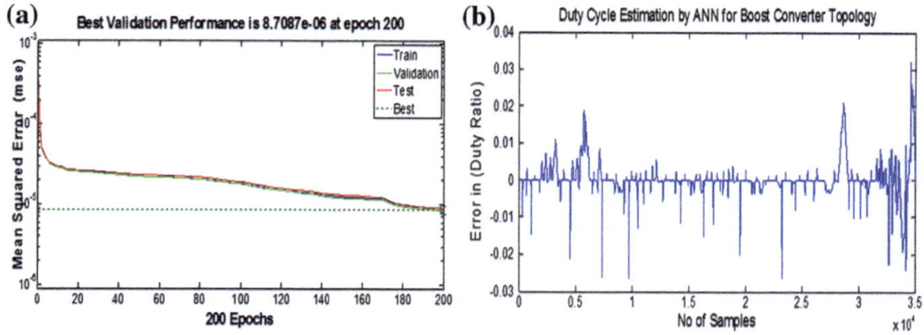

Fig. 5 **a** Convergence curve of MSE error of ANN-based MPPT controller for CBC **b** training error in duty cycle estimation for CBC

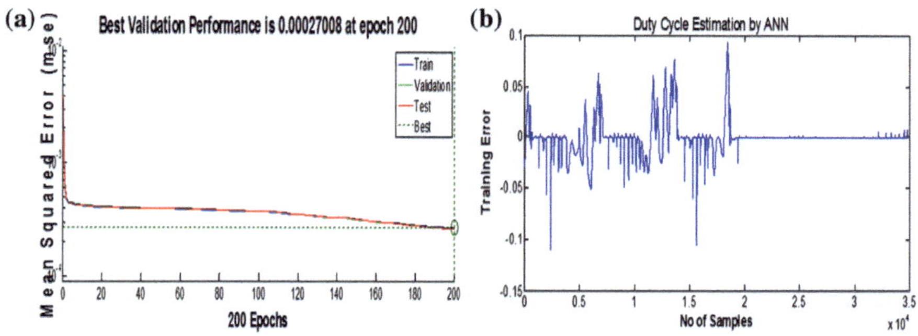

Fig. 6 **a** Convergence curve of MSE error of ANN-based MPPT controller for SEPIC converter **b** training error in duty cycle estimation for SEPIC converter

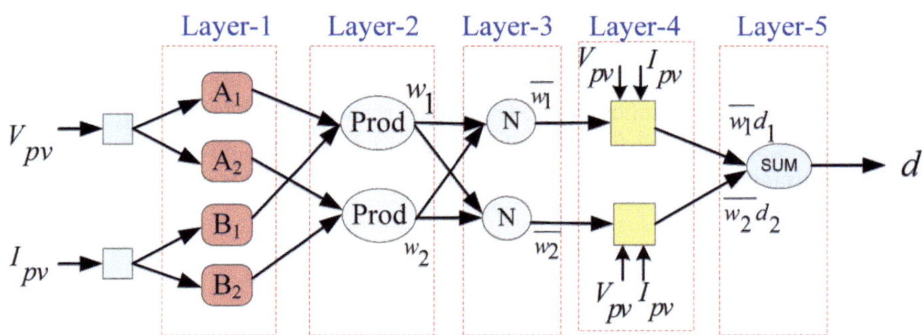

Fig. 7 ANFIS architecture of two-input Sugeno fuzzy model with two rules for MPPT

3.4.1 Architecture of ANFIS Controller

The ANFIS based MPPT algorithm has a better ability to track the MPP then conventional MPPT algorithms [33, 34]. Presently, photovoltaic panel voltage and current are considered as an input to ANFIS controller and one output is the duty cycle of the converter. These two inputs generate the control variable to control the conduction period of the semiconductor switch in the DC–DC converter, so that maximum power extraction from the solar panel can be achieved. Owing to its computational efficiency and interpretability, the Sugeno fuzzy inference model is highly preferred [35]. A first order Sugeno fuzzy model with two inputs can be expressed as

Rule1: If V_{pv} is A_1 and I_{pv} is B_1 THEN $d_1 = p_1 V_{pv} + q_1 I_{pv} + r_1$
Rule2: If V_{pv} is A_2 and I_{pv} is B_2 THEN $d_2 = p_2 V_{pv} + q_2 I_{pv} + r_2$

where, V_{pv} and I_{pv} are the input vector, d is the output function. The A_1, A_2, B_1 and B_2 are the fuzzy sets in antecedent and p_1, p_2, q_1, q_2, r_1 and r_2 are designed parameters that are estimated during the training process. The typical architecture of ANFIS based MPPT controller is shown in Fig. 7

Layer1 : Fuzzification Layer

Every nod in this layer is square/adaptive nodes that represents fuzzy clustering between input V_{pv} and I_{pv}. The output of this layer is expressed in Eq. (7).

$$\begin{aligned} O_i^1 &= \mu_{A_i}(V_{pv});\ i = 1, 2 \\ O_i^1 &= \mu_{B_{i=2}}(V_{pv});\ i = 3, 4 \end{aligned} \tag{7}$$

where, μ_{A_i} and $\mu_{B_{i=2}}$ can adopt any fuzzy membership function and O_i^1 is the membership value for the crisp inputs V_{pv} and I_{pv}. The subscripted '1' and 'i' represent the layer number and node number, respectively. Membership functions "μ" can be any shaped function like trapezoidal, triangle, and Gaussian. In this chapter triangular member ship function is considered and it is expressed as Eq. (8)

$$\mu_A(x) = \begin{cases} 0 & x < a_i \\ (x - a_i)/(b_i - a_i) & a_i \le x \le b_i \\ (c_i - x)/(c_i - b_i) & b_i \le x \le c_i \\ 0 & x > c_i \end{cases} \tag{8}$$

Layer2 : Rule interface Layer

Every node in this layer is fixed node (circles) and it transmit the input signal by applying t-norm operator these values are given in Eq. (9):

$$O_i^2 = w_i = \mu_{A_i}(V_{pv}) \times \mu_{B_i}(I_{pv});\quad i = 1, 2 \ldots 5 \tag{9}$$

The inference process is performed in this layer and output of each node shows the firing strength of the rules. In this study total number of rule bases is 25.

$$Layer3 : Normalization\ Layer$$

Every node in this layer denoted by circle tagged with N as shown in Fig. 7. The ith node in this layer estimated the ratio of rule 'i' firing level to the total firing level/strength as shown in (10). Where w_i is referred as the normalized firing strength:

$$O_i^3 = \overline{w_i} = \frac{w_i}{w_1 + w_2}; \quad i = 1, 2 \tag{10}$$

$$Layer4 : Consequent\ Layer$$

In this layer, the node function can be acknowledged with following function:

$$O_i^4 = \overline{w_i} d_i = \overline{w_i}\left(p_i V_{pv} + q_i I_{pv} + r_i\right); \quad i = 1, 2 \tag{11}$$

where, $\overline{w_i}$ is the output normalization layer-3 and p_i, q_i, r_i are the parameter set called as the result parameters and required to be optimized during training process. The purpose of this layer is to establish an adaptive correlation between normalized firing value and result function.

$$Layer5 : Output\ Layer$$

It has a single fixed node that computes the overall sum of the input signals to acquire final output and expressed as Eq. (12):

$$O_i^5 = \sum \overline{w_i} d_i = \frac{\sum_i \overline{w_i} d_i}{\sum_i \overline{w_i}}; \quad i = 1, 2 \tag{12}$$

The hybrid learning algorithm for controller design is used in [36]. An adaptive FIS typically consists of two different adjustable parts: (a) antecedent part (b) consequent part. These parts can be optimized by hybrid learning procedure such as combination of GD (gradient descent) and LSE (least-squares estimator). It has two passes, forward pass and backward pass. In the forward pass of hybrid learning algorithm, node output goes forward until layer-4 and the consequent parameters are identified by the sequential least square method. In backward pass, the error signal propagates backward and premise parameters are updated by gradient descent that is back propagation learning method.

3.4.2 Steps to ANFIS Based MPPT Implementation

The steps to realize ANFIS controller for maximum power extraction from the photovoltaic panel are as follows:

1. Collect the experimental data such as panel voltage, current and duty ratio of any conventional MPPT algorithm.
2. Load data in Workspace in MATLAB.
3. Use the 'anfisedit' command in command window to start the ANFIS GUI editor.
4. Load data from work space to ANFIS editor GUI.
5. Initialize and generate the FIS by choosing suitable fuzzy model.
6. Select the 'Train', FIS optimization methods (hybrid or back-propagation), error tolerance limit and number of epochs.
7. Check the structure and membership function.
8. Test the data against the trained FIS.

3.4.3 Design of ANFIS Controller Based MPP Controller

In the present work, an ANFIS controller has been proposed to predict the duty cycle for particular DC–DC converter based on solar module terminal voltage and current. Here, the fuzzy rules of controller are generated on the basis of Takagi-Sugeno inference model. The results of P&O algorithm such as solar panel voltage, current and corresponding duty ratio of the converter topology are captured through control desk next generation software of dSPACE DS1103. In this study, the effect of solar illumination and load variation is also taken into account. Figure 8 depicts the Simulink model of ANFIS controller for implementing on dSPACE DS1103. The ANFIS optimized parameters such as surface, rule base, membership of panel voltage current for CBC and SEPIC are shown in Figs. 9 and 10, respectively. The parameters and features of trained ANFIS controller are listed in Tables 4 and 5 for various converters under different load condition as well as during change in solar irradiance.

3.4.4 Experimental Setup Details and Procedure

In order to validate the real time effect of load line on the performance of MPPT, different algorithms such as ANN, ANFIS, and modified IC have been implemented on CBC and SEPIC converter topologies. The experimental setup for implementing these algorithms on different converter topologies is shown in Fig. 11.The Ecosense PV emulator (IGE-PV4C400-001 model) is used to emulate user defined solar panel of 'K-Trom series 150 W. The PV emulator is used for repeating the same environ-

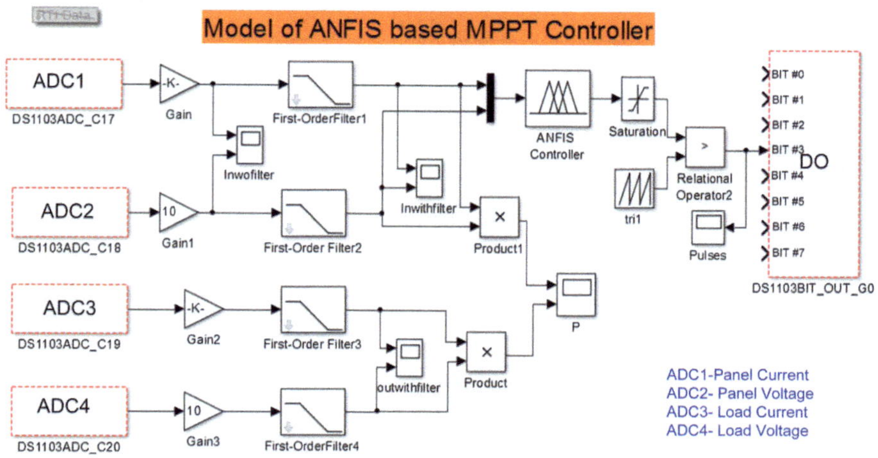

Fig. 8 Implementation of ANFIS-based MPPT control scheme on dSPACE DS1103 controller

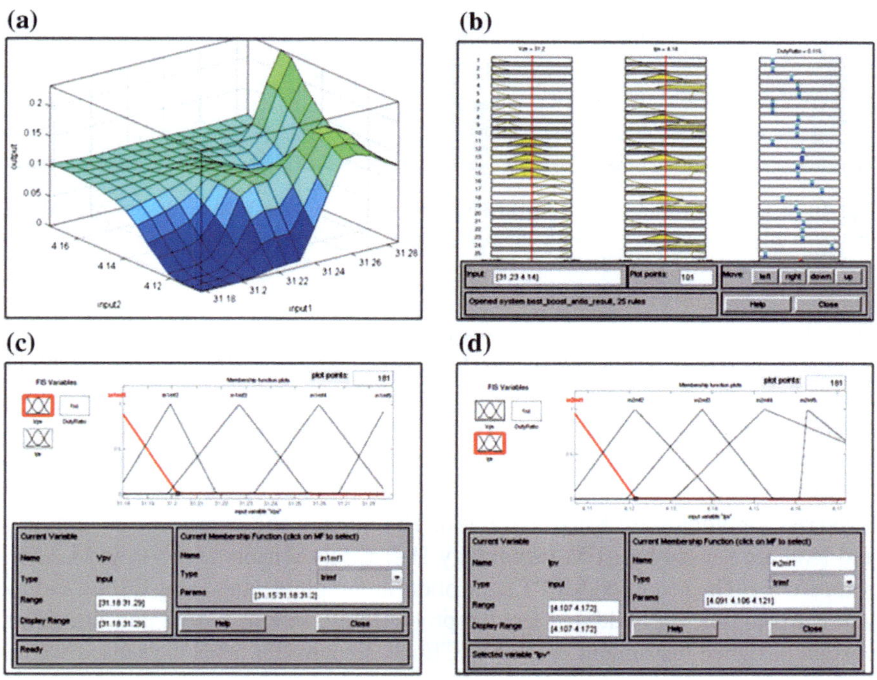

Fig. 9 ANFIS controller results for CBC **a** Surface **b** Rule Base **c** Membership functions of solar panel voltage **d** Membership function of solar panel current

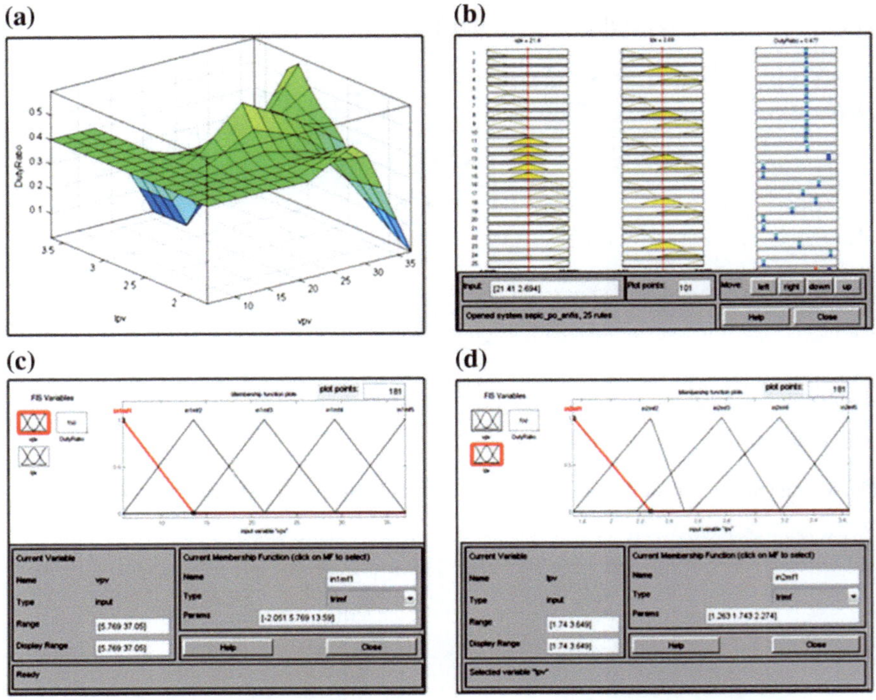

Fig. 10 ANFIS controller results for SEPIC converter **a** Surface **b** Rule Base **c** Membership functions of solar panel voltage **d** Membership function of solar panel current

mental test condition for comparing the performance of different MPPT algorithms. Also, it is used to experimentally determine the maximum extracted power from the panel. From the previous discussion it can be observed that, the maximum power from the solar panel can be extracted provided that the input impedance seen by the converter $R_i \approx R_{mp}$. This condition of maximum power point depends upon the converter configuration. The specification detail of 150 W solar panel and different converter topologies are listed in Table 6. The voltage and current of PV emulator and load are sensed by LEM technology Hall Effect sensors LV25-P and LA55-P respectively. The proposed MPPT controllers are implemented using dSPACE-1103 that helps to quickly build and test the proposed algorithms before converting into custom designed instrument. The dSPACE-1103 controller takes the sensed voltage and current signal through its Analog I/O port and it gene-rates the PWM pulse for controlling the duty cycle of DC–DC converter.

Table 4 Parameters of proposed ANFIS controller

Topology	Panel voltage range	Panel current range	Duty ratio range	No of rules	MF type
CBC	31.81–31.29	4.107–4.172	0.0–0.2	25	Triangular
SEPIC	5.77–37.05	1.742–3.649	0.1–0.6	25	Triangular

Table 5 The features of proposed ANFIS-based MPPT controller for CBC and SEPIC converter power conditioning units

Feature of ANFIS controller					
Type	Sugeno	AggMethod (aggregation method)	Sum	No. of output MF	25
And Method	Product	No. of input	2	Output MF type	Linear
Or method	probor (probabilistic or)	No. of input MF	5	No. of rules	25
DefuzzMethod	wtaver (weighted average)	Input MF type	Triangular	No of samples	69904
ImpMethod (implication method)	Product	No. of output	1	No of epochs	27

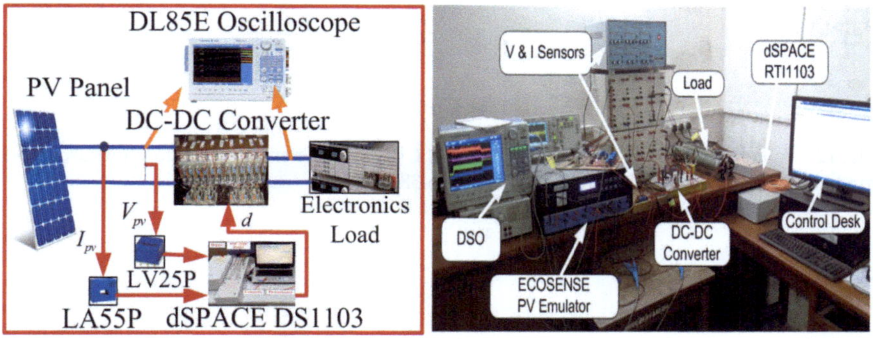

Fig. 11 Experimental setup for extracting maximum power from PV emulator

Table 6 Specifications of photovoltaic panel and converter

K-Trom150 W solar panel parameters	Voc = 43.2 V; Vmp = 34.5 V; Isc = 4.8 A; Imp = 4.37 A; Pmp = 150 W; αI = 0.00288 mA/oC; βv = −0.14256 mV/oC
CBC	L = 5 mH; I_{LSAT} = 10 A; CO = 1200 uF; f = 10 kHz; Diode = UF5402; SW = IRFP360
SEPIC converter	L1 = 5 mH; I_{LSAT} = 10 A; Co = 1200 uF; f = 10 kHz; L2 = 5 mH; C1 = 220 uF; Diode = UF5402; SW = IRFP360

Note: I_{LSAT} is the saturation current of the inductor

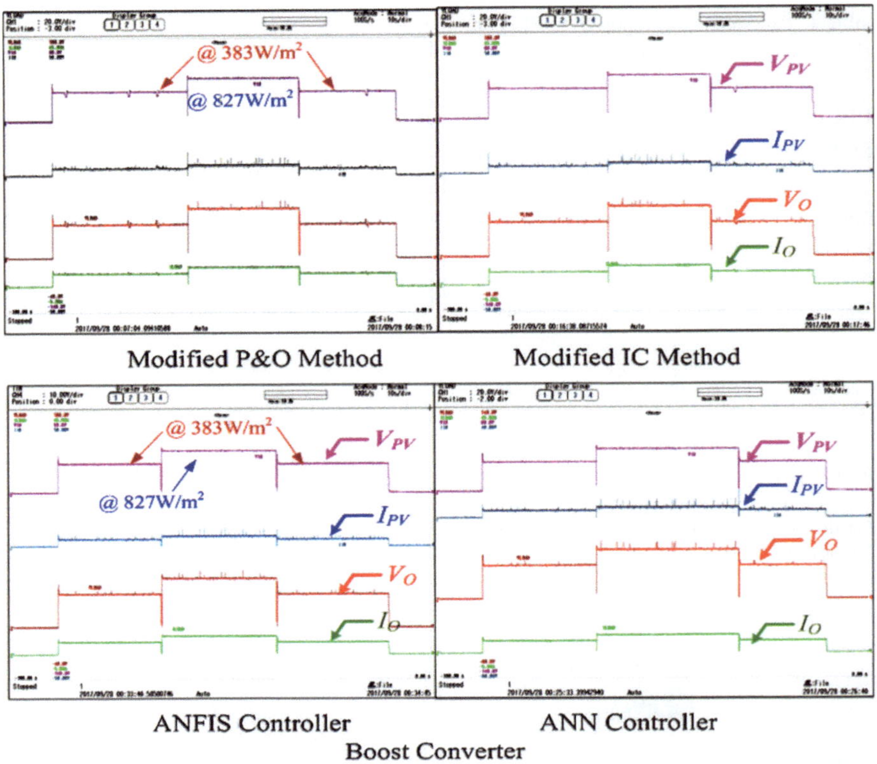

Fig. 12 Voltage, current of solar panel and load of CBC for different MPPT algorithm during solar irradiance change from 383 W/m^2 to 827 W/m^2 to 383 W/m^2

4 Result Analysis

The results obtained from the simulation study of various MPPT techniques are analyzed and validated from the experimental results taking into consideration the effect of solar isolation and load.

4.1 *Effect of Solar Isolation Change*

In order to emulate the solar isolation change, Nagpur city with latitude 21.145 °N and longitude 79.0882 °E on 15th April 2017 has been considered. In order to emulate the real environmental condition at particular day, the channel-1 of Ecosense PV emulator has been set to emulate the 150 W solar panel at 827 W/m^2 solar irradiance for 12.00Noon to 1.00PM and channel-2 is set to 383 W/m^2 for 9.00AM to 10.00AM. The step load change has been introduced at $t = 3$ s from 827 W/m^2

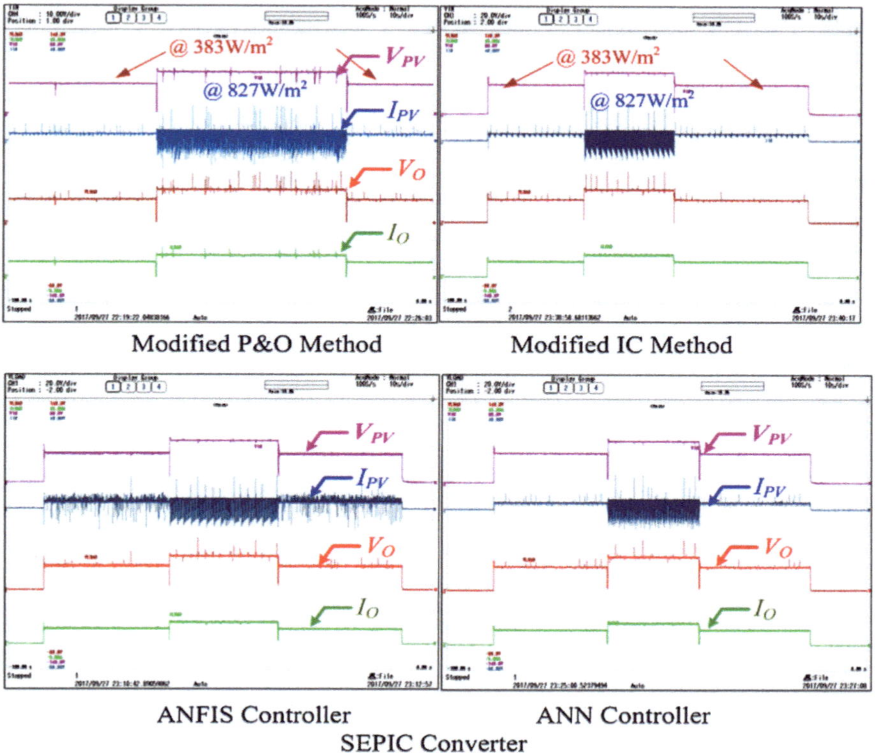

Fig. 13 Voltage, current of solar panel and load of SEPIC for different MPPT algorithm during solar irradiance change from 383 W/m^2 to 827 W/m^2 to 383 W/m^2

to 383 W/m^2. During this experiment, intelligence controller extracts more power than conventional methods. The results of CBC and SEPIC converter are depicted in Figs. 12 and 13 respectively. The CBC converter has highest efficiency due to lesser circuit components and current density for same power requirement. It is noteworthy that the CBC has limited operating region unlike SEPIC converter. Also the SEPIC converter has more current ripples during the isolation change as depicted in Fig. 13. The performance of MPPT algorithms during solar isolation change is reported in Table 7.

4.2 Power Extraction from PV Panel During of Load Variation

The performance of intelligent power extraction methods (ANN and ANFIS) and modified incremental conductance algorithm are investigated during load variation.

Table 7 Performance of MPPT algorithms during change in solar irradiance

MPPT →	CBC						SEPIC					
	Modified IC		ANFIS		ANN		Modified IC		ANFIS		ANN	
G (W/m^2)	G1	G2	G1	G2	G1	G2	G1	G2	G1	G2	G1	G2
V_{PV}	22.2	31.8	22	31.7	22	31.8	22.5	31.7	22.2	31.7	22.5	31.7
I_{PV}	2.5	3.74	2.52	3.77	2.52	3.75	2.5	3.88	2.77	3.83	2.5	3.83
P_{PV}	55.5	118.93	55.44	119.51	55.44	119.25	56.25	122.99	61.49	121.41	56.25	121.41
V_o	24.84	36.3	24.62	36.3	24.71	36.3	16.85	24.12	17.13	24.13	17	24.13
I_o	2.18	3.27	2.21	3.28	2.19	3.25	2.65	3.88	2.69	3.87	2.66	3.86
P_o	54.15	118.7	54.41	119.06	54.12	117.97	44.65	94.75	46.08	93.38	45.2	93.15
$\%\eta$	97.57	99.81	98.14	99.63	97.61	98.93	79.38	77.04	74.94	76.91	80.36	76.72

Where, G1:383 W/m^2; G2:827 W/m^2

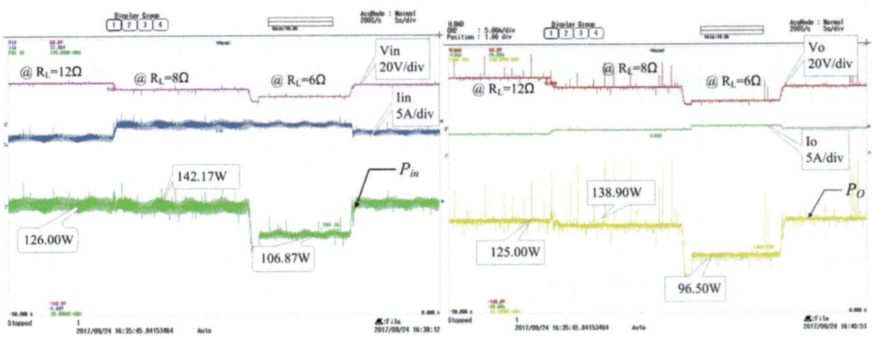

(a) Solar Panel (b) Load

Fig. 14 Voltage, current and power CBC with modified IC-based MPPT algorithm during change in load

The "Ecosense" PV emulator IGEPV4C400- 01 [36] is used to synthesize the 150 W solar panel and it is operated in simulated mode that uses stored current-voltage table of particular solar panel at STC. The details of emulated solar panel are $V_{oc} = 35$ V, $I_{sc} = 5.75$ V, $V_{mp} = 28.4$ V and $I_{mp} = 5.28$ A at STC. The converters are considered as inefficient that has parasitic resistance of 2 Ω, inductance approximately of 0.8 Ω for connecting wires of semiconductor switches. During this study, the PV emulator is operated in constant current region near MPP. The experimental results of a CBC and SEPIC converter during change in load for different MPPT algorithms are depicted in Figs, 14, 15, 16, 17, 18 and 19. The extracted power from various MPPT algorithms is reported in the Table 8. In CBC, for all considered MPPT algorithms, overall efficiency of system was found more than 90%. In ANN-based MPPT algorithm power extraction efficiency is also high compared to other method. Further, it could be observed that the ANFIS performance is better in term of efficiency compared to other techniques.

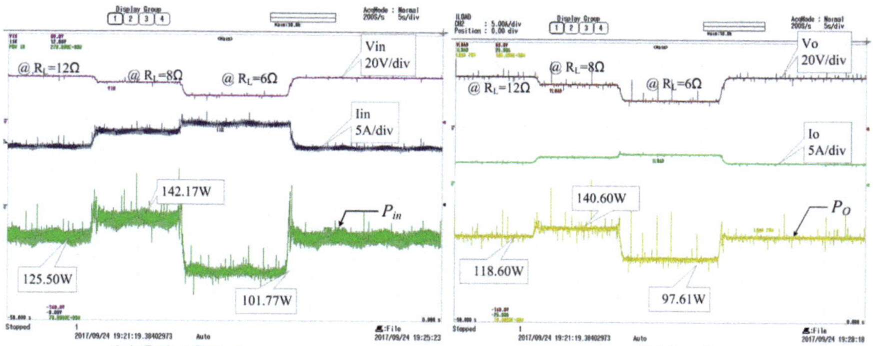

(a) Solar Panel (b) Load

Fig. 15 Voltage, current and power of CBC with ANFIS-based MPPT algorithm during change in load

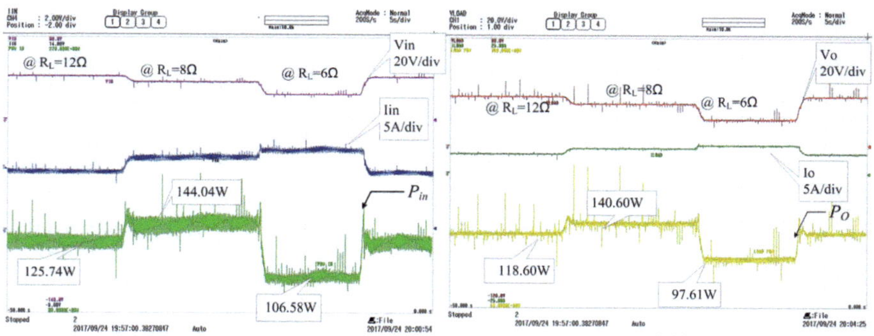

(a) Solar Panel (b) Load

Fig. 16 Voltage, current and power of CBC with ANN-based MPPT algorithm during change in load

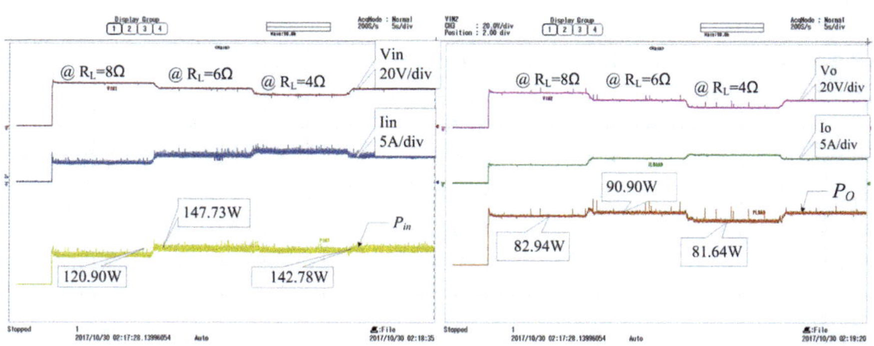

(a) Solar Panel (b) Load

Fig. 17 Voltage, current and power of SEPIC converter with modified IC-based MPPT algorithm during change in load

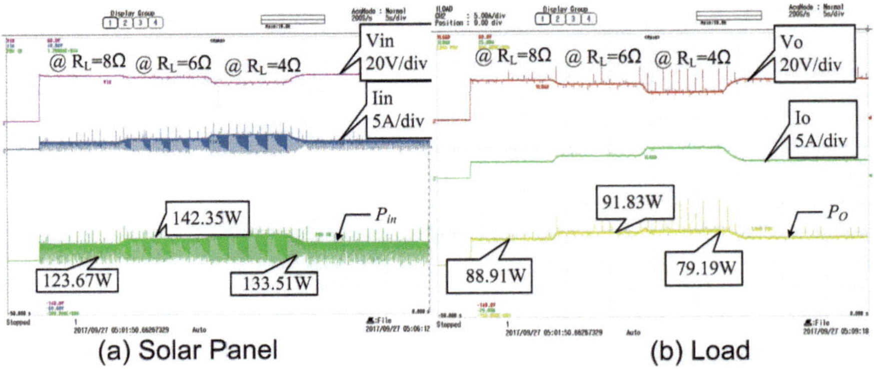

Fig. 18 Voltage, current and power of SEPIC converter with ANFIS-based MPPT algorithm during change in load

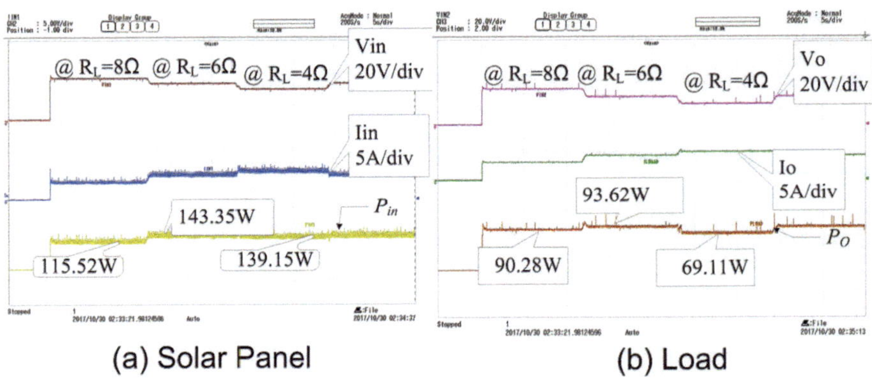

Fig. 19 Voltage, current and power of SEPIC converter with ANN-based MPPT algorithm during change in load

Table 8 Maximum extracted extraction by various MPPT algorithms

MPPT→	CBC			SEPIC		
	Modified IC	ANFIS	ANN	Modified IC	ANFIS	ANN
G (W/m^2)	8 Ω	8 Ω	8 Ω	6 Ω	6 Ω	6 Ω
V_{PV}	27.50	27.50	27.70	28.00	29.30	28.60
I_{PV}	5.17	5.17	5.20	5.07	4.92	5.03
P_{PV}	142.17	142.17	144.04	141.96	144.15	143.35
V_o	30.20	30.43	30.70	18.10	18.44	18.99
I_o	4.60	4.62	4.54	5.00	4.98	4.93
P_o	138.90	140.60	139.38	90.50	91.83	93.62
$\%\eta$	97.70	98.90	96.76	63.75	63.70	65.47

5 Conclusion

For improving the performance of photovoltaic system, several conventional and soft-computing-based MPPT methods have been used. This study presents the implementation of hybrid ANFIS controller for maximum power extraction from PV module. Further, the results are compared with well-liked Incremental Conductance method and ANN-based MPPT scheme. The hardware results are very promising and conclude that the ANFIS controller performance is better than other conventional methods in terms of efficiency, stability and precision. Also, ANFIS is suitable for the problems where prior knowledge of consequent parameters of the rules is unknown. Further, it learns the parameters and tunes the membership function according to it.

References

1. Catalão JPS (2015) Smart and sustainable power systems: operations, planning, and economics of insular electricity grids. CRC Press, Boca Raton
2. Committee RS (2016) Renewables 2016 global status report, REN21, renewable energy policy network for the 21st century, Athens
3. Panagiotis Karampelas LE (2016) Electrical distribution-intelligent solution for electricity transmission and distribution networks, 2016th edn. Springer, Berlin
4. Charles JP, Pierre M (1981) A practical method of analysis of the current-voltage characteristics of solar cells. Sol Cells 4:169–178
5. Kok BC, Goh HH, Chua HG (2012) Optimal power tracker for stand-alone photovoltaic system using artificial neural network (ANN) and particle swarm optimisation (PSO). In: International conference on renewable energies and power quality, pp 1–6
6. Kulaksiz AA, Akkaya R (2012) Training data optimization for ANNs using genetic algorithms to enhance MPPT efficiency of a stand-alone PV system. Turk J Elec Eng Comp Sci 20(2):241–254
7. Rizzo SA, Scelba G (2015) ANN based MPPT methods for rapidly variable shading conditions. Appl Energy 145:124–132
8. Negnevisky M (2005) Artificial intelligence: a guide to intelligent systems, 2nd edn. Pearson, Delhi
9. Rekioua D, Matagne E (2012) Modeling of solar irradiance and cells. Optimization of photovoltaic power systems, 2012th edn. Springer, London, pp 31–87
10. Sumathi S, Ashok KL, Surekha P (2015) Application of MATLAB/SIMULINK in solar PV systems. Solar PV and wind energy conversion systems, vol XXIV. Springer, Cham, pp 59–143
11. Kumari JS, Babu CS (2012) Mathematical modeling and simulation of photovoltaic cell using Matlab-Simulink environment. Int J Electr Comput Eng 2(1):26–34
12. el Tayyan AA (2013) A simple method to extract the parameters of the single-diode model of a PV system. Turk J Phys 37:121–131
13. King DL, Kratochvil JA, Boyson WE (1997) Temperature coefficients for PV modules and arrays: measurement methods, difficulties, and results. In: 26th IEEE photovoltaic specialists conference
14. Mohamed MA (2015) Solar irradiance estimation of photovoltaic module based on Thevenin equivalent circuit model. Int J Renew Energy Res 5(4):971–972
15. Haihong B, Weiping Z, Bing C (2016) Control simulation and experimental verification of maximum power point tracking based on RT-LAB. Int J Eng 29(10):1372–1379

16. Karanjkar DS, Chatterji S, Kumar A, Shimi SL (2014) Fuzzy adaptive proportional-integral-derivative controller with dynamic set-point adjustment for maximum power point tracking in solar photovoltaic system. Taylor Fr. Jr. Syst Sci Control Eng 2583
17. Haque A (2014) Maximum power point tracking (MPPT) scheme for solar photovoltaic system. Energy Technol Policy 1(1):115–122
18. Hua C, Lin J, Tzou H (2003) MPP control of a photovoltaic energy system. Eur Trans Electr Power 13(4):239–246
19. Taghvaee MH, Radzi MAM, Moosavain SM, Hizam H, Marhaban MH (2013) A current and future study on non-isolated DC-DC converters for photovoltaic applications. Renew Sustain Energy Rev 17:216–227
20. Walker GR, Sernia PC (2004) Cascaded DC-DC converter connection of photovoltaic modules. IEEE Trans Power Electron 19(4):1130–1139
21. Radjai T, Gaubert JP, Rahmani L, Mekhilef S (2015) Experimental verification of P&O MPPT algorithm with direct control based on fuzzy logic control using CUK converter. Int Trans Electr Energy Syst 25(12):3492–3508
22. Dixit TV (2018) Real time investigation on performance enhancement of power conditioing units for photo-voltic and fuel cell, NIT Raipur, (C.G.) India
23. Enrique JM, Duran E, Sidrach-de-Cardona M, Andujar JM (2007) Theoretical assessment of the maximum power point tracking efficiency of photovoltaic facilities with different converter topologies. Sol Energy 81:31–38
24. Ahmed J, Salam Z (2016) A modified P & O maximum power point tracking method with reduced steady-state oscillation and improved tracking efficiency. IEEE Trans Sustain Energy 7(4):1506–1515
25. Kok ST, Mekhilef S (2014) Modified incremental conductance algorithm for photovoltaic system under partial shading conditions and load variation. IEEE Trans Ind Electron 61(10):5384–5392
26. Chekired F, Mellit A, Kalogirou SA, Larbes C (2014) Intelligent maximum power point trackers for photovoltaic applications using FPGA chip: a comparative study. Sol Energy 101:83–99
27. Safari A, Mekhilef S (2011) Simulation and Hardware implementation of incremental conductance MPPT with direct control method using cuk converter. IEEE Trans Ind Electron 58(4):1154–1161
28. Thangavelu A, Vairakannu S, Parvathyshankar D (2017) Linear open circuit voltage variable step-size incremental conductance strategy-based hybrid MPPT controller for remote power applications. IET Power Electron 10(11):1363–1376
29. Zheng H, Li S (2016) Fast and robust maximum power point tracking for solar photovoltaic systems. Am J Eng Appl Sci 9:755–769
30. Liao T, Huang N (1999) Genetic algorithm-based self-learning fuzzy pi controller for buck converter. Eur Trans Electr Power 9(4):233–239
31. Jang JR (1993) ANFIS: adaptive-network-based fuzzy inference system. IEEE Trans Syst Man Cybern 23(3):665–685
32. Rezvani A, Izadbakhsh M, Gandomkar M, Vafaei S (2015) Implementing GA-ANFIS for maximum power point tracking in pv system. Indian J Sci Technol 8(May):982–991
33. Sheraz MAAM, Muhammed K (2015) An efficient ANFIS-based pi controller for maximum power point tracking of pv systems. Springer Trans Arab J Sci Eng 40:2641–2651
34. Tarek B, Said D, Benbouzid MEH (2013) Maximum power point tracking control for photovoltaic sys-tem using adaptive neuro- fuzzy 'ANFIS. In: Eighth international conference and exhibition on ecologi-cal vehicles and renewable energies, pp. 1–7
35. Acar M, Avci D (2010) An adaptive network-based fuzzy inference system (ANFIS) for the prediction of stock market return: the case of the Istanbul stock exchange. Expert Syst Appl 37:7908–7912
36. Ecosense, FOUR CHANNEL SOLAR PV EMULATOR:USER, IGE-PV4C400-001/Power: 1600 W (400 × 4)

An Online Self Recurrent Direct Adaptive Neuro-Fuzzy Wavelet Based Control of Photovoltaic Systems

Syed Zulqadar Hassan, Tariq Kamal, Sidra Mumtaz and Laiq Khan

Abstract Solar through photovoltaic is an inexhaustible energy source which contributes to enhance the sustainability of the society. Though, photovoltaic systems experience some fundamental problems such as low conversion efficiency particularly during high weather variations and the high nonlinearity between the photovoltaic output power and current. These problems involve in photovoltaic systems need the use of advanced intelligent control methods. This book chapter develops a new direct adaptive maximum power point tracking control for photovoltaic systems. The new proposed technique integrates a Chebyshev wavelet in the consequent part of existing neuro-fuzzy structure. The parameters of the proposed controller are tuned adaptively online using backpropagation algorithm. The performance of the proposed method is tested under high uncertainties appearing from solar irradiance, temperature and fluctuations in load. Finally, simulation results are provided to show that the proposed control method is better than other existing methods in terms of efficiency, load tracking and output power.

S. Z. Hassan (✉)
School of Electrical Engineering, Chongqing University, Chongqing, China
e-mail: syedzulqadar.hassan.pk@ieee.org

T. Kamal
Faculty of Engineering, Department of Electrical and Electronics Engineering, Sakarya University, Sakarya, Turkey
e-mail: tariq.kamal.pk@ieee.org

T. Kamal
Research Group in Electrical Technologies for Sustainable and Renewable Energy (PAIDI-TEP023), Department of Electrical Engineering, Higher Polytechnic School of Algeciras, University of Cadiz, Algeciras, Spain

S. Mumtaz · L. Khan
Department of Electrical Engineering, COMSATS University, Abbottabad Campus, Islamabad, Pakistan
e-mail: sidramumtaz@ciit.net.pk

L. Khan
e-mail: laiq@ciit.net.pk

R.-E. Precup et al. (eds.), *Solar Photovoltaic Power Plants*, Power Systems,
https://doi.org/10.1007/978-981-13-6151-7_11

1 Introduction

The continuous increase in energy demand, adversarial environmental impact, and high cost of fossil fuels used in conventional power stations have strengthen the utilization of sustainable energy resources, particularly solar energy which has exposed its importance as a sustainable energy source to the world energy consumption. Solar through photovoltaic provides unlimited amount of energy that can be processed directly to electrical power via Photovoltaic (PV) system. Nonetheless, the performance of PV system relies on weather forecasts which creates a high non-linearity between the PV output power and current as illustrated in Fig. 1. To get maximum benefits from solar PV, different conventional and intelligent techniques and approaches have been developed by many researchers in the literature. The most adapted conventional Maximum Power Point Tracking (MPPT) techniques are two, namely an Incremental Conductance (InC) [1, 2], and Perturb and Observe (PO) [3, 4]. Both these algorithms and their modification [5, 6] are highly populated in the academia. The drawback of PO is that the tracking is quicker for large perturb step, whereas slower for small perturb step. As a result, oscillations are present around Maximum Power Point (MPP). The oscillations problems in PO technique during rapidly varying environmental conditions were eliminated in InC. Nevertheless, the cost of InC method is high due to the requirements of high sampling compliance [7].

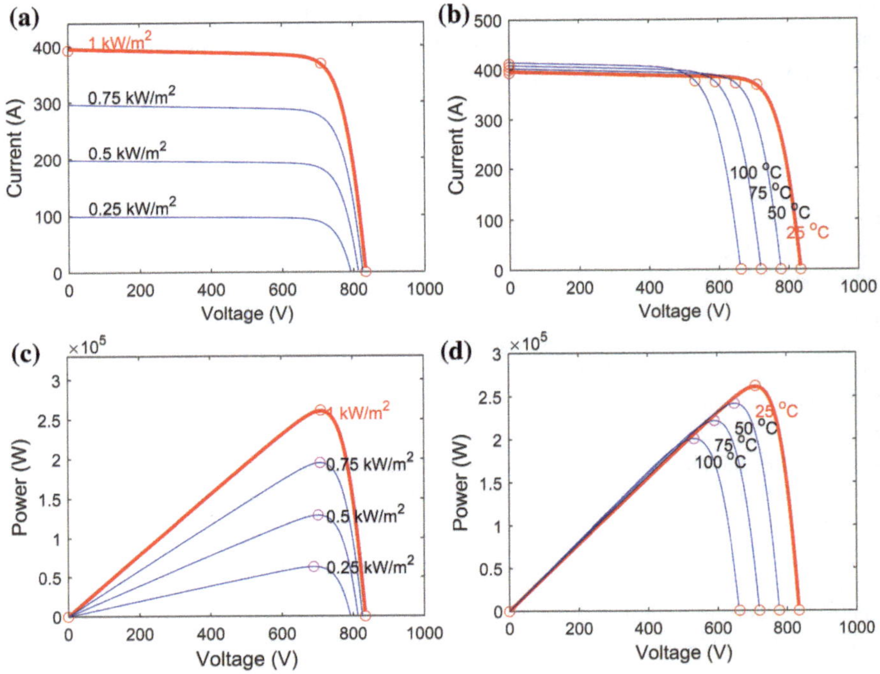

Fig. 1 *P-V* and *I-V* characteristics curves of PV system

Some authors have reported sliding mode [8], feedback linearization [9], and ripple correlation control [10] of PV. When properly designed sliding mode provides a robust solution to a nonlinear system, but chattering an undesirable phenomenon is associated with this method and if the switching is not appropriately selected, it may lead to limit cycle [11]. Similarly, feedback linearization transforms a nonlinear system to a linear system, but its big disadvantage is the lack of enough robustness in presence of system parameter's uncertainties such as different weather situations or rapid load variations. In ripple correlation technique, the optimum point is traced at a fast-tracking speed by changing the operating current according to its location. However, since the differentiators are highly sensitive to disturbance, the MPP accuracy is below than other MPPT methods [12]. Some authors have designed evolutionary algorithms for the control of PV such as Genetic Algorithm (GA) [13], cuckoo search [14], ant colony optimization [15], and particle swarm optimization [16].

PV equations can be treated as the fitness function for the GA algorithm, the algorithm provides solutions in a very less time, but it fails to find the most optimal solution in all cases. Similarly, cuckoo search responds fast and track Global Maximum Power Point (GMPP) even under certain partial shading states. Despite the partial shading problem can be solved effectively using bio-inspired techniques, however, the average efficiency of these methods entirely depends on initial conditions and the number and type of chosen parameters. In ant colony optimization method, the goal is to converge the ants to follow the shortest path possible. In case of solar PV, the control parameter is the current of the PV array. Increasing the number of ants increases the accuracy but slow down the process as many iterations are required to reach the MPP. Likewise, particle swarm optimization can find the MPP under varying environmental conditions, however, it is a complex algorithm and the selection of its parameter has an impact on the optimization process, i.e., it will take a long time if the wrong parameters are chosen.

Over the last few decades, artificial Intelligence based approaches such as fuzzy control, neural networks and their hybrid (neuro-fuzzy) have tremendously contributed to predict accurate generation of renewable energy sources and their impacts on the electric network such as maintaining system reliability, stability integrating hybrid solar, wind and energy storage systems, net load forecasting, and line loss predictions. For instance, in [17, 18], the authors have control PV via fuzzy in a grid integrated hybrid system. Similarly, the application of neural networks and neuro-fuzzy at different aspects such as power quality improvement, maximum power transfer, and solar radiation forecast are found in [19–22]. However, the accurate application of fuzzy needs a-priori-knowledge, while neural networks are computationally intensive to train, hence, their limitations are versatility [23]. Similarly, neuro-fuzzy controllers fail to find local minima of the search space [24].

In the literature, many authors have stated that the integration of wavelets in neuro-fuzzy structure can solve the concern of local minima [24]. The introduction of wavelets in neuro-fuzzy structure substantially enhances its computational speed [25]. Wavelet transform provides a time-frequency localization of the non-stationary signals. In wavelet transform, the basis functions are used to preserve the maximum energy of the signal [26]. All the above discussed control methods are modelled and developed considering certain operating and load conditions. In case of high nonlinearity in weather climate/load conditions where there are several MPPs in the P-V curve, the above techniques are failed in finding the global maximum power points MPPs among the local MPPs, therefore it is difficult to achieve maximum benefits from PV system at all time. An adaptive control paradigm is highly effective to handle non-linearity and uncertain fluctuations.

This chapter contributes to the work on an uncertain PV system to provide maximum power through an online full recurrent direct adaptive wavelet based neuro-fuzzy MPPT control. The modeling of the proposed control method starts from a set of fuzzy IF-THEN rules. Then a neural network is designed to provide the gradient information in the proposed algorithm. Then a Chebyshev wavelet is integrated in the exiting neuro-fuzzy structure to enhance the performance of proposed controller in terms of output power, and efficiency. MATLAB simulations are presented to show the superiority over other maximum power point tracking control methods.

The rest of the work is ordered as follows: Section 2 discusses the background regarding PV system. Section 3 is related to the development of the proposed controller. Comparison through results is covered in Sect. 4. Finally, conclusions are provided in Sect. 5.

2 Solar Photovoltaic Energy Conversion System (SPECS)

The entire layout of the SPECS is shown in Fig. 2. It consists of a PV array, a DC–DC boost converter and a load. The PV array contains solar cells, which generate electrical energy from the sunlight associated with solar radiations. The function of DC–DC converter is to increase the available output voltage level of PV at the desired current and/or voltage to achieve maximum benefits from the PV module. The dynamic model of PV system is characterized in (1)–(3).

$$\frac{dV_{pv}}{dt} = \frac{1}{C_x}(I_{pv} - i_1) \tag{1}$$

$$\frac{di_{L_1}}{dt} = \frac{1}{L_1}(V_{pv} - v_{C_z}(1 - d)) \tag{2}$$

$$\frac{dv_{C_z}}{dt} = \frac{1}{C_z}\left(i_{L_1}(1 - d) - \frac{v_{C_z}}{R_L}\right) \tag{3}$$

where I_{pv} is the current across PV array; i_1 is the current in inductor, L_1; V_{pv} is the voltage across PV array; v_x is the voltage across input capacitor, C_x; v_z is the voltage across output capacitor, C_z; R_L is the resistance of load; and d is the duty cycle. Inherently, the *P-V* power output and *I-V* curves are highly non-linear as given in (4).

$$i_{pv} = n_p I_{ph} - n_p I_d \left[\exp\left(\frac{V_{pv} + R_s I_{pv}}{n_s v_t} \right) - 1 \right] \quad (4)$$

where v_t calculates the terminal voltage of PV array; n_s gives the number of solar cells connected in series; n_p represents the number of solar cells in parallel; I_d is the current passing through diode, d; and R_s is the series resistance of the PV array.

2.1 DC–DC Boost Converter

DC–DC boost converter plays an important role in PV system. It acts as a switching-mode regulator and generates the required voltage levels. The impedance that seen at the input of the boost converter from PV side depends upon the d variations between 0 and 1. The converter uses Pulse Width Modulation (PWM) generator. The appear equivalent resistance of boost converter is given as:

$$R = R_i(1 - d)^2 \quad (5)$$

Using the maximum power transfer theory, maximum power is provided to the external load when the value of the equivalent resistance, R is equal to the output resistance of the PV system [27].

2.2 Smoothing Filter

The smoothing filter is an important part of any renewable energy system which serves to remove voltage ripples and consists of a series inductor attached in parallel with capacitor as illustrated in Fig. 3. The current through the inductor is written as:

$$i_1 = \frac{1}{L_1} \int (v_i - v_0) \quad (6)$$

$$v_L = L_1 \frac{di_1}{dt} \quad (7)$$

where v_L stands for inductor voltage. The output voltage v_o can be calculated as:

$$v_0 = \frac{1}{C_z} \int (i_c dt) \quad (8)$$

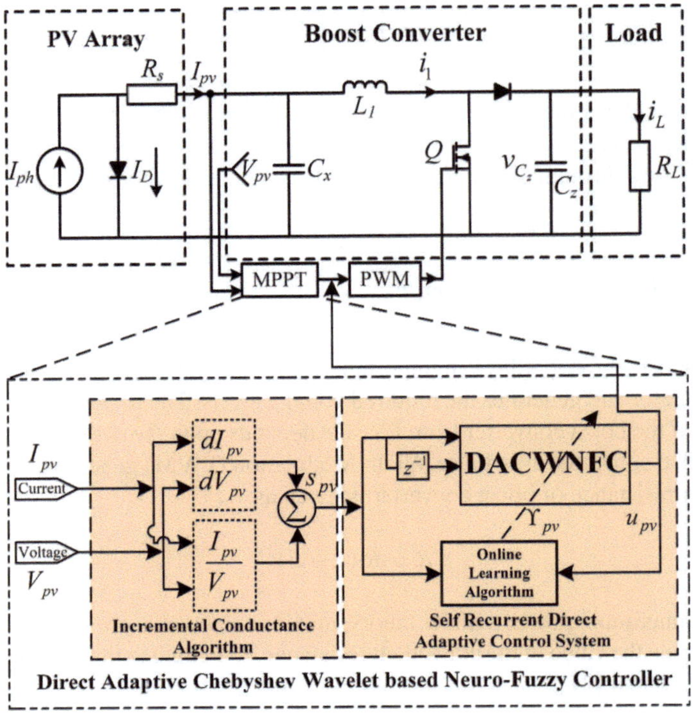

Fig. 2 SPECS system and proposed adaptive control model

3 Proposed Direct Adaptive Control Paradigm Design

In this section, a Direct Adaptive Chebyshev Wavelet based Neuro-Fuzzy Controller (DACWNFC) is designed to extract maximum power from PV system. The condition to obtain maximum benefits from PV system is defined in the following cost function.

$$\downarrow \Upsilon_{pv} = \frac{1}{2} s_{pv}^2(n) + \frac{\xi_{pv}}{2} u_{pv}^2(n) \tag{9}$$

s.t.

$$\begin{aligned} P_{pv} &\leq P_{peak} \\ V_{pv} &\leq V_{oc} \\ I_{pv} &\leq I_{sc} \\ T_{pv} &\leq T_{max} \\ \phi_{pv} &\leq \phi_{max} \end{aligned}$$

Fig. 3 Smoothing filter

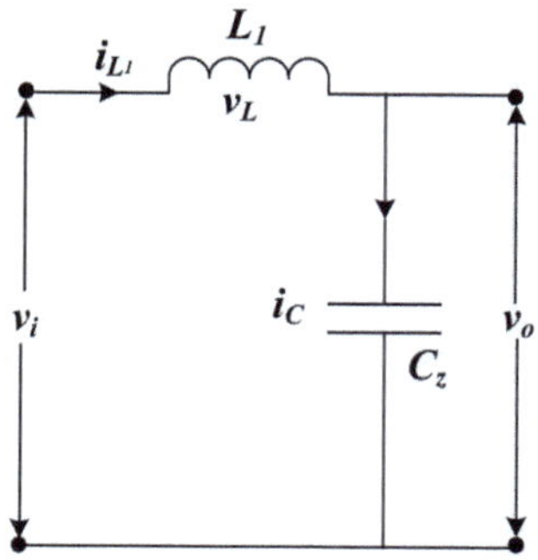

where P_{peak}, V_{oc} and I_{sc} are the peak power, open circuit voltage and short circuit current, respectively. Similarly, the T_{max} and ϕ_{max} are the maximum temperature and solar irradiance, respectively. When the change of PV power with respect to the operating voltage is zero, then the MPP is achieved as follows (Fig. 2):

$$s|_{mpp} = \frac{\partial P_{pv}}{\partial V_{pv}}|_{mpp} = \left[\frac{I_{pv}}{V_{pv}} + \frac{\partial I_{pv}}{\partial V_{pv}}\right]_{mpp} = 0 \tag{10}$$

where s is the slope of PV power with respect to the operating voltage. The change in d alters the value of impedance at the input of a DC/DC converter. For a boost converter, the load impedance (output impedance) is always high than the input impedance.

There are seven layers in the proposed structure as shown in the Fig. 4. IF-THEN rules are defined as follows:

$$R^i : \text{ if } x_1 \text{ is } \mu_1^i, x_2 \text{ is } \mu_2^i \text{ and } x_p^i \text{ is } \mu_p^i, \text{ then } y = \mathfrak{P}_{pq}$$

These seven layers are discussed as follows:

Layer 1: The input from the plant (PV) are collected in this layer. These inputs are directly sent to the fuzzification layer (i.e., layer 2) using nodes.

Layer 2: Here the linguistic terms and their degree of membership are allocated to each input. The linguistic terms using Gaussian membership function for each input as follows:

$$\mu_j^i = e^{\frac{-1}{2}\left[\frac{x_j - \mathfrak{m}_j^i}{\varsigma_j^i}\right]^2} \tag{11}$$

Layer 3: The firing intensity of each rule is determined using product T-norm in this layer.

$$\bar{\mu}_j^i = \prod_{i=1}^{q} \mu_j^i \tag{12}$$

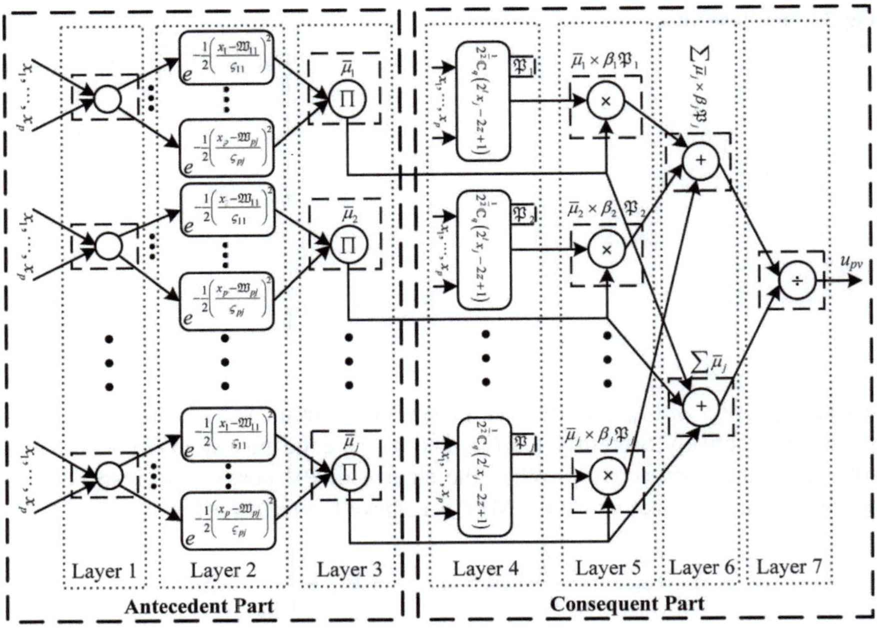

Fig. 4 Neurofuzzy structure of the proposed controller

Layer 4: Here Chebyshev wavelet functions are integrated in this layer. The weighted consequent value after applying Chebyshev wavelet of each rule is determined as follows:

$$\Xi^i = \sum_{j=1}^{2^{n-1}} \beta_j^i \mathfrak{P}_{pq}(x) \tag{13}$$

where, Ξ^i is the output of this layer and β represents weighting factor.

Layer 5: The output of layer 3 which is the last layer of antecedent part and layer 4 which is first layer of consequent part are multiplied for each input and then they are sum.

Layer 6: This layer provides the sum of output of layer 3.

Layer 7: The final output of neurofuzzy structure is calculated and given as:

$$u(n) = \frac{\sum_{j=1}^{p} \bar{\mu}_j^i \Xi^i}{\sum_{j=1}^{p} \bar{\mu}_j^i} \tag{14}$$

The above derived wavelet based neurofuzzy structure is proven as a universal approximator for continuous functions over compact sets. The Chebyshev wavelet which is taken from the interval, [0 1] as follows:

$$\mathfrak{P}_{pq}(x_j) = \begin{cases} 2^{\frac{l}{2}}\mathbb{C}_q(2^l x_j - 2z + 1), & \frac{z-1}{2^{l-1}} \leq x_j \leq \frac{z}{2^{l-1}} \\ 0 & \text{Otherwise} \end{cases} \tag{15}$$

where $z = 1, 2, \ldots, 2^{l-1}$ which is the translation parameter and $f = 0, 1, \ldots, F - 1$ is the order of the polynomial.

$$\mathbb{C}_f = \begin{cases} 1/\sqrt{\pi}, & f = 0 \\ \sqrt{2/\pi}\tilde{\mathbb{C}}_f & f > 0 \end{cases} \tag{16}$$

where $\tilde{\mathbb{C}}_f$ are Chebyshev polynomials which can be calculated as:

$$\tilde{\mathbb{C}}_0 = 1, \tilde{\mathbb{C}}_1 = x_j \text{ and } \tilde{\mathbb{C}}_{f+1} = 2x_j\tilde{\mathbb{C}}_z - \tilde{\mathbb{C}}_{z-1} \tag{17}$$

The cost function using (10) is given as:

$$\downarrow \Upsilon_{pv} = \frac{1}{2}s_{pv}^2(n) + \frac{\xi_{pv}}{2}u_{pv}^2(n) \tag{18}$$

The control law $u(n) = u_{pv}(n)$ for PV system is:

$$u_{pv}(n) = \frac{\sum_{j=1}^{p}\mu_j^i \Xi^i}{\sum_{j=1}^{p}\mu_j^i} = \frac{\sum_{j=1}^{p}\left(\left[\prod_{j=1}^{p}\exp\left(\frac{-1}{2}\left[\frac{x_j - \mathfrak{W}_j^i}{\varsigma_j^i}\right]^2\right)\right] \times [\beta_j^i \times \mathfrak{P}_{pq}(x_j)]\right)}{\sum_{j=1}^{p}\left[\prod_{j=1}^{p}\exp\left(\frac{-1}{2}\left[\frac{x_j - \mathfrak{W}_j^i}{\varsigma_j^i}\right]^2\right)\right]} \tag{19}$$

In general the update equations for control law $u_{pv}(n)$ are written as:

$$\mathfrak{T}_j^i(n+1) = \mathfrak{T}_j^i(n) + \xi_{pv}\frac{\partial \Upsilon_{pv}(n)}{\partial \mathfrak{T}_j^i(n)} \tag{20}$$

$$\mathfrak{T}_j^i(n+1) = \mathfrak{T}_j^i(n) + \xi_{pv}\frac{\partial}{\partial \mathfrak{T}_j^i(n)}\left[\frac{1}{2}s_{pv}^2(n) + \frac{\xi_{pv}}{2}u_{pv}^2(n)\right] \tag{21}$$

$$\mathfrak{T}_j^i(n+1) = \mathfrak{T}_j^i(n) + \xi_{pv}\left[s_{pv}\frac{\partial s_{pv}(n)}{\partial u_{pv}(n)} + \xi_{pv}u_{pv}(n)\right]\frac{\partial u_{pv}(n)}{\partial \mathfrak{T}_j^i(n)} \tag{22}$$

Now for simplification $\frac{\partial s_{pv}(n)}{\partial u_{pv}(n)} = 1$ and $\mathfrak{J} = s_{pv} + \xi_{pv}u_{pv}(n)$.

$$\mathfrak{T}_j^i(n+1) = \mathfrak{T}_j^i(n) + \xi_{pv}\mathfrak{J}\frac{\partial u_{pv}(n)}{\partial \mathfrak{T}_j^i(n)} \tag{23}$$

The s_{pv} is used to tune the parameters $\mathfrak{T}_j^i \in \{\mathfrak{W}_j^i, \varsigma_j^i, \beta_j^i\}$ of the controller. ξ_{pv} is the learning rate, $\mathfrak{W}_j^i$ is the mean of the Gaussian fuzzy function for jth input, ith rule, ς_j^i is the variance of the Gaussian fuzzy function for jth input, ith rule and β_j^i is the weight of the Gaussian membership function for jth input, ith rule. The update equations for individual parameters are given as:

$$\mathfrak{W}_j^i(n+1) = \mathfrak{W}_j^i(n) + \xi_{pv}\mathfrak{J}\frac{\partial u_{pv}(n)}{\partial \mathfrak{W}_j^i(n)} \tag{24}$$

$$\varsigma_j^i(n+1) = \varsigma_j^i(n) + \xi_{pv}\mathfrak{J}\frac{\partial u_{pv}(n)}{\partial \varsigma_j^i(n)} \tag{25}$$

$$\beta_j^i(n+1) = \beta_j^i(n) + \xi_{pv}\mathfrak{J}\frac{\partial u_{pv}(n)}{\partial \beta_j^i(n)} \tag{26}$$

The gradient $\frac{\partial u_{pv}(n)}{\partial \mathfrak{W}_j^i(n)}$ can be derived as:

$$\frac{\partial u_{pv}(n)}{\partial \mathfrak{W}_j^i(n)} = \frac{\partial u_{pv}(n)}{\partial \bar{\mu}_j^i}\frac{\partial \bar{\mu}_j^i}{\partial \mathfrak{W}_j^i(n)} = \left[\frac{\Xi_j - u_{pv}(n)}{\sum_{j=1}^{p}\mu_j}\right]\frac{\partial \bar{\mu}_j^i}{\partial \mathfrak{W}_j^i(n)} \tag{27}$$

$$\frac{\partial u_{pv}(n)}{\partial \mathfrak{W}_j^i(n)} = \left[\frac{\Xi_j - u_{pv}(n)}{\sum_{j=1}^{p}\mu_j}\right]\mu_i\frac{\left(x_j - \mathfrak{W}_j^i\right)}{\varsigma_j^{i^2}} \tag{28}$$

Similarly, the gradient $\frac{\partial u_{pv}(n)}{\partial \varsigma_j^i(n)}$ can be derived as:

$$\frac{\partial u_{pv}(n)}{\partial \varsigma_j^i(n)} = \frac{\partial u_{pv}(n)}{\partial \bar{\mu}_j^i}\frac{\partial \bar{\mu}_j^i}{\partial \varsigma_j^i(n)} = \left[\frac{\Xi_j - u_{pv}(n)}{\sum_{j=1}^{p}\mu_j}\right]\frac{\partial \bar{\mu}_j^i}{\partial \varsigma_j^i(n)} \tag{29}$$

$$\frac{\partial u_{pv}(n)}{\partial \varsigma_j^i(n)} = \left[\frac{\Xi_j - u_{pv}(n)}{\sum_{j=1}^{p}\mu_j}\right]\mu_i\frac{\left(x_j - \mathfrak{W}_j^i\right)^2}{\varsigma_j^{i^3}} \tag{30}$$

The gradient $\frac{\partial u_{pv}(n)}{\partial \beta_j^i(n)}$ can be derived as:

$$\frac{\partial u_{pv}(n)}{\partial \beta_j^i(n)} = \frac{\partial u_{pv}(n)}{\partial \mathfrak{P}_{pq}(x)}\frac{\partial \mathfrak{P}_{pq}(x)}{\partial \beta_j^i(n)} = \left[\frac{\mu_j}{\sum_{j=1}^{p}\mu_j}\mathfrak{P}_{pq}(x_j)\right] \tag{31}$$

By putting the values from (28), (30) and (31) into (24)–(26), we get:

$$\mathfrak{W}_j^i(n+1) = \mathfrak{W}_j^i(n) + \xi_{pv}\mathfrak{J}\left[\frac{\Xi_j - u_{pv}(n)}{\sum_{j=1}^{p}\mu_j}\right]\mu_i\frac{\left(x_j - \mathfrak{W}_j^i\right)}{\varsigma_j^{i^2}} \tag{32}$$

$$\varsigma_j^i(n+1) = \varsigma_j^i(n) + \xi_{pv}\mathfrak{J}\left[\frac{\Xi_j - u_{pv}(n)}{\sum_{j=1}^{p}\mu_j}\right]\mu_i\frac{\left(x_j - \mathfrak{W}_j^i\right)^2}{\varsigma_j^{i^3}} \tag{33}$$

$$\beta_j^i(n+1) = \beta_j^i(n) + \xi_{pv}\mathfrak{J}\left[\frac{\mu_j}{\sum_{j=1}^{p}\mu_j}\mathfrak{P}_{pq}(x_j)\right] \tag{34}$$

4 Simulation Results

The proposed DACWNFC, PV system with rated output power of 261 kW were modeled and designed in MATLAB. The initial parameters in the updating laws were taken as:

$j = 5, \xi = 0.224, [\varsigma_{1,0}(0), \varsigma_{1,1}(0), \ldots, \varsigma_{1,5}(0), \varsigma_{2,0}(0), \varsigma_{2,1}(0), \ldots, \varsigma_{2,5}(0)] =$
$[0.01, 0.014, 0.017, 0.019, 0.025, 0.035, 0.04, 0.045, 0.055, 0.062]$
$[\mathfrak{W}_{1,0}(0), \mathfrak{W}_{1,1}(0), \ldots, \mathfrak{W}_{1,5}(0), \mathfrak{W}_{2,0}(0), \mathfrak{W}_{2,1}(0), \ldots, \mathfrak{W}_{2,5}(0)] =$
$[0.024, 0.028, 0.033, 0.36, 0.41, 0.46, 0.5, 0.56, 0.61, 0.7]$
$[\beta_{1,0}(0), \beta_{1,1}(0), \ldots, \beta_{1,5}(0), \beta_{2,0}(0), \beta_{2,1}(0), \ldots, \beta_{2,5}(0)] =$
$[0.12, 0.16, 0.20, 0.25, 0.27, 0.30, 0.39, 0.45, 0.57, 0.62]$

To demonstrate the excellent behavior of the proposed DACWNFC, three other controllers, i.e., fuzzy based MPPT, and InC Proportional Integral Derivative (PID) based MPPT are designed in MATLAB/SIMULINK and simulations were performed for all controllers under same uncertainties coming from load fluctuations and weather situations. In this chapter, the weather data which were used during simulation was taken from [28]. Solar irradiance W/m^2 and ambient temperature (°C) levels were measured after every minute for three summer days and as illustrated in Figs. 5 and 6. The irradiance level shown high variations during day time due to the rise and set of sun. For instance, at 6:13 h (433 min) the sun starts to appear and sets at at 19:18 h (1158 min) as given in Fig. 5. The average irradiance level goes to 1000 W/m^2 during day 1. Due to high level of cloud cover, the solar irradiance level is quite less with respect to other days. The irradiance level is considered as partial shading. The average temperature during day 2 rises to 44.28 °C and at night it falls to 15 °C as shown in Fig. 6.

The PV system output power comparison for day 1 is shown in Fig. 7. The reference power is represented with red dashed line. It is clearly observed, that the proposed DACWNFC output power is very close to the reference power followed by Fuzzy Logic Controller (FLC) and PID. The peak power obtained from PV system is 365 kW (at 820 min). Similarly, the PV system output power comparison for day 2 is shown in Fig. 8. In this case, due the sudden variation in irradiance level, the PV output power also suddenly changes and performance of different controller is observed. From Fig. 8, the PID controller performance is very poor as compared to FLC and DACWNFC. The maximum power obtained on this day is 218 kW

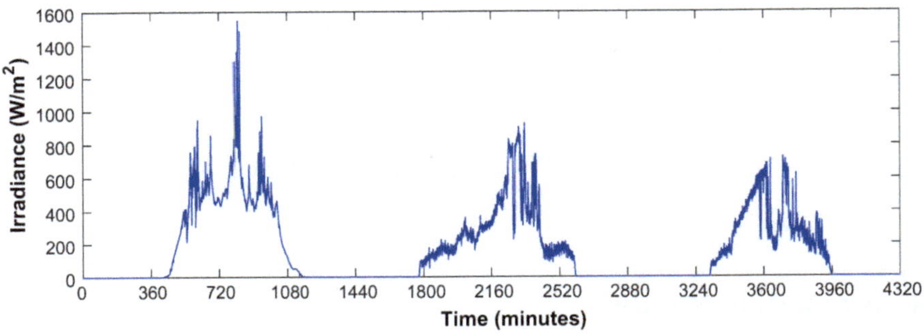

Fig. 5 Solar irradiance

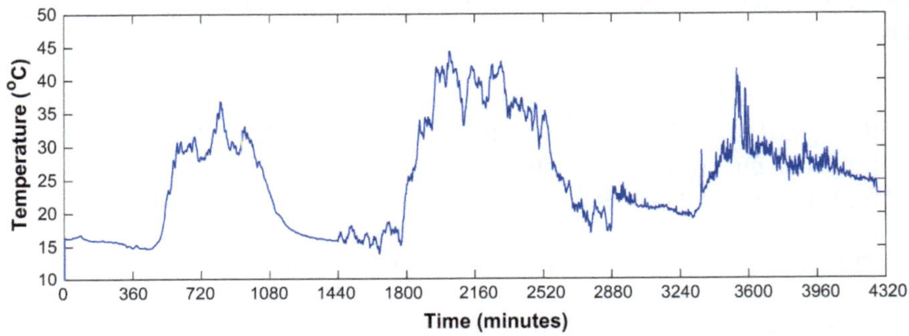

Fig. 6 Ambient temperature

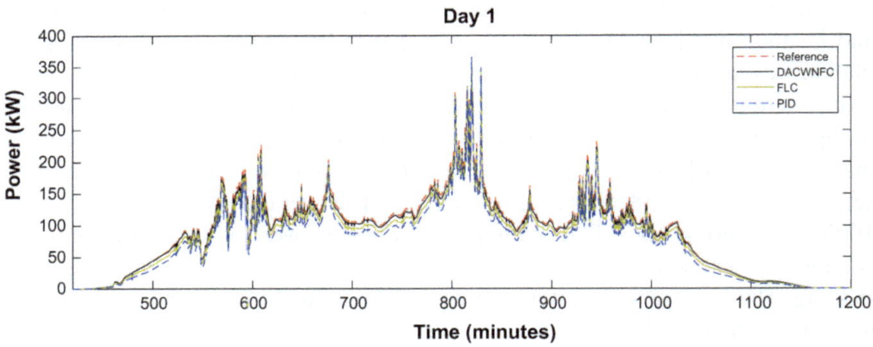

Fig. 7 PV system output power on day 1

(at 2236 min). Similarly, due to partial shading condition, the maximum PV output power obtained is only 166 kW (at 3703 min) as shown in Fig. 9. Throughout, the proposed DACWNFC is superior than both fuzzy based MPPT and InC PID MPPT. The obvious reason for it is the integration of wavelets in the neuro-fuzzy controller.

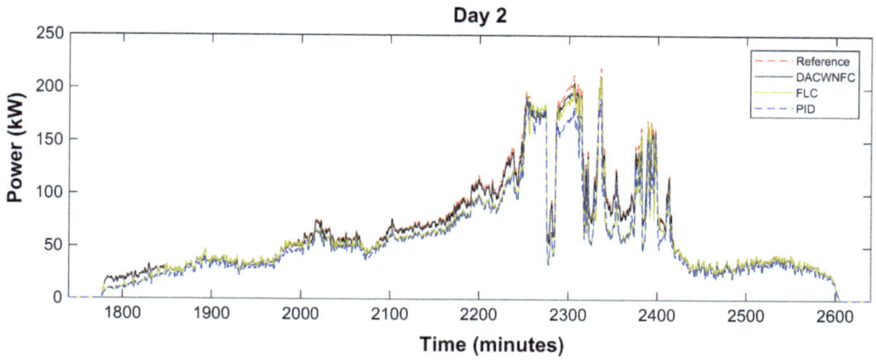

Fig. 8 PV system output power on day 2

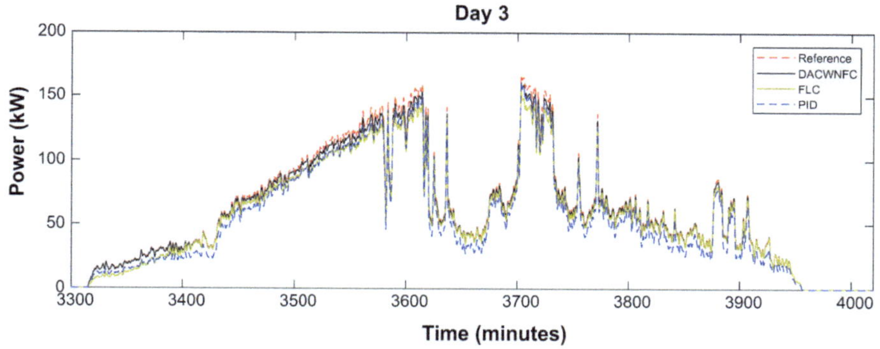

Fig. 9 PV system output power on day 3

The proposed controller tries to keep the slope zero while tracking the MPP of solar PV. The scatter plot of different controllers, including DACWNFC is shown in Fig. 10. Due to sudden irradiance variations, the MPP goes to a another new point which in turn changes the value of s (an input error for the controller). For any new operating point, the controller minimizes the input error through changing the d. From Fig. 10, the InC PID based approach, the maximum slope observed is above 1000, while majority of the point lies between ±500. Likewise, the maximum slope for a fuzzy based MPPT method ranges to ±500, whereas the proposed DACWNFC accomplishes the MPP very quickly and the maximum slope points does not excess ±30.

The PV overall efficiency (η_{pv}) and PV average efficiency (η_{pv-av}) are also determined in case of all controllers and are given as follows:

$$\eta_{pv} = \frac{\int_0^t P_{pv}(t)dt}{\int_0^t P_{ref}(t)dt} \times 100\% \tag{35}$$

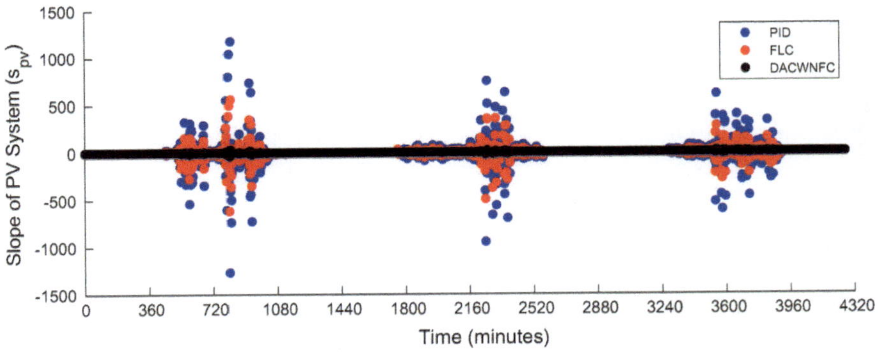

Fig. 10 Scatter plot of MPPT error using various controllers

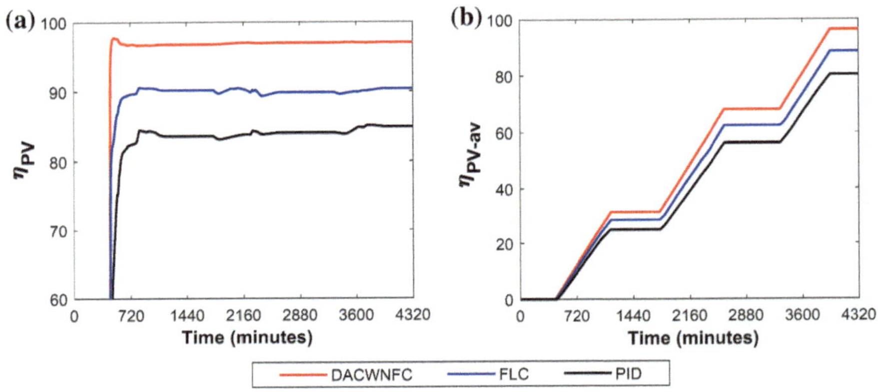

Fig. 11 PV MPP tracking efficiency comparison: **a** η_{pv}, **b** η_{pv-av}

$$\eta_{pv-av} = \frac{1}{T}\eta_{pv} \tag{36}$$

From Fig. 11, the PV efficiency for DACWNFC is observed as 97.07%, while other controllers efficiencies varies with time. Correspondingly, the mean efficiency for DACWNFC linearly rises with a certain slope as compare to others. Similarly, various performance indexes, i.e., Integral Squared Error (ISE), Integral Time-weighted Absolute Error (ITAE), Integral Absolute Error (IAE), and Integral Time-weighted Squared Error (ITSE) are calculated in (37)–(40) and are shown in Fig. 12.

$$IAE = \int_0^t |e(t)|dt \tag{37}$$

$$ITAE = \int_0^t t|e(t)|dt \tag{38}$$

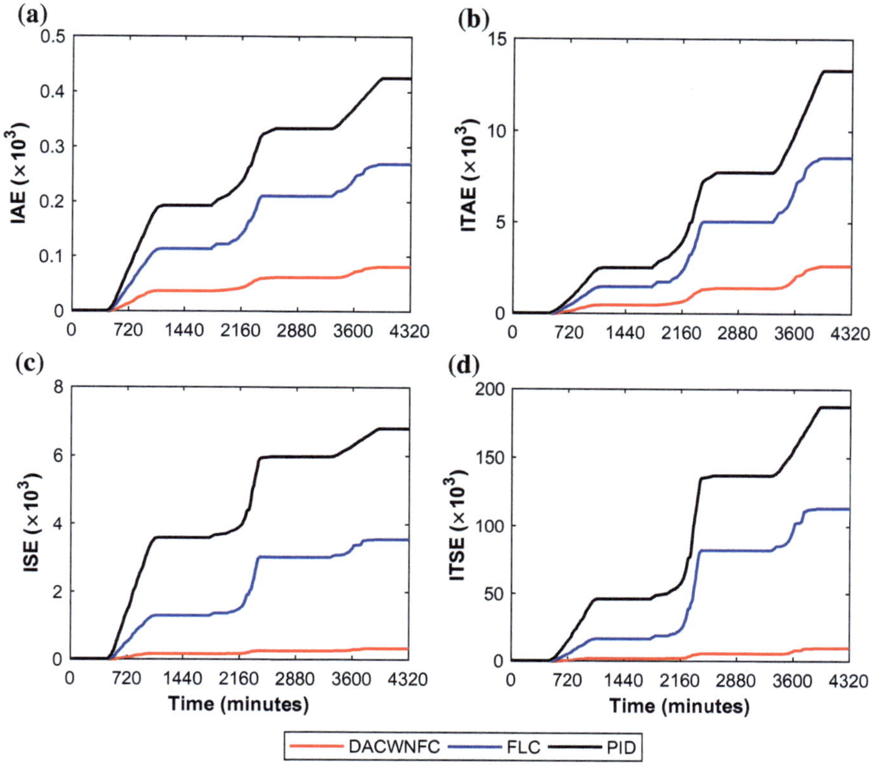

Fig. 12 Dynamic performance indexes: **a** IAE **b** ITAE **c** ISE **d** ITSE

$$ISE = \int_0^t e^2(t)dt \tag{39}$$

$$ITSE = \int_0^t te^2(t)dt \tag{40}$$

where $e(t) = P_{ref}(t) - P_{pv}(t)$. It is clear from the Fig. 12 that all index values are quite small and better with DACWNFC as compared to other two control methods. Table 1 presents the values of calculated indexes and efficiencies based on a 4320 min simulation with a sampling time of 10^{-6} s. The efficiency of PV is seen as high as compared to other controllers. Likewise, the error indexes are also observed below than those of FLC based MPPT and InC based PID MPPT. From the results, the proposed DACWNFC shows a better tracking and efficiency with respect to other controllers.

Table 1 PV efficiencies and performance parameters. DACWNFC, FLC and InC PID

Controllers	η_{pv} (%age)	$\eta_{pv\text{-}av}$ (%age)	IAE (10^3)	ITAE (10^3)	ISE (10^3)	ITSE (10^3)
DACWNFC	97.07	96.36	0.082	2.644	0.338	10.06
FLC	90.43	88.64	0.270	8.59	3.56	113.94
PID	84.92	80.33	0.425	13.29	6.80	187.12

5 Conclusions

In this chapter, a PV system was controlled under real load and long weather parameters through four different controllers, i.e., a new improved direct self-recurrent direct adaptive neuro-fuzzy wavelet embedded MPPT controller, fuzzy based MPPT controller, and InC PID based MPPT controller. The improvement in the new proposed controller was performed by an integrating a Chebyshev wavelet in the exiting neuro-fuzzy architecture. The proposed controller was better under real disturbances appearing from load change and long weather pattern fluctuations. Unlike many existing MPPT methods, the proposed technique has high self-adaptation ability for a new operating condition at any time. Many simulation results and comparison have shown that the new direct adaptive controller is better than fuzzy MPPT, and InC PID based MPPT in terms of efficiency, power transfer and dynamic response. This chapter will be of high interest for those who are looking for research in renewable energy using wavelets.

References

1. Putri RI, Wibowo S, Rifa'i M (2015) Maximum power point tracking for photovoltaic using incremental conductance method. Energy Procedia 68:22–30. https://doi.org/10.1016/j.egypro.2015.03.228
2. Sera D, Mathe L, Kerekes T, Spataru SV, Teodorescu R (2013) On the perturb-and-observe and incremental conductance MPPT methods for PV systems. IEEE J Photovolt 3(3):1070–1078. https://doi.org/10.1109/JPHOTOV.2013.2261118
3. Kjær SB (2012) Evaluation of the hill climbing and the incremental conductance maximum power point trackers for photovoltaic power systems. IEEE Trans Energy Convers 27(4):922–929. https://doi.org/10.1109/TEC.2012.2218816
4. Ishaque K, Salam Z, Lauss G (2014) The performance of perturb and observe and incremental conductance maximum power point tracking method under dynamic weather conditions. Appl Energy 119:228–236. https://doi.org/10.1016/j.apenergy.2013.12.054
5. Tey KS, Mekhilef S (2014) Modified incremental conductance algorithm for photovoltaic system under partial shading conditions and load variation. IEEE Trans Ind Electron 61(10):5384–5392. https://doi.org/10.1109/TIE.2014.2304921
6. Alik R, Jusoh A (2017) Modified perturb and observe (P&O) with checking algorithm under various solar irradiation. Sol Energy 148:128–139. https://doi.org/10.1016/j.solener.2017.03.064

7. Libo W, Zhengming Z, Jianzheng L (2007) A single-stage three-phase grid-connected photovoltaic system with modified MPPT method and reactive power compensation. IEEE Trans Energy Convers 22(4):881–886. https://doi.org/10.1109/TEC.2007.895461
8. Chu C-C, Chen C-L (2009) Robust maximum power point tracking method for photovoltaic cells: a sliding mode control approach. Sol Energy 83(8):1370–1378. https://doi.org/10.1016/j.solener.2009.03.005
9. Lalili D, Mellit A, Lourci N, Medjahed B, Berkouk EM (2011) Input output feedback linearization control and variable step size MPPT algorithm of a grid-connected photovoltaic inverter. Renew Energy 36(12):3282–3291. https://doi.org/10.1016/j.renene.2011.04.027
10. Esram T, Kimball JW, Krein PT, Chapman PL, Midya P (2006) Dynamic maximum power point tracking of photovoltaic arrays using ripple correlation control. IEEE Trans Power Electron 21(5):1282–1291. https://doi.org/10.1109/TPEL.2006.880242
11. Shah I, ur Rehman F (2017) Smooth higher-order sliding mode control of a class of underactuated mechanical systems. Arab J Sci Eng 42(12):5147–5164. https://doi.org/10.1007/s13369-017-2617-9
12. Alajmi BN, Ahmed KH, Finney SJ, Williams BW (2011) Fuzzy-logic-control approach of a modified hill-climbing method for maximum power point in microgrid standalone photovoltaic system. IEEE Trans Power Electron 26(4):1022–1030. https://doi.org/10.1109/TPEL.2010.2090903
13. Shaiek Y, Smida MB, Sakly A, Mimouni MF (2013) Comparison between conventional methods and GA approach for maximum power point tracking of shaded solar PV generators. Sol Energy 90:107–122. https://doi.org/10.1016/j.solener.2013.01.005
14. Ahmed J, Salam Z (2014) A maximum power point tracking (MPPT) for PV system using cuckoo search with partial shading capability. Appl Energy 119:118–130. https://doi.org/10.1016/j.apenergy.2013.12.062
15. Titri S, Larbes C, Toumi KY, Benatchba K (2017) A new MPPT controller based on the ant colony optimization algorithm for photovoltaic systems under partial shading conditions. Appl Soft Comput 58:465–479. https://doi.org/10.1016/j.asoc.2017.05.017
16. Ishaque K, Salam Z, Shamsudin A, Amjad M (2012) A direct control based maximum power point tracking method for photovoltaic system under partial shading conditions using particle swarm optimization algorithm. Appl Energy 99:414–422. https://doi.org/10.1016/j.apenergy.2012.05.026
17. Hassan SZ, Li H, Kamal T, Nadarajah M, Mehmood F (2016) Fuzzy embedded MPPT modeling and control of PV system in a hybrid power system. In: 2016 international conference on emerging technologies (ICET). IEEE, pp. 1–6. https://doi.org/10.1109/ICET.2016.7813236
18. Rahmani B, Li W (2017) A current-based fuzzy controller for MPPT of grid-connected PV systems. J Renew Sustain Energy 9(2):23503. https://doi.org/10.1063/1.4977826
19. Khaldi N, Mahmoudi H, Zazi M, Barradi Y (2014) The MPPT control of PV system by using neural networks based on Newton Raphson method. In: 2014 international renewable and sustainable energy conference (IRSEC). IEEE, pp. 19–24. https://doi.org/10.1109/IRSEC.2014.7059894
20. Sedaghati F, Nahavandi A, Badamchizadeh MA, Ghaemi S, Abedinpour Fallah M (2012) PV maximum power-point tracking by using artificial neural network. Math Probl Eng 2012:1–10. https://doi.org/10.1155/2012/506709
21. Vazquez JR, Martin AD, Herrera RS (2013) Neuro-Fuzzy control of a grid-connected photovoltaic system with power quality improvement. In: 2013 IEEE EUROCON. IEEE, pp. 850–857. https://doi.org/10.1109/EUROCON.2013.6625082
22. Chen SX, Gooi HB, Wang MQ (2013) Solar radiation forecast based on fuzzy logic and neural networks. Renew Energy 60:195–201. https://doi.org/10.1016/j.renene.2013.05.011
23. Kamal T, Karabacak M, Hassan SZ, Li H, Fernandez LM (2018) A robust online adaptive B-spline MPPT control of three-phase grid-coupled photovoltaic systems under real partial shading condition. IEEE Trans Energy Convers, p. 1. https://doi.org/10.1109/TEC.2018.2878358
24. Hassan SZ, Li H, Kamal T, Arifoğlu U, Mumtaz S, Khan L (2017) Neuro-fuzzy wavelet based adaptive MPPT algorithm for photovoltaic systems. Energies 10(3):394 . https://doi.org/10.3390/en10030394

25. Abiyev RH, Kaynak O (2008) Identification and control of dynamic plants using fuzzy wavelet neural networks. In: 2008 IEEE international symposium on intelligent control. IEEE, pp. 1295–1301
26. Badar R, Khan L (2013) Hybrid neuro-fuzzy legendre-based adaptive control algorithm for static synchronous series compensator. Electr Power Compon Syst 41(9):845–867. https://doi.org/10.1080/15325008.2013.792882
27. Mukerjee AK, Dasgupta N (2007) DC power supply used as photovoltaic simulator for testing MPPT algorithms. Renew Energy 32(4):587–592. https://doi.org/10.1016/j.renene.2006.02.010
28. Pakistan (2018) Pakistan meteorological department solar radiation data

MIX
Papier aus verantwortungsvollen Quellen
Paper from responsible sources
FSC® C105338

FSC
www.fsc.org

If you have any concerns about our products,
you can contact us on
ProductSafety@springernature.com

In case Publisher is established outside the EU,
the EU authorized representative is:
Springer Nature Customer Service Center GmbH
Europaplatz 3, 69115 Heidelberg, Germany

Printed by Libri Plureos GmbH
in Hamburg, Germany